Setting the Agenda
for American Archaeology

Classics in Southeastern Archaeology
Stephen Williams, Series Editor

Publication of this work has been supported in part by
the Dan Josselyn Memorial Fund

Setting the Agenda for American Archaeology:

The National Research Council Archaeological Conferences of 1929, 1932, and 1935

Edited and with an Introduction by
Michael J. O'Brien and R. Lee Lyman

THE UNIVERSITY OF ALABAMA PRESS
Tuscaloosa and London

Copyright © 2001
The University of Alabama Press
Tuscaloosa, Alabama 35487-0380
All rights reserved
Manufactured in the United States of America

2 4 6 8 9 7 5 3 1
02 04 06 08 09 07 05 03 01

Typeface: Adobe Garamond

∞
The paper on which this book is printed meets the minimum requirements of American National Standard for Information Science–Permanence of Paper for Printed Library Materials, ANSI Z39.48–1984.

Library of Congress Cataloging-in-Publication Data

Setting the agenda for American archaeology : the National Research Council archaeological conferences of 1929, 1932, and 1935 / edited and with an introduction by Michael J. O'Brien and R. Lee Lyman. p. cm. — (Classics in southeastern archaeology)
Includes bibliographical references and index. ISBN 0-8173-1084-3 (alk. paper)
1. Archaeology—Research—United States—History—20th century. 2. National Research Council (U.S.). Committee on State Archaeological Surveys. I. O'Brien, Michael J. (Michael John), 1950- II. Lyman, R. Lee. III. Series.
CC95 .S48 2001
930.1—dc21
00-012564

British Library Cataloguing-in-Publication Data available

Contents

List of Illustrations   vii

Preface and Acknowledgments   ix

Introduction   1
Michael J. O'Brien and R. Lee Lyman

1. Report of the Conference on Midwestern Archaeology, St. Louis, 1929   85

2. Conference on Southern Pre-History, Birmingham, 1932   207

3. The Indianapolis Archaeological Conference, 1935   333

Appendix 1. State Archaeological Surveys: Suggestions in Method and Technique   429

Appendix 2. Guide Leaflet for Amateur Archaeologists   455

Index   469

Illustrations

Clark Wissler   5

Roland B. Dixon   5

Arthur C. Parker   13

A. V. Kidder with Alfred M. Tozzer and Carl E. Guthe   19

Frederick W. Hodge   27

Matthew W. Stirling alongside Olmec colossal head 4 at La Venta   29

Fay-Cooper Cole   38

John R. Swanton   42

Henry B. Collins   43

James B. Griffin examining pottery from the Kincaid site   49

W. C. McKern   51

The analytical relations among traits, components, and foci in the Midwestern Taxonomic Method   56

The Serpent Mound, Adams County, Ohio   86

Monk's Mound, Cahokia Group, in East St. Louis, Illinois   203

The "Red" Mound at Cahokia   203

Pottery vessels from the mounds near Lewiston, Illinois   203

Example of methods of excavation at Lewiston, Illinois   204

Post-molds uncovered by excavation of a mound near Pembroke, Kentucky   205

Specimen of pottery used by the "Hopewell" people of Ohio   205

Examples of handiwork found in a "Hopewell" mound north of La Crosse, Wisconsin   206

Mound B at Moundville   208

Conferees attending the 1932 Conference on Southern Pre-History in Birmingham, Alabama   209

Indian tribes and linguistic stocks in the eastern part of the United States about 1700   324

Relation of Indian tribes in the Southeast to the physical divisions   325

Hypothetical route of De Soto   327

Distribution of certain cultural traits in the Southeast   328

Tribal movements of the eastern Indians   329

Significant tribal locations   330

Archaeological culture areas   331

Tentative Archaeological Culture Classification for Upper Mississippi and Great Lakes areas   414

Preface and Acknowledgments

One of the lesser-known aspects of Americanist archaeology is the substantial role played by the National Research Council during the 1920s and 1930s. Our use of the term "lesser-known" is not meant to imply that the role of the National Research Council (NRC) has gone unreported (see Griffin 1976a, 1985; Guthe 1952, 1967) but rather that its critical importance in shaping the course and complexion of Americanist archaeology has perhaps not been given the place in the history of the discipline that it deserves. And yet, it really wasn't the NRC itself that played the critical role but rather the archaeologists affiliated with it—persons such as Roland B. Dixon, A. V. Kidder, Frederick W. Hodge, Clark Wissler, and Carl E. Guthe. Their vehicle for plotting the future course of archaeology in the United States, especially in the Midwest and Southeast, was the Committee on State Archaeological Surveys (CSAS), which was organized in 1920 under the newly created Division of Anthropology and Psychology within the NRC. For 17 years, until it was abolished in June 1937, the CSAS labored to bring a sense of professionalism to the manner in which archaeology was being conducted in the United States.

Few graduate programs in archaeology existed in the 1920s and 1930s, and those that did—first at Harvard, Chicago, Pennsylvania, California, and Columbia and eventually at Yale and Michigan—had not turned out enough archaeologists to meet the curricular needs of private and public colleges and universities, which increasingly were offering courses not only in prehistory but also in such practical aspects as excavation and artifact analysis. For example, at our own institution, the University of Missouri, archaeology was taught by a historian of the ancient world and by a sociologist. Of even more urgency was the fact that a nation's fascination with the past was leading to a rapid destruction of archaeological sites and the commercialization of antiquities. Neither phenomenon was entirely new in the 1920s, but the escalation of wholesale looting that occurred in the eastern United States during the early decades of the twentieth century, together with the upward spiral in commercial value of archaeological objects, finally reached such a level that the problem could no longer be ignored by those with a professional commitment to understanding the past. Something had to be done to awaken in the public a sense for preserving the archaeological record.

But professional archaeologists had a bigger problem than simply educating the general public about the scientific value of the archaeological record and why it should be protected. By the late 1920s every state had at least one museum or historical society engaged in some form of archaeological exploration, and countless towns and cities had societies of one kind or another that had been organized ostensibly for "scientific" purposes. Societies in the larger cities such as Boston, Philadelphia, New York, and Washington, D.C., counted professional archaeologists among their ranks, but those in the smaller cities, although composed of learned individuals, had no guiding voices on how to explore the past without destroying the very record being examined. As a result, fieldwork was oftentimes little more than a pot-hunting expedition used to fill the shelves and curio cabinets of a society's members. For the same reason—lack of expertise—much of the fieldwork carried out by local museums and historical societies was not much better.

Faced with this growing problem, professional archaeologists had several options, one of which was to do nothing and hope the problem would go away. There was, however, little likelihood of that happening. Another option was to point out deficiencies in the work being done and hope that people were wise enough to listen to those who supposedly were experts on the matter. This would not have worked either. It didn't help matters that most professional archaeologists of the period had been trained in the Northeast and that the majority of the destruction of the archaeological record was taking place in the South and Midwest. The Civil War was but a distant memory in the Northeast, but this was not true in the South. The last thing members of a local southern archaeological society wanted to see was a carpetbagger archaeologist from a northeastern institution telling them how to excavate a site or to keep track of the artifacts they were collecting.

There was only one solution that might work, and that was to attempt to draw in not only the lay public but also amateur prehistorians and to make them feel as if they were part of something important—something that everyone, regardless of educational grounding or geographic area, could work toward. Archaeologists had used the public before—the federal Bureau of American Ethnology had long sent circulars to various parts of the East and Midwest asking for the help of locals in locating sites and collections of antiquities—but there had never been a well-designed plan to educate the literally thousands of potential correspondents out in the hinterland. Nor had any concerted effort been made to elevate the quality of fieldwork being done by local and state societies and museums.

One means of doing that was to coordinate efforts among the various state museums and societies, especially with respect to site surveys, and to publish brief résumés of their work. By the early 1920s several states either had begun statewide surveys or were anticipating such action, and one of the early roles of the Committee on State Archaeological Surveys, as the name implies, was to encourage and assist the states in their efforts. As results became known, summary statements were published annually in the *American Anthropologist.* Encouraging cooperation among various institutions was one thing, but it did not really address the overriding concern of the CSAS, which was the deplorable lack of standards that existed in fieldwork and analysis. The committee's solution to this nagging problem comprised two parts. One was to print and circulate brief pamphlets on archaeological method, the first in 1923 and a second in 1930. The other was to sponsor a series of regional seminars aimed at elevating the consciousness of midwesterners and southerners who were interested in the past. The seminars took place in St. Louis in 1929; in Birmingham, Alabama, in 1932; and in Indianapolis in 1935. The reports that issued from those meetings are the focus of this book. Taken in the aggregate the reports and the pamphlets on method offer a rare opportunity to see not only some of the problems that Americanist archaeologists faced in the early decades of the discipline but also the steps they took to overcome them.

When we first had the idea of assembling such a volume, we considered whether it would make more sense to publish only the volumes from the Birmingham and Indianapolis conferences. The tenor of those two meetings was more methodological than that of the St. Louis meeting, which was aimed more at how to preserve the archaeological record rather than at how to study it. But then we realized that deleting the St. Louis proceedings would distort the picture in terms of the role played by the CSAS in its efforts to call attention to a dwindling resource base and to train a cadre of laypersons to assist in gathering useful information about the past. Based on the ever-growing reports of activities at the state level that appeared in the *American Anthropologist*—the last year the reports appeared, 1932, they occupied 29 pages—the St. Louis conference was a success. The CSAS followed up on that success the following year by issuing the second pamphlet on archaeological method, and in 1932 by organizing a second conference aimed specifically at bringing southeastern museums and societies into the fold.

Unlike the St. Louis conference, the Birmingham conference focused very little on site destruction or on preserving the archaeological record. As they had in St. Louis, organizers of the Birmingham conference dedicated several sessions to discussions of findings from various geographical areas, but they dedicated even more time to methodological and technical topics such as field and laboratory methods, museum work, and publication. Whereas the St. Louis conference was a demonstration of what archaeology *could* do, the one in Birmingham was a nuts-and-bolts demonstration of *how* to do it. In our minds, that conference played a pivotal role in Americanist archaeology, because it was there, perhaps for the first time, that professional anthropologists and archaeologists from a wide range of institutions, including the Bureau of American Ethnology and the U.S. National Museum, laid out a research agenda that both professionals *and* nonprofessionals could follow. In reading the report of the Birmingham conference, one cannot escape the feeling that from an intellectual standpoint it was as profitable for professionals as it was for nonprofessionals. The legacy of the conference was the unity with which fieldwork and analysis were carried out in the Southeast during the 1930s and 1940s, much of it at the hands of James A. Ford, who at the age of 21 was in attendance at Birmingham.

From our vantage point, the most important aspect of the Birmingham conference is what it tells us about Americanist archaeology generally, especially in terms of how archaeologists were coming to grips with the study of time. A key item on the agenda at Birmingham was developing a chronological ordering of archaeological remains—not a mean feat, it was thought, in an area that lacked standing prehistoric architecture, perishable artifacts, and thick, stratified deposits. Chronological control was viewed as being so central to the pursuit of archaeology in the Southeast that all other studies were held to be of secondary importance. In the

Southwest, where there had been concerted archaeological effort for well over two decades, helped along by the founding of the Pecos Conference in 1927, chronology building was old hat. To be sure, new pieces were constantly being hung on the chronological framework, but by 1932 southwesternists were satisfied that the ordering was more or less complete, and they began to pursue other aspects of prehistory. As their knowledge of the prehistoric Southwest grew, prehistorians began to look increasingly to the East for comparative data, but all they saw were personnel from local museums and amateur societies out collecting artifacts. The Birmingham conference was organized to address this deficiency, and in the report that was issued we can see in clear form the blueprint that archaeologists working outside the Southwest were beginning to follow in their pursuit of chronological ordering.

When the Indianapolis conference was held three years later, the CSAS focused almost exclusively on professional archaeologists in terms of who was invited to the meeting, and the number of delegates and guests was less than half the number in attendance in Birmingham. By 1935 the upper Mississippi Valley and the Great Lakes region had witnessed considerable archaeological survey and excavation, but missing was any kind of meaningful synthesis that tied together the various pieces of site- or region-specific work. Comparisons across the upper Midwest were difficult to make in part because of a lack of standardized terminology, but even more significantly because of a lack of an analytical framework on which to hang the various taxonomic units that were being proposed in an effort to keep track of time, space, and form. The framework that came out of the Indianapolis conference was termed "A Tentative Archaeological Culture Classification for Upper Mississippi and Great Lakes Areas." It soon became known as the Midwestern Taxonomic Method, or, in honor of its chief architect, W. C. McKern, simply the McKern Method. Our experience has been that although modern archaeologists have heard of the method, they are not too sure of what it attempted to do or of the importance it assumed, at least for a while, in Americanist archaeology. Part of the problem from a modern perspective is that McKern's exposition of the method, publication of which occurred subsequent to the Indianapolis conference, is not particularly clear. Another part of the problem is that the method in its purest form ignored time and placed analytical emphasis on the presence or absence of traits. Today this sounds almost counterintuitive, and reading only published versions (e.g., McKern 1939, 1940) tells us nothing about why a method that was not built around time was ever developed in the first place. Thus the report that emanated from the Indianapolis conference is so important: it outlines the reasons why an integrative framework was needed and gives us some insight into the thinking of archaeologists working in the upper Midwest at the time.

The Midwestern Taxonomic Method was only one of many analytical schemes developed during what is commonly referred to as the culture-history period of Americanist archaeology—an appropriate characterization of that time span from about 1915 to 1960, when those in the discipline saw documenting the development of prehistoric cultures in the Americas as their highest calling. Although it fell from favor in the 1960s, many of the central tenets of culture history were carried over to newer paradigms and thus continue to be fundamental within Americanist archaeology. Many of the methods contemporary archaeologists use to examine formal variation in artifacts and the distribution of that variation across space and through time were formulated between about 1915 and 1935, and if one is at all interested in how and why those methods came to be part of the standard archaeological tool kit, then reports such as those issued by the CSAS, and especially those emanating from the Birmingham and Indianapolis conferences, become invaluable sources of information.

The report on the St. Louis conference was originally published in typeset format in the *National Research Council Bulletin* series, and the other two reports were mimeographed for distribution. To enhance the present volume as a reference and research tool, we have retained the original pagination from all three reports, with original page numbers appearing in brackets. All three reports are reprinted in full, including illustrations and lists of attendees. Instead of breaking the flow of the reports by inserting editorial notes, we cover those in the introduction. Page numbers accompanying quotes from the reports that we use in the introduction refer to the original pagination.

Upon finding in the National Research Council Archives copies of the two pamphlets on archaeological method issued by the CSAS, we realized that the intent and hopes of the committee could only be fully appreciated if those pamphlets were included in this volume. The pamphlet issued in 1923 is reproduced in Appendix 1, and the 1930 pamphlet, issued as number 93 in the NRC's *Reprint and Circular Series,* is reproduced in Appendix 2.

As with the production of any book, several people need to be thanked for their generous contributions to the process. We greatly appreciate the help and encouragement we received from the University of Alabama Press, specifically from Judith Knight and Suzette Griffith. We also thank Stephen Williams and Charles McNutt for comments on an early draft. Over the years a number of individuals have given us numerous insights into what went on during the culture-history period of Americanist archaeology, and we thank especially Gordon R. Willey, the late William G. Haag, and the late James B. Griffin. We also thank Dan Glover for helping us with various aspects of manuscript production, including drafting the figures, obtaining various photographs, and scanning the original reports into a usable format; Daniel Barbiero, archivist with the NRC, who arranged for us to use the archives; and Michael P. Hoffman and Trish Kaufmann for answering our queries about certain individuals. Finally, we thank E. J. O'Brien, who made numerous suggestions for improving the manuscript, and Kristin Harpster, who did an excellent job of copyediting the final draft.

Setting the Agenda
for American Archaeology

Introduction
Michael J. O'Brien and R. Lee Lyman

The creation of the National Research Council (NRC) in 1916 reflected a growing concern that the United States was ill-prepared to enter a war into which it was inexorably being pulled. The council's express purpose was to assist the National Academy of Sciences (NAS), which had been signed into existence by President Abraham Lincoln in 1863, in advancing the cause of knowledge and advising the federal government on matters of science and technology. From its inception, the NAS had undertaken a wide variety of studies for different branches of government, but by the second decade of the twentieth century it was obvious that the body was too small to deal effectively with the exponential growth of science and technology taking place not only in the United States but also in Europe and Russia. Members of the NAS, including the outspoken astrophysicist George E. Hale, who served as the organization's foreign secretary, saw this scientific and technological explosion as a potential threat to the security of the United States. At Hale's instigation, members urged President Woodrow Wilson to create a body that could broaden the scope of the NAS and coordinate efforts among government, industrial, and educational organizations to strengthen not only national defense but the security of American industry as well (Cochrane 1978; Hale 1916, 1919). Hale was made the first chairman of the newly created council, which drew its membership from universities, private research institutions, and various branches of government. After the war, the NRC was made a permanent body when President Wilson signed Executive Order No. 2859 on May 11, 1918.

This was how Vernon L. Kellogg, permanent secretary of the NRC, saw the charter of the organization:

> The council is neither a large operating scientific laboratory nor a repository of large funds to be given away to scattered scientific workers or institutions. It is rather an organization which, while clearly recognizing the unique value of individual work, hopes especially to help bring together scattered work and workers and to assist in coordinating in some measure scientific attack in America on large problems in any and all lines of scientific activity, especially, perhaps, on those problems which depend for successful solution on the cooperation of several or many workers and laboratories, either within the realms of a single science or representing different realms in which various parts of a single problem may lie. It particularly intends not to duplicate or in the slightest degree to interfere with work already under way; to such work it only hopes to offer encouragement and support where needed and possible to be given. It hopes to help maintain the morale of devoted isolated investigators and to stimulate renewed effort among groups willing but halted by obstacles. (NRC 1921:6)

Until 1943 the NRC was divided into two broad sections, one concerned with relationships with the government and other bodies, and the second representing specific scientific disciplines. Each section was subdivided into divisions, with the membership composed of representatives of scientific societies and various government departments. One division on the scientific side of the house was the Division of Anthropology and Psychology, which during its lifetime oversaw the creation of 55 committees, each charged with specific tasks dictated by members of the division's executive board. As one might expect given the diversity of subject matter subsumed under the broad rubric of anthropology and psychology, the committees were diverse in terms of purpose. For example, a Committee on Accurate Publicity for Anthropology was established in 1928, a Committee on Pelvic Structure in 1926, a Committee on Psychology of Highway in 1922, and a Committee on Vestibular Research in 1921. There was even a proposal in 1919 to create a Committee on Morality, but that idea was soon abandoned. For modern students attending their first American Anthropological Association meeting and feeling that the discipline has lost its focus, it might be comforting to know that things were not completely different in the 1920s.

The Committee on State Archaeological Surveys

One of the first committees created within the Division of Anthropology and Psychology was the Committee on State Archaeological Surveys (CSAS) in 1920. Clark Wissler (Figure 1), curator of anthropology at the American Museum of Natural History in New York City and chairman of the Division of Anthropology and Psychology, reported on the formation of the committee:

> A committee was appointed to encourage and assist the several States in the organization of State archaeological surveys similar to the surveys conducted by the States of Ohio, New York, and Wisconsin. The chairman of this committee is R. B. Dixon, of Harvard University. The plan contemplates the coordination of all the agencies within those States, enlisting the cooperation of local students and interested citizens so that an effective appeal may be made to the various State legislatures for special appropriations for these surveys. (NRC 1921:53)

Roland B. Dixon (Figure 2), a hybrid ethnologist-archaeologist who was on the faculty at Harvard and was curator of ethnology at the Peabody Museum (Harvard), and who had served as president of the American Anthropological Association in 1913, was joined on the CSAS by Berthold Laufer, an expert on Chinese art and material culture who was on the staff of the Field Museum of Natural History in Chicago, and by C. E. Seashore, a neuropsychologist at the University of Iowa and the man who would succeed Wissler as chairman of the Division of Anthropology and Psychology. Subcommittees were created in four states—Illinois, Indiana, Iowa, and Missouri (Indiana Academy of Science 1921:79; NRC 1921:54). In July 1921 Dixon resigned as chairman—apparently the "correspondence involved was distasteful to him"[1]—and the CSAS was reorganized, with Wissler, having completed his term as chairman of the Division of Anthropology and Psychology, serving as committee chairman and Dixon, Laufer, Frederick W. Hodge of the Museum of the American Indian (Heye Foundation) in New York City, and Amos W. Butler of Indianapolis serving as members. Hodge was an ethnologist-archaeologist who before assuming the directorship of the Museum of the American Indian had worked at the Bureau of American Ethnology, where he was appointed ethnologist-in-charge in 1910 (Lonergan 1991:294). Butler was an ornithologist of considerable reputation, having founded the Indiana

Academy of Science in 1885, and at the time of his appointment was serving as secretary of the State Board of Charities in his home state. The committee was expanded to 7 members a year later and, to obtain better geographic coverage, to 11 members in 1924.

One might well ask why such an important entity as the NRC was involved with state archaeological surveys when there were other, seemingly more important, scientific and technological issues facing postwar America—serious issues of national welfare and defense, not the examination of pelvic structure or the psychology of highway. The answer, we think, lies in the credentials and political acumen of several key anthropologists involved with the NRC from the start. Dixon was influential from his post at Harvard, having served not only as president of the American Anthropological Association but as a trainer of a generation of archaeologists and ethnologists. Wissler was a powerful force in Americanist archaeology and ethnology from his dual positions as curator of anthropology at the American Museum of Natural History in New York City and later as professor of anthropology at Yale.[2] Working under Wissler at the museum were some of the leading figures in southwestern anthropology—Leslie Spier and Nels Nelson, for example—and Wissler was friends with Alfred L. Kroeber, the most influential anthropologist in the western half of the United States from his position at the University of California and the person who on July 1, 1921, assumed the vice chairmanship of the Division of Anthropology and Psychology. As division chairman, Wissler had the respect of the discipline and could guide the unit's direction, and one of the first things he did was to create the CSAS.

One impetus for forming the committee was the destruction of archaeological sites that was occurring with increasing frequency across the eastern United States, much of it the result of indiscriminate fieldwork by amateur societies. Making matters worse was the absence of any baseline data against which to judge the magnitude of destruction. In other words, site surveys had never been conducted in most states, and hence there was no way to gauge the percentage of sites being destroyed. A few states, such as Ohio, New York, and Wisconsin, had conducted state surveys, and Wissler was determined to see similar surveys established in other states.[3] The best means of accomplishing that objective was through an arm of the Division of Anthropology and Psychology, which, following Kellogg's vision, would act as both an organizing body and a clearing house for information.

The CSAS's decision to focus first on Illinois, Indiana, Iowa, and Mis-

Figure 1. Clark Wissler, longtime ethnologist with the American Museum of Natural History and the first chairman of the National Research Council's Division of Anthropology and Psychology, ca. 1940. (Reproduced by permission of the Society for American Antiquity from *American Antiquity* 13, no. 3 [1948])

Figure 2. Roland B. Dixon, longtime member of the Peabody Museum (Harvard) staff and faculty member in the Harvard anthropology department, ca. 1910. (From Coon and Andrews 1943; reprinted courtesy Peabody Museum, Harvard University)

souri was not entirely accidental. All four states had strong statewide support for science and history as well as active historical and scientific societies. The decision to include Indiana was certainly no surprise given that Wissler grew up there and had received all his degrees, including one in law, from the University of Indiana (Guthe 1940). Carl Guthe of the University of Michigan, who assumed the chairmanship of the CSAS in 1927, reported a year later at the International Congress of Americanists meeting in New York that

> Before the end of 1920, interest had been awakened in Illinois and Indiana. A discussion of the Illinois project constituted a part of the meeting of Section H [Anthropology] of the American Association for the Advancement of Science at Chicago in December of that year. A few days prior to this, the plans for a similar project for Indiana had been presented to the Indiana Academy of Science and the Indiana Historical Conference, both of which organizations appointed committees to further the work. During 1921, W. K. Moorehead began his excavations at the great Cahokia mound group in East St. Louis, Illinois, working in cooperation with the University of Illinois, and in Indiana the State Historical Commission and the State Department of Conservation, through the State Geologist, jointly developed a survey of the State by counties, recording all facts of an archaeological nature obtained either by field parties or questionnaires. (Guthe 1930b:52)[4]

With respect to the composition of the four state committees, Wissler (1922:233) reported soon after their formation that

> In Indiana the State Academy of Sciences and the Historical Society appointed a State committee to cooperate, viz., Dr. Frank B. Wynn, Dr. Stanley Coulter, Judge R. W. McBride; for Illinois and Iowa similar State committees; Illinois, Dr. Berthold Laufer, Dr. Otto L. Schmidt, Dr. Charles L. Owen; Iowa, Prof. B. F. Shambaugh, Dr. E. R. Harlan, E. K. Putnam. The Missouri survey was initiated by the Anthropological Society of St. Louis and is under the direction of the following committee representing a number of societies and institutions: Dr. R. J. Terry, Leslie Dana, B. M. Duggar, R. A. Holland, George S. Mepham, Dr. H. M. Whelpley, J. M. Wulfing, Dr. C. H. Danforth. Satisfactory progress has been made in each of these States. The Indiana Survey is by the State under the direction of the State Geologist. In Iowa the work has begun under a grant from the [State Historical Society of Iowa]; in Missouri under a fund raised by the

above-mentioned committee. As the results of all these surveys will be published, the outlook is stimulating.

Anatomy of a State Committee

For a perspective on the goals and methods of the state organizations that were brought about under the CSAS, we focus on the Anthropological Society of St. Louis, which in many ways was typical of the kinds of organizations that the committee was attempting to assist. Like societies in some of the other states, it grew out of an amalgam of earlier organizations, or more precisely, out of a recombination of members from different societies, some of which had long histories. The earliest scientific society in Missouri was the Academy of Science of St. Louis, which was formed in March 1856. Fifteen members—seven medical doctors, three lawyers, and five professors—attended the first meeting, and the constitution and bylaws they adopted spelled out the objectives of the fledgling society:

> Section 1. It shall have for its object the promotion of Science: it shall embrace Zoology, Botany, Geology, Mineralogy, Palaeontology, Ethnology (especially that of the Aboriginal Tribes of North America), Chemistry, Physics, Mathematics, Meteorology, and Comparative Anatomy and Physiology.
>
> Sec. 2. It shall furthermore be the object of this Academy to collect and treasure Specimens illustrative of the various departments of Science above enumerated; to procure a Library of works relating to the same, with the Instruments necessary to facilitate their study, and to procure original Papers on them.
>
> Sec. 3. It shall also be the object of this Academy to establish correspondence with scientific men, both in America and other parts of the world. (cited in O'Brien 1996:42)

The Academy of Science of St. Louis was small, but it was anything but dormant. In terms of topics that were pursued by the members, there was little in the realm of science that did not fall under the academy's purview. Understandably, topics that fell broadly under the rubric of natural history, including ethnology, archaeology, and paleontology, enjoyed keen interest. From the beginning, the academy reached out to eastern societies and institutions, perhaps as a means of gaining recognition but probably also because of an insatiable thirst for knowledge on the part of its highly educated members. The academy had established two types of membership: associate and corresponding. The former was for members living in St. Louis County and who were thus able to attend meetings, and the latter was for persons living elsewhere. It is obvious from examining the *Journal of Proceedings*[5] for the first month and a half of the society's existence that members were interested in adding to the corresponding membership some of the most well-known names in science —men such as Joseph Henry, first secretary of the Smithsonian Institution; Ferdinand V. Hayden, geologist with the U.S. Geological Survey; and Joseph Leidy of Philadelphia, arguably the top vertebrate paleontologist of the time.

The Academy of Science of St. Louis was active in archaeological fieldwork through the late 1800s (O'Brien 1996), but by the turn of the century it had been eclipsed in prominence by the Missouri Historical Society, which had been founded in 1866, and within a few more years by the St. Louis Society of the Archaeological Institute of America (AIA), which was organized in 1906. From its beginning, the historical society maintained an active interest in prehistory and as early as 1880 proposed a statewide survey of known archaeological sites (Broadhead 1880). The society was also intensely interested in obtaining collections of artifacts from Missouri sites, as remarks made by Frank Hilder in 1880 made clear:

> [Hilder] spoke to the disjointed efforts made to collect relics of the people who once dwelt in these lands. It was certainly most discreditable that one had to resort to the Smithsonian Institut[ion], the Peabody Museum, and the Blackmore Museum in Salisbury, England, to find proper collections of our prehistoric remains. He hoped to see the time when St. Louis would possess a collection in which the ancient history of the race can be studied.
>
> The spirit in which the work had been begun by the Historical Society gave promise that it would be the agency to bring together a collection which would not only rival, but surpass, any similar archaeological and historical collection.[6]

The society wasted no time in following up on Hilder's plea, as is evidenced by an advertisement placed in the January 23, 1881, edition of the *Missouri Republican* (cited in Trubowitz 1993): "The Society particularly wishes to procure archaeological specimens, popularly known as Indian curiosities or stones, flint arrow and spear heads, chisels, discoidal stones, stone axes, pottery from mounds, etc., and will be thankful for every object of this class." The historical society was extremely successful in acquiring various collections; an inventory made in 1903 showed that the organization had at least 11,000 artifacts in storage and almost 14,000 on display in 42 cases (Trubowitz 1993:3).

The other St. Louis organization that was gaining prestige in the early twentieth century was the local affiliate of the AIA known as the St. Louis Society. More than 100 people attended the organizational meeting on February 8, 1906. The AIA had been formed in 1879 with the goals of "promoting and directing archaeological investigation and research, —by the sending out of expeditions for special investigation, by aiding the efforts of independent explorers, by publication of reports of the results of the expeditions which the Institute may undertake or promote, and by any other means which may from time to time appear desirable" (AIA 1880:6).

One activity sponsored by the St. Louis Society was Gerard Fowke's (1910) survey and excavation of sites in southeastern and central Missouri. Fowke was a peripatetic journeyman connected at times with the Bureau of American Ethnology (O'Brien 1996), but apparently it was an unsalaried connection. The research proposal that was drawn up in advance of Fowke's work in Missouri has a modern ring to it (Pool 1989): records of the November 1906 meeting state that "it was unanimously agreed from the outset that, in the archaeological investigations that were proposed, the object should be scientific results, whether negative or positive, rather than in the making of large finds of relics; and that a district should be selected and worked systematically, regardless of whether the finds were great or small, so that the archaeological record might be complete."[7] Interestingly, it was the national organization, the AIA, and not the local chapter that transmitted the final report of the work to the Bureau of American Ethnology for publication in its *Bulletin* (Fowke 1910).

To note that Fowke aligned himself with the St. Louis chapter of the AIA does not do justice to an important episode in the history of Missouri archaeology, because the relationship that apparently developed between Fowke and the chapter was much more productive than is evident on the surface. In many respects the relationship between Fowke and the professional and the nonprofessional chapter members foreshadowed what was to come several decades later with the founding of the Missouri Archaeological Society in 1935, and it was the kind of relationship that Wissler hoped to foster through the CSAS. To understand the relationship between Fowke and the St. Louis chapter of the AIA requires a brief discussion of a group with the bizarre name of the "Knockers" that was formed by members of three local St. Louis organizations: the AIA, the Academy of Science of St. Louis, and the Missouri Historical Society. The spark of this group was a father-and-son combination—David I. Bushnell, Sr. and Jr.—the latter of whom had begun to make a name for himself in archaeology. Never formally trained as an anthropologist, Bushnell, Jr., worked as an assistant for the University of California in the 1890s, for the Peabody Museum (Harvard) between 1901 and 1904 (e.g., Bushnell 1904), and as an employee of the Bureau of American Ethnology beginning in 1907.

David Browman (1978:2) suggests, and we think he is correct, that it was the younger Bushnell's emerging prominence in the field and his ties to institutions such as the Peabody Museum that acted as a magnet in attracting other professional archaeologists such as Fowke to St. Louis. And it wasn't only Fowke who began keeping company with the Knockers. Browman notes that leading archaeologists of the day—men such as Earl Morris, who went on to have a distinguished career as a southwestern archaeologist with the Carnegie Foundation, and Edgar Hewett, a southwesternist and Mayanist who in 1906 became director of American research with the AIA—regularly turned up at various meetings of the Knockers.

By at least the early 1920s, another group, the Anthropological Society of St. Louis, had been formed. It was organized "chiefly by members of the Medical School in Washington University" and had the purpose of "bringing together all the institutions in St. Louis interested in historical and archaeological work."[8] This was the group identified by the Committee on State Archaeological Surveys as its contact point in Missouri—a status reflected in the organization's listing in the "Notes on State Archaeological Surveys" (Wissler 1922). At least two members of the Knockers—George S. Mepham and J. M. Wulfing—are listed as members of the committee, along with Dr. Henry M. Whelpley, a pharmacist, and Dr. R. J. Terry and Dr. Charles H. Danforth, both of whom were on the staff of the Washington University School of Medicine. Whelpley, an avid artifact collector (Blake and Houser 1978), was one of the driving forces behind the local chapter of the AIA and had long advocated a statewide survey of Missouri (Pool 1989). Thus the group that was put together to represent the state to the CSAS might have been short on professional archaeological expertise, but few states had a more respected body to coordinate such a survey than Missouri.

Wissler, as he would do with other state groups, offered the St. Louis group advice on what the goals of a statewide survey should be. In a long letter to Danforth, secretary of the St. Louis society, Wissler noted that

> As I see the problem in such State surveys, there are, in the main, two alternatives: First, to project a rapid comprehensive survey of the State to be carried out in a year or two and to be supported by a specific appropriation. Second, the inauguration of a modest program under the auspices of a society or some existing State agency which can be counted upon to continue the work of the survey indefinitely.
>
> To my way of thinking, the second is preferable. For one thing, an appropriation of sufficient magnitude for quick comprehensive survey is not likely to materialize. On the other hand, if fortune did favor us and the appropriation were made by the State, we could not look forward with confidence to a continuation of the work in the future.[9]

Wissler also made sure Danforth understood that support for the state survey "must necessarily be found within the State concerned. No direct financial support from the outside can be expected." Apparently Wissler's warning did not dampen the St. Louis group's enthusiasm, because Danforth soon wrote him back, noting that "Several hundred dollars are already in sight and it is proposed to make an immediate (next spring) survey of St. Louis County and, if possible, inspire simultaneous surveys of several other favorable counties."[10] Despite this enthusiasm, there is no evidence that the planned surveys were ever carried out, and by 1928 the Anthropological Society of St. Louis was listed as "inactive."[11]

Annual Reports of the State Committees

Yearly reports of state surveys appeared in the *American Anthropologist* from 1922 to 1934, at which point that journal decided to stop carrying them. The reports then moved to the newly created journal *American Antiquity*.[12] Some of the later summaries also appeared in various volumes of the *Pan American Union Bulletin*. As Guthe (1930b:56) pointed out, the initial survey summaries were so well received that

> the Committee several years later sought and secured the cooperation of those institutions which were not State agencies, but were likewise engaged in archaeological research. Their cordial response made possible the expansion of these summaries to record nearly all of the archaeological fieldwork in North America. That this aspect of the Committee's activities has been a popular one is evidenced by the growth of the summaries. That for the year 1921 contains reports from thirteen State agencies [Wissler 1922], that for the year 1924, the first of the expanded summaries, gives reports from eleven State agencies and eight other institutions [Kidder 1925], that

for 1927 reports on the work of nineteen State agencies, and thirteen other institutions [Guthe 1928]. Today the Committee conducts a correspondence with representatives of fifty-one institutions.

Summaries were prepared by contributing correspondents from whichever state organization was sponsoring the work. Twelve states—Alabama, Arizona, California, Colorado, Illinois, Indiana, Kansas, Nebraska, New York, Ohio, Tennessee, and Wisconsin—plus New England were represented in the first set of summaries, which were for work carried out in 1921. If one were to read only what was published, it might sound as if the CSAS was making significant inroads into establishing statewide surveys across the Midwest and East, but this was not the case. For example, although Nebraska and Kansas submitted summaries for 1921, Wissler made it clear in a letter to C. E. Seashore, who had succeeded Wissler as chairman of the Division of Anthropology and Psychology, that all was not well in those states: "The situation in Nebraska is such as to render inadvisable any effort to launch a survey. The organizations that should be interested in the project are not working in harmony, chiefly because of the questionable scientific character of some of the men.... There is considerable interest in the subject in Kansas, but no live leadership at present."[13]

As might be expected given the long history of work in those states, the 1921 summaries from New York, Ohio, and Wisconsin were the lengthiest. Ohio's mounds had long figured prominently in Americanist archaeology, receiving in-depth treatment early in the nineteenth century, especially through the work of Ephraim G. Squier and Edwin H. Davis (1848), and Frederic Ward Putnam (e.g., 1887), and continuing through the opening decades of the twentieth century with the work of Warren K. Moorehead (1892a, 1892b, 1897, 1899, 1922) of Phillips Academy in Andover, Massachusetts, and Henry C. Shetrone (1920) and William C. Mills (1906, 1907) of the Ohio State Archaeological and Historical Society. Likewise, the mounds and petroglyphs of Wisconsin had been well documented through the work of Theodore H. Lewis (e.g., 1883) and others.

The most interesting of the reports published in 1922 is the one on New York, written by Arthur C. Parker (Figure 3) of the New York State Museum and the man who would become, in 1935, the first president of the Society for American Archaeology. Parker's paternal great-grandfather was a Seneca, and Parker himself had been president of the Society of American Indians in 1914–1915. The New York survey began in 1905, and the

Figure 3. Arthur C. Parker, director of archaeology at the New York State Museum and first president of the Society for American Archaeology, ca. 1950. (Reproduced by permission of the Society for American Antiquity from *American Antiquity* 21, no. 3 [1956])

office of the Archaeologist of the State Museum was created a year later, with Parker as the director. We say the New York report is the most interesting of the 13 that were published that year for several reasons. First, Parker made the objectives of the survey clear. Although the operation was referred to as a "survey," the term was not used in exactly the same way as it is today. Whereas modern archaeologists tend to think of site survey as an analytical exercise in its own right, prehistorians in the early part of the twentieth century saw it almost solely as an immediate prelude to excavation. Granted, in some states preservation efforts grew out of surveys—Mills made this clear in his 1921 summary report on Ohio, as Charles E. Brown did in his report on Wisconsin—but it seems undeniable that many statewide surveys were little more than prospecting exercises.

This rationale is apparent in Parker's three-part plan for New York: "The Museum began by exploring and excavating important sites without regard to culture. If a site seemed of special interest and likely to yield information and artifacts it received the attention of the season" (Wissler 1922:239). The CSAS had faced this problem early on, as Guthe (1930b:55) later noted: "The first problem before the Committee was that of defining what constituted an archaeological survey. It is immediately apparent that a survey consists of exploration and excavation." But there was more to it than that, and Guthe (1930b:55) summarized what had been the position of the committee from the start:

> Since it is always advisable to take a careful inventory of the assets of a given project before paying special attention to a detailed aspect of it, and since the scientific excavation of an archaeological site requires at least the supervision of a technically trained man, the Committee has always recommended, upon the inauguration of a survey, that emphasis be placed on a somewhat detailed exploration of the State, covering a period of several years, if necessary, before a program of excavation is undertaken.
>
> The most patent aspect of such an exploratory survey is a compilation and description of the many kinds of archaeological sites found in the State. . . . With this must be coupled an examination of the literature of the subject. . . . Moreover, each State contains a number of amateurs, who have become interested through the discovery of "Indian relics" in their immediate vicinity. The director of the State survey must not overlook the latent possibilities of these enthusiasts, but must become acquainted with them, enlist their support, and record and evaluate the material found in their private collections.

It was a good plan, but by the end of the decade it was becoming obvious to the CSAS that the real interest of nonprofessionals lay in excavation—a phenomenon not uncommon in modern times. Had this not been the case, there probably would have been no need for the NRC-sponsored Conference on Midwestern Archaeology in 1929.

Returning to Parker's summary of activities in New York, the second reason for our interest in the report resides in his third paragraph, where he laid out a suspected chronological ordering of what he termed "occupations" (Wissler 1922:239–240):

> As a result of this work the survey has determined the general localities and the chief characteristics of several occupations. The latest is the historic Iroquois in central and western New York and the Algonquian along the coast. Using these as datum we have been able to chart the successive oc-

cupations of the several areas within the State. By general areas, these are broadly as follows. (In reading these lists note that the higher the number the earlier the date.)

Western New York
1. Historic Iroquois (Seneca), tributary Algonquian peoples
2. Seneca and others who followed the Erie and Neutral
3. Erie, Neutral, Seneca, Iroquoian indeterminate
4. Algonquian, various tribes
5. Earth-work builders with pottery between Algonquian and Iroquoian
6. Mound-Builder-like sites
7. Algonquian (?)
8. Early Algonquian (?)
9. Indeterminate

Central New York (south to the Pennsylvania line)
1. Historic Cayuga, Onondaga, and Oneida
2. Andaste in the south along the Susquehanna and tributaries
3. Algonquian about the Finger Lakes
4. Mound-Builder-like
5. Algonquian
6. Early Algonquian
7. Algonquian (?)
8. Eskimoan (?)
9. "Red Paint" (?)
10. Indeterminate

Northern New York and Mohawk Valley
1. Iroquoian (in Jefferson County, early Onondaga)
2. Algonquian
3. Early Algonquian
4. "Red Paint" (?)
5. Eskimoan (?)
6. Indeterminate

(Contemporaneous with 3, in the Mohawk Valley there were "stone grave" people.)

Southern New York and Coast
1. Algonquian tribes
2. Iroquoian influence
3. Pre-Colonial Algonquian (Iroquoian traces)

4. Early Algonquian, certain Eskimo-like traces (?)
5. Indeterminate

Questions of chronology were not new in 1921, although it had not been that long since Americanist archaeologists began to think seriously about how to measure the passage of time in anything more than crude fashion. William Henry Holmes and his colleagues at the Bureau of American Ethnology had finally succeeded in demonstrating that purported evidence of glacial-age humans in North America was suspect (e.g., Holmes 1892, 1893a, 1893b, 1897; Hrdlička 1907, 1918), and for several years the notion of a relatively shallow time depth to the archaeological record was more or less axiomatic (Meltzer 1983, 1985). When Parker penned his summary of occupations in New York, the landmark chronological work undertaken in the Southwest by Nels Nelson (1916), A. V. Kidder (1916; Kidder and Kidder 1917), A. L. Kroeber (1916a, 1916b), and Leslie Spier (1917) was no more than five years old (Lyman et al. 1997; O'Brien and Lyman 1999a). This work had little effect on efforts in the East, although some archaeologists working there (e.g., Mills 1907) were beginning to wonder if there was more time depth to the archaeological record than had previously been proposed. What makes Parker's scheme so remarkable is that it implies *considerable* time depth to the archaeological record of New York.

Do not be misled into thinking that the perceived shallowness to the archaeological record in the East means that archaeologists were not interested in marking the passage of time. Simply because there was no incontrovertible evidence of glacial-age humans in the East did not imply that there was *no* time depth, and eastern archaeologists were interested in measuring whatever time depth there was. They knew very well the law of superposition—that artifacts at the bottom of a stratigraphic sequence were deposited before those on top—and they often used that positioning as a proxy for age differences among sets of artifacts recovered from different vertical positions (Lyman and O'Brien 1999). The problem was in figuring out how *much* older one set of artifacts was than another. For example, H. C. Mercer, a Harvard-trained curator of archaeology at the University of Pennsylvania Museum, excavated a trench through a mound adjacent to the Delaware River in New Jersey and upon making some postfacto stratigraphic observations concluded that the mound was stratified and contained the remains of "two village sites, set one upon the other,—an upper and a lower" (Mercer 1897:72). Mercer knew the superposed "village sites" were different in age, but he lamented that "the upper site might have been inhabited one or five hundred years after the lower was overwhelmed. If, therefore, we sought for inference as to the relative age of the two sites, we could only hope to find it in a comparison of the relics discovered. Realizing this, the depth, position, and association of all the specimens found, and particularly their occurrence above or below the lines of stratification, was carefully noted" (Mercer 1897:74). As we point out later, this early interest in stratigraphic relations preceded the so-called stratigraphic revolution in the Southwest (e.g., Browman and Givens 1996) by some 20 years and can be traced back to Frederic Ward Putnam, second director of the Peabody Museum (Harvard), and the training he provided students and staff.

Carl E. Guthe: The Quintessential Committee Chairman

By the summer of 1922, local committees in several states were so well run that the national committee discontinued its official connection with them (Guthe 1930b:53). At the same time, several midwestern states were asking for support, and in 1923 Wissler's committee began developing a plan for surveys in the Mississippi Valley, which led to the circulation of a pamphlet on suggestions regarding the aims and methods of statewide surveys (Wissler et al. 1923; see Appendix 1). Costs associated with the production of the pamphlet were subsidized by the State Historical Society of Iowa.[14] Interestingly, the secondary message of the pamphlet was that all work should be done by or under the supervision of professionally trained individuals, a foreshadowing of the tone of the 1929 Conference on Midwestern Archaeology.

Wissler retired as chairman in July 1924 and was replaced by A. V. Kidder, a Harvard-trained archaeologist working in the Southwest under the aegis of Phillips Academy in Andover, Massachusetts. The CSAS was again enlarged and consisted of holdovers Wissler, Roland B. Dixon, Frederick W. Hodge, Amos W. Butler, Marshall Saville, Charles E. Brown, and Peter A. Brannon, and new members W. C. Mills of the Ohio State Archaeological and Historical Society, Henry M. Whelpley of the Anthropological Society of St. Louis, and Charles R. Keyes of the State Historical Society of Iowa. Guthe (1930b:53–54) noted that during Kidder's service as chairman, the CSAS "continued to extend its contacts, particularly in the southern and western portions of the country and its function as an advisory board was thereby strengthened and expanded."[15] Under both Wissler's and Kidder's chairmanship the committee continued to hold formal and informal meetings—for example, at the American Association for the Advancement of Science meeting in Cincinnati in 1923 and at the Central Section (later the Central States Branch) of the American Anthropological Association meeting in Columbus, Ohio, in 1926—but it was at the 1927 Central Section meeting in Chicago that the CSAS took its boldest step to date, proposing that the NRC establish a cooperative laboratory for the study of pottery from the eastern United States.[16] The University of Michigan offered to maintain the laboratory in its Museum of Anthropology,[17] and it became known officially as the Ceramic Repository for the Eastern United States, with "Eastern" referring to anything east of the Rocky Mountains.

Kidder resigned his position as chairman of the CSAS in the fall of 1927, and the person who succeeded him was the same man who had responsibility for the daily operation of the Michigan repository, Carl E. Guthe. Under Guthe's chairmanship, which lasted until 1937, the committee stabilized and became the organizing force for which it had been designed. In fact, despite the rotating nature of NRC committee membership, the makeup of the CSAS was remarkably stable from 1924 on, with Brannon, Brown, Butler, and Keyes serving until the CSAS was abolished in 1937.

Guthe received his Ph.D at Harvard in 1917 and worked with another Harvard graduate, Frederick H. Sterns, in Nebraska in 1915 and then with Kidder at Pecos Pueblo, New Mexico, from 1916 until 1921 (Figure 4). He became associate director of the Pecos project in 1917. From 1920 to 1922 he also worked as a research associate for the Carnegie Institution's Tayasal, Guatemala, project that was directed by Sylvanus G. Morley (Griffin 1976b). Guthe joined the staff of the University of Michigan in 1922 and became associate director of the newly created Museum of Anthropology, eventually assuming the directorship in 1929. Given both his position and training, as well as his enthusiasm, Guthe was a natural choice to head the CSAS.

Guthe went to great lengths to increase the effectiveness of the committee in its relations with nonprofessionals, even taking an extended trip in the summer of 1928 to visit coordinating offices in 15 states in the Mississippi Valley. He later recalled, "Impressed by the attitudes and accomplishments of these earnest amateurs, I felt they deserved to be helped rather than censured" (Guthe 1967:434). Perhaps, but he did not mince words in the report that summarized what he found during his trip. For example, with respect to Arkansas he stated that archaeology there was

Figure 4. A. V. Kidder (*center*) with Alfred M. Tozzer (*left*) and Carl E. Guthe, Kidder's assistant at Pecos Pueblo, New Mexico, 1916. (From Woodbury 1973; photo courtesy Columbia University Press)

"the hobby of [Samuel C.] Dellinger, a biologist at the State University who has seen fit to leave our letters unanswered.... The 'Arkansas Museum of Natural History and Antiquities' is a newly formed group, with a big paper organization. The situation here is pathetic because of the well-intentioned but blissfully ignorant enthusiasm of the promoters. A quantity of extremely obvious frauds have been purchased by them."[18] Amusingly, in Mississippi he found "[t]wo inadequately trained young men... conducting excavations" for the director of the Mississippi Department of Archives and History. Those two "inadequately trained young men" were Moreau B. Chambers and James A. Ford (O'Brien and Lyman 1998, 1999b).

Despite the decade-long effort of the committee to foster cooperation among various state organizations and to channel local energies into less commercially motivated activities, the outlook was still bleak in 1929, as Guthe (1967:435) recalled almost four decades later:

> In 1929 ... archaeological explorations were under way in about half of the states of the Union, many of them carried out by lay students of the subject. The lack of communication between groups was enormous. State political boundaries served as corral fences, preventing archaeologists in one state from communicating with their colleagues in adjacent and neigh-

boring states. Nor were the channels of communication between the professional and the serious-minded laymen as broad and open as they should have been.

The professionals were outspoken in their condemnation of Indian-relic collectors and dealers who destroyed irreplaceable archaeological evidence. . . . Equally objectionable, because of the resulting destruction of evidence, were the activities of well-intentioned amateurs who did not understand the dangers of careless excavation and neglected to keep adequate records.

The only possible solution to the problem resided where it had for the previous decade: "the cultivation and friendly education of another type of amateur," namely, the "[s]erious-minded, thoughtful collectors, [who,] intrigued by the conditions and associations under which the relics were found, sought information on their origins and functions by consulting libraries, fellow collectors, and, when possible, professional archaeologists" (Guthe 1967:435). By 1929 this approach had paid dividends but certainly not big ones. How could the CSAS change the situation? The answer, it seemed, was to hold a large conference and attack the issue head on. Not simply a conference such as had been held at the annual meetings of the American Anthropological Association and the American Association for the Advancement of Science—those were attended only by professionals—but a large gathering of both amateurs and professionals, where the former could listen to recommendations offered by the latter, and the latter could listen to the concerns of the former.

This is how Knight Dunlap, chairman of the Division of Anthropology and Psychology from 1927 to 1929, pitched the conference to Edmund Day, director of the Laura Spelman Rockefeller Memorial, the foundation Dunlap approached for funding to offset the estimated $3,000–4,500 needed to host such a meeting:

> The Conference on American Archaeology seems to be the most important thing to be done for the anthropologists at the present time. Some of the mid-western states are "sold" on the idea of comparative work, and realize that institutions working, or wishing to work on their mounds, etc., do not wish to "rob" them, or to interfere with "States Rights." Other states are still on the defensive. It is believed that in this Conference the officials of states already favorable would help with the other states. . . . The most favorable place in which to call this Conference, seems at present to be Indianapolis.[19]

The Conference on Midwestern Archaeology

Given the model nature of the statewide survey of Indiana, its capital was a logical venue for such a meeting, but Indianapolis was passed over in favor of St. Louis. Fifty-three people, including 9 of the 11 members of the CSAS, attended the two-day conference, which was held at the Hotel Coronado on May 17–18, 1929. Among them were Dunlap; Henry S. Caulfield, governor of Missouri; W. E. Freeland, majority leader in the Missouri House of Representatives; G. R. Throop, chancellor of Washington University; Thomas M. Knapp, chancellor of St. Louis University; John C. Futrall, president of the University of Arkansas;[20] Rufus Dawes, president of the Chicago World's Fair Centennial Celebration; and William J. Cooper, U.S. commissioner of education. Those on the professional-anthropological side included Matthew W. Stirling, chief of the Bureau of American Ethnology; Fay-Cooper Cole of the University of Chicago, chairman of the Division of Anthropology and Psychology from 1929 to 1930; William S. Webb, head of the newly created Department of Anthropology and Archaeology at the University of Kentucky; Frans Blom, director of the Department of Middle American Research at Tulane University; J. Alden Mason of the University of Pennsylvania; and Clark Wissler and Nels Nelson of the American Museum of Natural History. Several state archaeologists and geologists also attended the meeting, including Calvin S. Brown, an archaeologist with the Mississippi Geological Survey, and M. M. Leighton, chief of the Illinois State Geological Survey.

Three fairly high-profile amateurs also attended—Don F. Dickson of Lewiston, Illinois; Harry J. Lemley of Hope, Arkansas; and Jay L. B. Taylor of Pineville, Missouri. Dickson had earned a reputation as a preservationist by erecting a structure over human skeletons he unearthed on his property in Fulton County, Illinois (Harn 1980), and Lemley was a collector of Caddoan artifacts, although he had contact with professionals throughout part of his life (O'Brien and Lyman 1998) and would go on to publish articles on his excavations (e.g., Lemley 1936; Lemley and Dickinson 1937). Taylor had assisted Warren K. Moorehead in his excavations at Cahokia, located across the Mississippi River from St. Louis in Collinsville, Illinois (Moorehead 1929a), and he knew Wissler and Nelson very well—in the case of Nelson all too well, for it was Nelson (1928) who in a very clear and concise argument shredded Taylor's (1921a, 1921b) claims of authenticity of a bone with an engraving of a mastodon that

Taylor had ostensibly found in a cave in southwestern Missouri (O'Brien 1996).

The conference consisted of three parts: (1) an open meeting of the CSAS on Friday morning, followed by a trip to Cahokia mounds guided by Moorehead and an evening lecture by Henry C. Shetrone, director of the Ohio State Museum; (2) the main conference on Saturday morning and afternoon; and (3) Saturday evening dinner and presentations, which were broadcast on radio station KMOX. The conference proceedings reveal the striking disparity of topics that were addressed. As one might expect, given the ink that had been spilled up to that point, numerous presenters, from Governor Caulfield on down, spoke of preserving archaeological sites for the future. There were polemical statements on the need for preservation, which is not unexpected given the political nature of the meeting and the fact that some of the presentations were being broadcast to the public, but there also were presentations that dealt with specific advantages that accrued from preservation, such as increased tourism. Several presenters excoriated vandals and relic hunters for the catastrophic damage done to an irreplaceable resource—exactly the problem the committee had been working for a decade to solve, with little visible success. Arthur C. Parker made persuasive arguments in this direction in his paper, "The Value to the State of Archaeological Surveys." He laid out four reasons for surveying and preserving archaeological remains:

> 1. *Archaeology explains the prehistory of the state.*—The recoveries from ancient sites constitute visual exhibits of the people who occupied the state before the coming of a population of European origin....
>
> 2. *Archaeological remains constitute a vast reservoir of valuable knowledge.*—Judged by every moral standard the state is bound to conserve and protect its resources. The aboriginal sites within each state constitute unique and fundamental sources of archaeological facts, highly valued by the scientific world....
>
> 3. *Archaeological remains are monumental exhibits.*—The marking of prehistoric Indian sites and their protection from promiscuous digging would not only attract the attention of the sight-seeing public, but would stimulate the investigation by scientists....
>
> 4. *Archaeological collections are exhibits of lasting worth.*—Wherever archaeological collections have been made by trained students of prehistory the resulting exhibits and publications describing them have constituted genuine contributions to knowledge. (Parker 1929:33–34)

Parker railed against unskilled collectors and the effect they were having on the archaeological record:

> The relic-hunter digs only to destroy and his recoveries are often abortive things with undetermined parentage.... Whether the relic-hunter will continue to ruin the field, or whether state-supported agencies shall preserve the field and draw from it the information that an enlightened age demands, depends very largely upon the citizens of each state; but it depends most of all upon how thoroughly archaeologists who understand the importance of their quest are able to present it to the public. Archaeology must advertise and it must seek thereby to stimulate such a desire to know more of prehistory that support will follow. (Parker 1929:37–38)

It was one thing to say that states should take control of preserving their archaeological resources, but there was a catch, and Parker knew it: *Which* organization within a state was best suited to carry out a survey and to spearhead preservation efforts? Parker (1929:34) pointed out that it "matters little what institution or agency promotes the survey so long as its operating force is composed of trained archaeologists familiar with the problems to be met or capable of meeting these problems when they occur." To him the ideal institution, "other things being equal, is a state museum, for then there will be a centralized repository for the specimens, and at least a certain amount of clerical and professional help." He then noted—an understatement if there ever was one—that "A specially constituted commission cooperating with local groups may have difficulty in meeting the problem of distributing the recoveries, especially when it has invited the aid of numerous local historical and scientific societies" (Parker 1929:34–35). In other words, if a loose amalgam of persons constitutes the committee, how are they going to maintain control of the artifacts that result from field exercises, especially when their field crews consist of collectors? Even when a solid organization such as a state museum acts as the coordinating body, local organizations and municipalities will want to maintain control over artifacts, and, as Parker noted, the organizing body is going to have to educate them about the dangers in doing so. Modern readers may be struck by how little things have changed since 1929 when it comes to civic pride and private ownership of artifacts.

Although the topic of preservation and statewide committees dominated the Conference on Midwestern Archaeology, close reading of the proceedings turns up a few passages that give us some idea of the state of archaeological method in 1929 and that were preludes to topics that dominated much of the NRC-sponsored conference in Birmingham, Alabama, in 1932. Three papers and a prepared set of remarks on one of those papers furnish useful examples.

Emerson F. Greenman on Artifact Classification

The title of the paper by Emerson F. Greenman, curator of archaeology at the Ohio State Archaeological and Historical Society, was "A Form for Collection Inventories." Greenman received his Ph.D. from the University of Michigan in 1927 and had worked in Guthe's Museum of Anthropology; thus he was no novice when it came to artifact analysis. He began his paper by noting that "In view of the increasing activity in state archaeological survey work, some attempt should be made to bring about uniformity in the use of terms, and in the methods of describing archaeological objects, in order that the work done in one state may be compared with that in adjoining states. . . . Distributions [of artifacts] common to more than one state can only be worked out by the use of a uniform terminology" (Greenman 1929:82–83).

The classification system Greenman (1929:83–84) proposed was fairly rigorous and obviously had been well thought out. The system revolved around the identification of types, which Greenman defined as "the frequent linking together of a number of features on the same specimen" (83). He identified 11 projectile-point characteristics useful for defining types and 21 other characteristics as providing a means of narrowing the type definitions. He also listed four other sets of characteristics—those related to the overall shape of a projectile point—which were used as initial sorting criteria. Thus a point could have wide, shallow notches, or it could be angular- or side-notched or wide-stemmed. Greenman even devised a shorthand notation for his system; in our example just cited, such a specimen would be listed as an A/A, 5, 13. In several respects the system exhibited characteristics of some modern approaches to classification, including paradigmatic classification (Dunnell 1971; O'Brien and Lyman 2000).

Classification theoretically serves two functions—to structure observations so that they can be explained and to provide a set of terminological conventions that allows communication. In the United States, early classification systems were developed solely as a way to enhance communication between researchers who had multiple specimens they wanted to describe (see Dunnell's [1986:156–159] discussion of Rau [1876] and Wilson

[1899]). The "Report of the Committee on Archeological Nomenclature" (Wright et al. 1909), which was commissioned early in the twentieth century by the American Anthropological Association, exemplified this kind of approach. Since the intent of the persons devising the classification schemes was to standardize terminology, most systems were based on readily perceived differences and similarities among specimens. This meant that form received the greatest attention. However, despite the best efforts of the classifiers, form and function were often conflated. Certainly this was the case with the system devised by the Committee on Archaeological Nomenclature, headed by Charles Peabody of Phillips Academy. Despite the statement that "it has been the particular aim of the Committee to avoid or to get rid of those classes and names that are based on uses assumed but not universally proved for certain specimens" (Wright et al. 1909:114), many of the committee's unit names—such as vessels, knives, and projectile points—have functional connotations in English.

Piles of more or less similar-looking specimens that late-nineteenth- and early-twentieth-century classifiers were forever creating lacked any archaeological meaning: "In an effort to make categorization more systematic and scientific, these early workers had arbitrarily focused on formal criteria that lacked any archaeological or ethnographic rationale" (Dunnell 1986:159). Further, variation in artifact form within each pile—and to some extent between piles—of specimens had no perceived explanatory value and was simply conceived of as noise resulting from different levels of skill in manufacturing or from raw-material quality.

Greenman's scheme was different because he emphasized the identification of variation and established a concise set of criteria to be used in the identification. Using precise language, Greenman (1929:84) explained the rationale behind his classification system: "It is the intentional forms whose distributions are significant, and for that reason stress is laid upon the *types*." In other words, the classification system was developed to create groups that had spatial (and perhaps temporal) meaning; haphazard or idiosyncratic classification couldn't produce such groups. Greenman obviously believed that types were reflections of what the original makers of the projectile points had in mind—hence his use of the term "intentional forms." The epistemological significance of types would be an issue with which Americanist archaeologists would wrestle for decades after Greenman presented his system (e.g., Ford 1954a, 1954b, 1954c; Spaulding 1953, 1954a, 1954b).

Frederick W. Hodge and Warren K. Moorehead on Cultural Complexity

The paper by Frederick W. Hodge (Figure 5), which was read by Roland B. Dixon, was titled "The Importance of Systematic and Accurate Methods in Archaeological Investigation." It was a primer on select topics in archaeological method, including analytical uses to which certain artifacts can be put. For example, Hodge (1929:20) pointed out that pottery was "the most important means of cultural determination" available to the archaeologist. It was "the master-key, above everything else made by primitive man, to the determination of multiple occupancy through stratification, and by its usual fragile character it commonly did not find its way very far from the place of manufacture. It stands to reason therefore that it is of the greatest importance" (Hodge 1929:21). Being a product of the late nineteenth century and the stagelike evolutionism of Edward B. Tylor (1871), Lewis Henry Morgan (1877), and others, Hodge (1929:21) went on to note that "Not all Indians made pottery, to be sure, for some were low indeed in the culture scale, subsisting on the products afforded by a not too prodigal nature and making little in the way of utilitarian, ceremonial, or esthetic objects that have survived to the present time."

Warren K. Moorehead picked up on the notion of cultural complexity in his paper, noting that in the eastern United States there existed a large territory "in which mound art . . . is rather highly developed. Surrounding it in the greater area, mounds and their contents indicate less complex cultures" (Moorehead 1929b:74–75). Moorehead, whose view of archaeology was greatly colored by his work on Ohio mounds (Moorehead 1892a, 1892b, 1897) and by his ongoing work at Cahokia (Moorehead 1927, 1929a; see Kelly 2000), developed a 19-point scale for measuring the culture status of mound-building peoples. The "famous Hopewell culture of the lower Scioto valley [Ohio]" received 13 points, and "the high Etowah culture of north Georgia and of the Tennessee-Cumberland valleys of Tennessee" received 11 points. Fort Ancient—a term originally coined by W. C. Mills (1906) to refer to non-Hopewellian culture in southern Ohio and surrounding regions—was lower still. To Moorehead (1929b:75), Fort Ancient meant "neither high mound builder art nor yet an exceeding low status but might be roughly compared with the term middle class, commonly employed to differentiate the bulk of individuals from those who are extremely well to do or very poor." Illinois Hopewell

Figure 5. Frederick W. Hodge, at various times ethnologist-in-charge at the Bureau of American Ethnology, employee of the Museum of the American Indian, and director of the Southwest Museum in Los Angeles, ca. 1935. (Photo courtesy National Anthropological Archives, Smithsonian Institution)

groups fared less favorably, receiving eight points, but they outscored groups in southern Georgia and Florida, which received only four or five points, despite the fact that "there are an enormous number of shell mounds, platforms for houses or temples, and indications of a very heavy and industrious population" (Moorehead 1929b:75).

Moorehead (1929b:74) admitted that there was "overlap" between the "distinct mound builder cultures" and that archaeologists "have gone entirely too far in extending the boundaries of certain of these cultures." Related to this problem was the origin of the various mound-building groups, and in his paper Moorehead focused specifically on the southern-Ohio Hopewell. We bring up this topic because in the early 1930s it would consume the attention of several archaeologists working in the lower Mississippi River valley, in particular Frank M. Setzler and James A. Ford.

Moorehead, never one to pass up an opportunity to engage in fanciful flights of fantasy, believed that southern-Ohio Hopewell peoples originated in eastern Iowa and at some point migrated eastward. On reaching the Scioto River valley, "where conditions were extremely favorable for their development, they remained, became sedentary, and attained the culmination of their wonderful development" (Moorehead 1929b:77). He indicated that trade items found at Ohio Hopewell sites were evidence that the Hopewellians had a knowledge of the South, but his objection to a southern point of origin of Hopewell was that the Ohio mounds did not contain the kind of ceramic art that was so prominent in the South. Moorehead was apparently unfamiliar with Gerard Fowke's (1928) excavations at the Marksville site in Avoyelles Parish, Louisiana, where he recovered several vessels similar in form and design to vessels from Ohio Hopewell sites. Ironically, Fowke himself failed to note the similarities, even though he had spent considerable time working in Ohio. The similarities, however, would not be lost on Setzler, who in 1933 began a reexamination of the Marksville site. In assessing the resemblances between vessels from Marksville and those from southern Ohio, Setzler (1933a, 1933b) came down decidedly on a south-to-north migration of Hopewell peoples. Setzler would have more to say about this at the Indianapolis conference in 1935. Ford, Setzler's field assistant at Marksville, would have *much* more to say on the subject two decades later (Ford et al. 1955; Ford and Webb 1956).

Matthew W. Stirling and Historical Continuity

In our opinion, the most interesting remarks made at the St. Louis meeting were not in a prepared paper but in comments made by Matthew Stirling (Figure 6) in his discussion of Hodge's presentation. Stirling received his undergraduate degree from the University of California in 1920 and his master's degree from George Washington University in 1922. He joined the U.S. National Museum in 1921, and in 1928 was named chief of the Bureau of American Ethnology. We focus specifically on two points he made, each of which symbolizes where Americanist archaeology was headed in the late 1920s. First, Stirling (1929a:28) noted that "One cannot be a competent archaeologist without ethnological training. Archaeology is not merely a matter of digging and careful observation, but it requires an ability to interpret these observations accurately." This sentiment was not something that Stirling alone felt but rather was an implicit notion that had been present from the earliest days of Americanist archaeology.

Figure 6. Matthew W. Stirling alongside Olmec colossal head 4 at La Venta, Mexico, ca. 1940.

In the United States, degrees were not granted in archaeology but in anthropology—a phenomenon that holds true today. Any professional archaeologist in attendance at St. Louis would probably have agreed with Stirling's remarks, having spent several years taking courses in general ethnology as well as courses focused on the ethnology of particular groups or regions. As we discuss elsewhere (e.g., Lyman et al. 1997; O'Brien and Lyman 1998, 1999c), much of what passed as archaeological theory during the culture-history period was grounded in ethnological theory. Thus the archaeological record was viewed in ethnological terms, and it became commonplace to equate such things as artifact assemblages with particular "cultures."

The second point Stirling made was related to the first, and it concerned the tracking of ethnohistorically known groups back in time. Stirling (1929a:25) saw two extremes in archaeology: "On the one hand is the tying up of archaeological research with the historical period concerning which we have definite information, and on the other hand the projecting of it backwards to that period of which we may be able definitely to say that there was no human occupancy of this continent." There was, however, a means of linking these two extremes, and Stirling (1929a:25) laid it out in clear terms for his audience:

> It is possible to determine rather definitely the dates of the introduction of certain types of articles of European manufacture which may have been found in an archaeological site. We know when and where certain varieties of trade beads were made; we know rather definitely the period during which certain smoking pipes were manufactured and introduced as trade articles among the Indians, and there are innumerable other examples of the same sort which may aid greatly in giving us something definite from which to project backwards a chronological sequence.

Why, Stirling asked, should an archaeologist be depressed upon discovering a silver ornament or a string of glass beads alongside articles of native origin? To the contrary, "There is no justification for such a reaction, and in most instances the archaeologist should feel rather a sense of elation. Where an association of this sort is discovered it becomes possible by a process of overlapping to carry a native culture throughout its successive stages of development well back into the prehistoric period" (Stirling 1929a:25).

Stirling was advocating what his Smithsonian colleague Waldo Wedel (1938) would refer to a decade later as the *direct historical approach*. No one can legitimately argue with the logic of the approach, which was not new in the 1930s but, as we discuss in more detail in the next section, had been the strategy adopted in the 1880s by John Wesley Powell and Cyrus Thomas (1894) for the Division of Mound Exploration in its quest to destroy the myth that a race of people separate from Native Americans had constructed the thousands of mounds evident across the eastern United States: First, document similarities in cultural materials between those evident from ethnographic and ethnohistorical research and those evident archaeologically. Second, assume similar materials are temporally and phyletically related and construct a continuous thread, or cultural lineage, from the past to the present (Lyman and O'Brien 2000; O'Brien and Lyman 1999a). Roland B. Dixon (1913) had espoused just such a strategy in his presidential address to the American Anthropological Association in 1913.

In "An Introduction to Nebraska Archeology," Bureau of American Ethnology archaeologist William Duncan Strong (1935; see also Strong

1936), who received his Ph.D. under A. L. Kroeber at the University of California in 1926, noted the importance of the direct historical approach:

> It is the firm belief of the author that the possibilities of historic archeology in North America are not fully realized by the majority of anthropologists at the present time. . . . It seems surprising, therefore, that even today there are archaeologists more interested in segregating obscure early cultures of unknown periods and affiliations than they are in determining the historic cultures and sequences represented in the regions to be worked. Obviously, in such work the historic cultures need not be an end in themselves, but they do seem to represent the threads that give most promise of untangling the complex skein of prehistory. (Strong 1935:296)

There are two critical aspects of the direct historical approach. First, it provides "a fixed datum point to which sequences may be tied" (Steward 1942:337); that is, it provides a chronological anchor in the historical period to which archaeological materials of otherwise unknown relative age can be linked. Second, the more similar prehistoric materials are to the historically documented materials, the more recent they are; conversely, materials that are less similar to historically documented materials come from further back in time. Thus the direct historical approach demands the study of homologous similarity, a point generally unrecognized at the time (Lyman and O'Brien 1997, 2000). Without a chronological anchor, sequences cannot be established, and assemblages of artifacts have the unsavory characteristic of floating in time and thus being of minimal utility in determining the development of historically documented cultures. This is the point Stirling was making in his comments on Hodge's paper, and it was the same point made by Neil Judd, curator of archaeology in the U.S. National Museum, in a paper published in the *American Anthropologist* that same year. Judd (1929) lamented that archaeologists knew little about the late prehistoric remains of more than 200 historically known tribes and noted that a "relative chronology for each culture area is one of the surpassing needs of archaeology in the United States today" (Judd 1929:418). As we will see, the NRC-sponsored Conference on Southern Pre-History addressed this issue head on.

The Bureau of American Ethnology

In closing the St. Louis meeting, Fay-Cooper Cole (1929:112) expressed the feeling that "we will all leave here, much more assured of the future of archaeology than when we came here two days ago." There may have been some reason for such optimism, but it is apparent that the field was still plagued with difficulties. Nowhere was this more apparent than in the Southeast, where state and local institutions were for the most part working in an intellectual vacuum. Compared to the Midwest and Northeast, there was a dearth of trained archaeologists, which meant there was little or no hope of introducing current methodological advances to the amateur societies that seemed to crop up everywhere. This situation did not escape the notice of professional archaeologists working in other regions, and it was the major reason the Conference on Southern Pre-History was held in Birmingham, Alabama, late in 1932. Although the CSAS was instrumental in organizing that conference, we need to look a little deeper at the few professionals who were working in the Southeast just prior to that meeting to determine what their influence on the field was. Stirling's closing comments at the conference in St. Louis provide a starting point for examining that topic. After mentioning the myriad issues that participants had addressed, Stirling (1929b:109–112) added,

> there is one topic on which I might profitably add a few words, and that is something concerning the history and the nature of the institutions which I represent: The Smithsonian Institution and the Bureau of American Ethnology, which is a part of that great institution. . . .
>
> The Bureau of American Ethnology at the present time has, among its duties, not only the pursuit of field work in various parts of the country, but it has also become, in a way, a court of appeal for the population throughout the country who are interested in matters pertaining to anthropology. . . .
>
> There is probably no organization in the country that has published as many pages or as many volumes dealing with the American Indian and with the subject of Archaeology as has our Bureau. . . . We stand ready to assist at any time, to the best of our ability, any of you who are interested or professionally engaged in the study of archaeology.

This assistance showed up in a significant way in Birmingham just a few days before Christmas 1932. Although the conference was attended by professional archaeologists from a number of institutions, it was personnel from the Bureau of American Ethnology and its sister institution, the U.S. National Museum—individuals who, as Stirling put it, stood "ready to assist at any time"—who had by far the most impact on the group.

The Bureau of American Ethnology was founded in 1879, and its involvement in the Southeast dates to the formation of the Division of

Mound Exploration within the bureau in 1881 and the mandate that bureau director John Wesley Powell received from Congress to decisively answer the question of which group or groups constructed the thousands of earthen mounds so evident across the eastern United States. By the time he was appointed to head the Division of Mound Exploration, Cyrus Thomas, like most other prehistorians, was convinced of the equation of the mound builders with the American Indians. For example, in 1884 he asked and then answered the question, "'Who were the mound-builders?' We answer unhesitatingly, Indians—the ancestors of some, perhaps of several of the tribes of modern or historic times" (Thomas 1884:90). Thomas published the "Report on the Mound Explorations of the Bureau of Ethnology" in the *Twelfth Annual Report of the Bureau of Ethnology, 1890–1891* (Thomas 1894), and in it he discussed in detail the mound explorations carried out by members of his crews as they worked their way over two dozen eastern states, including all the southern states, with the exception of Texas and Virginia.

Thomas continually referred to historical records of the sixteenth through eighteenth centuries, where it was documented that the post-Columbian Indians were sufficiently "culturally advanced" (being sedentary agriculturists) to have built the mounds. In some cases Indians had actually been observed building them. Documenting typological similarity of artifacts from the historical and prehistoric periods (e.g., Holmes 1886, 1903) merely completed the evolutionary, ethnic, and cultural linkages on which the direct historical approach was founded (Meltzer and Dunnell 1992; O'Brien and Lyman 1999c). Thomas noted that there was no logical reason to suspect that the mound builders were of Mexican origin or that later Indian groups had pushed the mound builders south into Mexico. In other words, the archaeological record demonstrated to Thomas's satisfaction that a high degree of cultural continuity had existed for an untold number of millennia and that such threads of continuity showed no major disruptions. Undoubtedly, change had occurred—that much was indicated by the myriad forms of earthworks recorded and the different kinds of artifacts found within them—but such change was an orderly, continuous progression as opposed to a punctuated, disruptive progression of cultural epochs such as was evident in the European Paleolithic-Neolithic sequence (Lyman and O'Brien 1999; Meltzer 1983, 1985). To Thomas, continuity had ruled throughout human tenure in the East, and it is clear that he favored tribal differences to explain the immense variation evident in the archaeological record.

Jesse Jennings (1974:39) once noted that the publication of Thomas's report could be thought of as "marking the birth of modern American archeology," although as we have noted elsewhere (O'Brien and Lyman 1999c) we consider this to be an overstatement. Prior to the founding of the Division of Mound Exploration in 1881, archaeology was primarily an antiquarian activity, meaning that interest centered on artifacts and earthen monuments themselves rather than on using such things as a means to other ends. The work summarized by Thomas (1894) was superior in many ways to what had come before, primarily because he demanded rigor in how materials and information were gathered (Smith 1990), but it was not particularly revolutionary. Further, to use 1894, the date of publication of Thomas's final report, as marking the birth of modern American archaeology overlooks the excellent work done by Harvard's Frederic Ward Putnam and those he trained.

The case can be made that it was through Putnam's example (e.g., Putnam 1887), and certainly through the training he provided, that Americanist archaeologists began excavating stratigraphically and keeping track of artifacts by stratum. By the time the so-called stratigraphic revolution (Browman and Givens 1996; Willey and Sabloff 1993) occurred in New Mexico a decade and a half into the twentieth century (e.g., Nelson 1916, Kidder 1916), those trained or influenced by Putnam—Henry Mercer and Charles Peabody, for example—had been digging stratigraphically in the East since the late 1800s (Browman 2000; Lyman and O'Brien 1999; O'Brien and Lyman 2000). Call him what you will—the "Father of American Anthropology" (Phillips 1973), the "father of American archaeology" (Dexter 1966), or the "professionalizer of American archaeology" (Mark 1980; Willey and Sabloff 1993)—Putnam played as large a role in the birth and subsequent growth of Americanist archaeology as Cyrus Thomas did.[21]

With the death of the mound-builder myth in the closing decade of the nineteenth century, Bureau of American Ethnology archaeologists turned their attention to other matters, some of which had been of considerable concern to them for some time. The one that has received the lion's share of attention from historians of archaeology (e.g., Meltzer 1983, 1985) was the great debate over the antiquity of humans in North America. Southeastern prehistorians, with rare exceptions, did not figure into this debate, but they were active nonetheless, and their activities did not go unnoticed. Over time, both the Bureau of American Ethnology and the National Museum began turning their attention to the Southeast as their interest became piqued by what prehistorians were uncovering there. One such individual was Clarence B. Moore, yet another Harvard-trained prehistorian, who spent a quarter of a century, from roughly 1892 to 1917, exploring mounds along the major waterways of the southern states, in the process excavating several thousand skeletons and recovering countless ceramic vessels and other artifacts. Although his work was not sponsored by a federal agency, it would be important background material for research by later archaeologists. He underwrote not only the costs of his projects but also the expense of producing 20 reports dealing with the excavations, which appeared in the *Journal of the Academy of Natural Sciences of Philadelphia*. The reports are rather sketchy, but the accompanying field photographs and artifact illustrations are excellent. Moore's work (e.g., 1892, 1894, 1896, 1902, 1905, 1907, 1908, 1909, 1910, 1911, 1912, 1913), and especially the artifacts it produced, spurred a resurgence of interest in the Southeast, especially by small state organizations and regional museums—precisely the groups at which Matthew Stirling and the Bureau of American Ethnology took aim in the early 1930s.

The Conference on Southern Pre-History

It was into these intellectually rather shallow waters of southeastern archaeology that the CSAS waded in 1932 when it hosted its second regional meeting designed to facilitate communication among archaeologists.[22] Organizers, again led by Carl Guthe, were careful not to give the impression that a group of outsiders, all from the North, was telling southerners not only how to do archaeology but also how to organize a meeting. Neil Judd expressed this concern to Guthe in a letter written in September 1932: "As you well know, the South is most conservative and sectional in its attitude; in general it resents northern advice and aid however altruistic" (cited in Lyon 1996:54).

The three-day conference, which was, as Jon Gibson (1982:258) pointed out, "without doubt one of the most influential professional meetings ever held on Southeastern archaeology," convened at the Hotel Tutwiler in Birmingham, Alabama, on December 18, 1932. The report that was issued after the meeting[23] carried the text of the papers presented, along with comments made by session chairmen. The report makes it obvious that Guthe took Judd's concern seriously when he drew up the program, because although the major papers were by nationally recognized archaeologists and anthropologists from northern institutions—in addition to Guthe, Judd, Wissler, Cole, and Moorehead, presenters were Ralph Linton of the University of Wisconsin and John R. Swanton, Matthew W. Stirling, and William Duncan Strong of the Bureau of American Ethnology—their papers were interspersed among summaries of the archaeological records of individual states, presented for the most part by southern prehistorians familiar with those records. Peter A. Brannon of the Alabama Anthropological Society and long-term member of the CSAS[24] chaired the session "Recent Field Work in Southern Archaeology," in which Samuel C. Dellinger of the University of Arkansas spoke on Arkansas, Walter B. Jones of the Alabama Museum of Natural History spoke on Moundville cultures, Charles K. Peacock of the East Tennessee Archaeological Society spoke on Tennessee, and James E. Pearce of the University of Texas spoke on eastern Texas. In addition, Winslow Walker of the Bureau of American Ethnology spoke on Louisiana, and Henry B. Collins of the National Museum spoke on Mississippi.

Although it carried no byline, the short introduction to the conference volume was authored by Guthe. In it he stated the purpose of the conference:

> The Conference on Southern Pre-History ... was called for the purposes of reviewing the available information on the pre-history of the southeastern states, discussing the best methods of approach to archaeology in this region, and to its general problems, and the developing of closer cooperation through the personal contacts of the members of the conference. During the past few years, the interest in Indian pre-history of the lower Mississippi Valley and the southern Atlantic states has been increasing steadily, and a number of institutions have undertaken research work in this field. Developments from studies of the same period in the northern part of the Mississippi Valley and from work on certain Southwestern problems indicate that as the knowledge of the pre-historic cultures of the southeast increases, the problems of the neighboring areas will be more clearly understood. It was for the purpose of fostering more rapid increase of this knowledge that this conference of experts in the study of pre-history from all over the United States was called to meet with interested students of the South. (Guthe 1932b:1)

Guthe selected his words carefully because he was really saying that nowhere in the Southeast were approaches that were routinely employed in the Southwest being incorporated into fieldwork and analysis. Part of the problem lay in the attraction the Southwest had long held for prehistorians—archaeological brainpower had been drained into that region at the expense of other regions (O'Brien and Lyman 1999c)—and part of it lay in the fact that southern universities were not producing students trained in archaeology. In states such as Alabama the majority of work was undertaken by museums, often in conjunction with local archaeological societies. In other states, amateur-based societies were left to their own devices. In some cases the quality of work was credible for the time period, but in others it was deplorable.

In language a bit stronger than Guthe's, Collins, who was then assistant curator in the ethnology division of the National Museum, summed up the state of affairs in the Southeast. He was speaking specifically of one state, but his remarks were applicable to the region as a whole: "Although Mississippi is rich in aboriginal remains and a considerable number of these have been investigated, it cannot be said that the work has clarified to any great extent the archaeological problems involved. The early investigators, in accordance with the unfortunate tendency of the time, too often proceeded on the assumption that the accumulation of specimens was an end in itself rather than a means toward the elucidation of archaeological problems" (Collins 1932:38).

Ensuring that everyone was on the same page meant that the regional experts—the ones actually doing much of the work in the Southeast—either had to be trained in proper procedure or, failing that, had to be made aware of what proper procedure was. To that end, the last day of the conference was dedicated to three topics—"exploration and excavation," "laboratory and museum work," and "comparative research and publication"—with the morning devoted to presentations by Cole, Judd, and Wissler and the afternoon to discussions led by Moorehead, Strong, and Webb, who was soon to head much of the federal-relief archaeology that took place in the South (Griffin 1974; Haag 1985; Lyon 1996). The sessions were geared toward imparting information on the proper methods of excavating a site, of analyzing artifacts, of preserving those artifacts, and of presenting the results of the work. These were critical topics to members of the Committee on State Archaeological Surveys, as evidenced by their publishing the suggestions on field methods early in the history of the committee (Wissler et al. 1923). The publication by the committee of a second pamphlet on field methods (CSAS 1930; see Appendix 2) took place only two years before the Conference on Southern Pre-History.

Figure 7. Fay-Cooper Cole, of the University of Chicago, ca. 1950. (Photo courtesy University of Chicago Press)

Fay-Cooper Cole (Figure 7), who received his doctorate from Columbia in 1914 and had assumed the chairmanship of the anthropology department at the University of Chicago in 1929, discussed proper procedure for excavating a mound, using a procedure we elsewhere (Lyman and O'Brien 1999) refer to as the bread-loaf technique, after Gordon Willey's (1936) notation that excavating in such a manner was like slicing a loaf of bread:

> If [the site] is a mound it is staked out in squares (five foot squares are usually most convenient). A trench is started at right angles to the axis of the mound and is carried down at least two feet below the base. The face of the trench is now carried forward into the mound itself by cutting

thin strips from top to bottom. At the same time the top is cut back horizontally for the distance of a foot or more. If this procedure is followed it is possible to see successive humus layers as well as to note all evidences of intrusions....

A village site is best uncovered by a series of trenches much like those used in mound work. A cut is made down to undisturbed soil and the earth is thrown backward as the excavation proceeds. Horizontal and vertical cutting should be employed in hopes of revealing successive periods of occupancy. The worker should never come in from the top. He should never be on top of his trench, otherwise lines of stratification will almost certainly be lost. (Cole 1932:76, 78)

This method has a long history in Americanist archaeology, dating back to the late nineteenth century and the influence of Frederic Ward Putnam, but it is quite evident from reading the literature that what Cole had to say in 1932 must have appeared revolutionary to most southeastern archaeologists.

It is unclear how much of a result the methodological presentations by Cole, Judd, Strong, and others actually had on southeastern archaeology, but the same cannot be said of some of the papers presented in the sessions of December 19, especially those by Walker on Louisiana, Swanton on southeastern Indian groups, and Collins on Mississippi. The intellectual tradition of the Southeast was in large part set in motion by what they had to say.

Winslow Walker and Louisiana Prehistory

Walker's point was simple: everything that an archaeologist wanted to do necessarily hinged on the ability to order remains chronologically. By 1932, seriation and superposition had been used as ordering methods in the American Southwest for almost two decades, but this was not the case in the Southeast. In fact, seriation never caught on there, despite statements to the contrary (e.g., Ford 1962), and it was stratigraphic excavation and the accompanying use of sherds as index markers that would form the backbone of archaeological dating (O'Brien and Lyman 1998, 1999b; O'Brien et al. 2000). Walker (1932:48), however, had a different strategy in mind when he noted that "it is futile to attempt a classification of pre-historic mound cultures in the lower Mississippi Valley until we know more definitely whether or not they have any connection with the principal [historical] tribes found there.... Some of these Indians we know were builders of mounds, but just which ones, and through what stages of development they may have passed, are problems requiring further attention."

The link between peoples living during the prehistoric period and those occupying the region during historical times was what Walker referred to as the "proto-historic" period—a temporal unit about which, Walker (1932:48) admitted, "we are completely in the dark archaeologically." How did one deal with the protohistoric period? Walker (1932:48) had the answer—one that had long been apparent to Smithsonian Institution archaeologists working in the Southeast: "The clue to this phase is the identification of sites visited by the Spaniards in 1542 and by the French in 1682. Special investigations should be made of all relics purporting to date back to either of these periods of exploration." Walker (1932:48) also addressed the investigation of prehistoric remains: "Sites known to contain only prehistoric material should not, of course, be neglected, as there is much work to be done in determining the relationships of the northern and southern mound cultures. But it is more important to establish first the succession of historic and proto-historic cultures, before attempting to say positively just what cultures belonged strictly to prehistoric times." Walker was advocating the use of what his colleagues in the Bureau of American Ethnology and National Museum had been using for years: the direct historical approach.

As Gibson (1982:259) noted, what Walker had to say about the promise of Louisiana's archaeological record and the future directions that should be taken in an effort to understand that record apparently had a profound effect on two young men in attendance—James A. Ford and Fred B. Kniffen, the latter a newly appointed faculty member at Louisiana State University who had trained under Kroeber and geographer Carl Sauer at the University of California. Both Ford and Kniffen immediately began orienting their work in some of the directions in which Walker was pointing (O'Brien and Lyman 1998, 1999b), one direction being the correlation between archaeological-site location and river channels—or more precisely, using the history of river channels to date archaeological sites. Kniffen had already begun exploring the relation between site location and geomorphic features in southern Louisiana, especially relative to land subsidence, as part of Richard Russell's coastal-environments program at Louisiana State University, but he would soon develop several other innovative techniques (Kniffen 1936, 1938), in part because of Walker's influence (O'Brien and Lyman 1998, 1999b).

John R. Swanton and Southeastern Ethnohistory

The success that Walker and other archaeologists working in the Southeast had in applying the direct historical approach was based in large part on the work of John R. Swanton (Figure 8), a Harvard-trained archaeologist-turned-ethnologist who spent his career with the Bureau of American Ethnology. Swanton's early work was on North American Indian languages, and although he continued to produce linguistical texts throughout his career (e.g., Dorsey and Swanton 1912; Gatschet and Swanton 1932; Swanton 1919, 1940; Swanton and Halbert 1915; Thomas and Swanton 1911), he became better known for his ethnohistorical work, especially as it related to the route Hernán de Soto took during his southeastern entrada. Swanton was an archaeologist's dream—someone who both spoke the language and was sympathetic to the goals of prehistory. More importantly, Swanton was someone who could place individual Indian groups in particular places at particular times. This was no small feat in the Southeast, where Indian tribes had experienced centuries of contact with a succession of white groups—Spanish, French, British, and American—resulting in the constant movement of aboriginal groups from one locality to another. It took someone like Swanton, who Kroeber (1940a:3) characterized as "exhibit[ing] a streak of historical genius," to sift through the myriad historical documents on the Southeast and to figure out where particular aboriginal groups were at different times in the past. It was because of the perceived importance of Swanton's work to archaeology that Albert L. Barrows, assistant secretary of the National Research Council, asked Swanton not only to look over the preliminary program for the conference in Birmingham well in advance of the meeting but also to brief council chairman W. H. Howell on his thoughts—all in an effort to make the conference "as useful an occasion as possible in advancing the interests of archaeological research in the southeastern part of the United States."[25]

Swanton addressed the broad issue of southeastern prehistory in two papers he presented in Birmingham, one titled "Southeastern Indians of History" and the other "The Relation of the Southeast to General Culture Problems of American Pre-History." Neither was particularly earthshaking but rather a synopsis of what he had been advocating to archae-

Figure 8. John R. Swanton, longtime ethnologist with the Bureau of American Ethnology, ca. 1945. (From Biographical Memoirs *National Academy of Sciences* 34:328–29 [1960], Columbia University Press, used by permission)

ologists for years: use the ethnohistorical record as a starting point—the chronological anchor—for the reconstruction of prehistory in the Southeast.

Henry B. Collins and Southeastern Culture History

Henry B. Collins (Figure 9) paid homage to Swanton in his paper on historical-period sites in Mississippi: "Our knowledge of the ethnology of the Mississippi Indians is based almost entirely upon the work of Dr. John R. Swanton, whose careful researches have thrown much light on the linguistic and cultural affinities of the Muskhogean and other southern stocks" (Collins 1932:37). However, Collins (1932:37–38) also noted that "There yet remains the task of determining the limits of various groups

Figure 9. Henry B. Collins, assistant curator of ethnology, U.S. National Museum, Smithsonian Institution, ca. 1930. (Negative no. 11,033-A; photo courtesy National Anthropological Archives, Smithsonian Institution)

in pre-historic times [and] their relations one to another and to other southeastern groups, an undertaking that as yet has been hardly begun." Collins (1932:38) believed the most immediate problem facing southeastern archaeologists was the lack of a "basis for chronology," and like his colleagues at the National Museum and Bureau of American Ethnology, he advocated using the direct historical approach. Collins had done the same in an earlier paper on Choctaw village sites in Mississippi, in which he stressed how important it was for southern archaeologists "to seize upon every available source of tribal identification of the cultures represented, and [that] to accomplish this end there is probably no safer beginning than to locate the historic Indian village sites and to study their type of cultural remains for comparison with other sites of unknown age" (Collins 1927:259–260).

By the time of the Birmingham conference, Collins was convinced that of all the "available source[s] of tribal identification," pottery held the most hope for developing chronological ordering:

potsherds are of decided value as chronological determinants and, if present in sufficient quantities to show the entire pottery range of the site, are of far more significance than a number of complete vessels which might not happen to show such a range. In fact, the obliterating effect of white civilization has reached such a point that at many aboriginal sites potsherds are the only really useful material that the archaeologist is able to salvage. The lowly potsherd thus seems destined to bear much of the weight of the chronology that we all hope may sometime be established for Southern archaeology. (Collins 1932:38)

As we discuss in detail elsewhere (O'Brien and Lyman 1998, 1999b, 1999c), Collins (1927) also believed that a pottery type designates an ethnic group, such as a tribe, that ethnic groups have histories, and that a pottery type designates a specific period in the history of an ethnic group. These were common assumptions among southwesternists (Lyman et al. 1997), but they were novel thoughts from someone working in the Southeast. In short, they provided the epistemological warrant for application of the direct historical approach (Lyman and O'Brien 2000).

Of all the federal archaeologists working in the Southeast, Collins would have the most significant and lasting impact. His work in Louisiana and Mississippi during the 1920s is of particular interest because of the impact it had on succeeding generations of archaeologists—an intellectual genealogy that can be traced from Collins through Ford, who from the late 1930s to the middle 1950s was the dominant force in southeastern archaeology. Collins trained Ford in the late 1920s when the latter was still a high-school student, and Ford later used what he learned while working in western Mississippi as he set about the arduous task of carving up prehistoric time in the lower Mississippi River valley (O'Brien and Lyman 1998, 1999b; O'Brien et al. 2000).

The Legacy of the Conference on Southern Pre-History

Ralph Linton (1932:3), in the remarks that opened the second day of the conference, stated explicitly the research questions that would soon guide much of southeastern archaeology: "The worker in any of the surrounding regions finds evidences not merely of diffusion, but of actual migrations coming into his particular area from the southeast, but until the history of that region is better known, it is impossible for him to tell when such migrants left the southeast, what part of it they came from, what their cultural or racial affiliations may have been, or how they are linked to other cultures marginal to the same area."

Stirling (1932:20–21) reiterated Linton's remarks, thereby reinforcing them in the minds of those in attendance. He also specified the procedure for addressing the issues Linton raised: "The first problem in developing the archaeology of the given locality is to isolate the known historic cultures leaving a residue of unknown pre-historic, should such exist. Both vertical and horizontal stratigraphy can usually be applied. . . . From our knowledge of the pottery used by the historic tribes, many significant hints are offered regarding pre-historic movements of peoples." This procedure was nothing more than the direct historical approach. Stirling (1932: 22) also offered the important caution that "the inter-relationship of cultures [is] a flow rather than a series of static jumps." The significance of that caution was lost not only on archaeologists working in the Southeast but on those working in the Americas generally (Lyman and O'Brien 1997; Lyman et al. 1997; O'Brien and Lyman 1998, 2000).

Collins (1932:37) indicated, for example, that one could determine "the limits of the various [ethnic or tribal] groups in pre-historic times," and he stated that typological differences in pottery denoted "cultural differences" (Collins 1932:40). This was, in short, a way of saying that his understanding of the archaeological record was derived from ethnological theory and ethnographic data. Tribes were viewed as discrete chunks of humanity that bore distinct cultural traits and had particular locations in time and space. Assuming that it was possible to identify cultures in the archaeological record (usually on the basis of some typologically distinctive artifacts), when such an identification was made, each prehistoric culture must, it was thought, represent a discontinuous ethnic unit, such as a tribe. This way of thinking was simply the notion of culture areas, popularized in the earlier work of Clark Wissler (1914, 1916, 1917, 1923b, 1924) and having its roots in the culture-classification work of Otis T. Mason (1896, 1905), in Cyrus Thomas's (1894) regional groupings of mound forms, and in William Henry Holmes's (1886, 1903) regional groupings of pottery. This approach was already coming under close scrutiny by several midwestern archaeologists, and its replacement would form the central focus of the third regional conference organized by the CSAS.

With the benefit of hindsight, the ontological parallels between the concept of biological species and the concept of prehistoric cultures are remarkable (Lyman and O'Brien 1997; Lyman et al. 1997). The analytical problem is one of identifying the historically antecedent species or cultures that were also ancestral (in an evolutionary sense) to historically or ethnohistorically documented species or cultures, respectively. In other words, between about 1910 and 1970 phyletic histories of cultures were determined in precisely the same sense that prehistoric *Homo ergaster* is today conceived of as having evolved (perhaps) into *Homo sapiens*. The procedure for determining these phyletic histories was introduced in the Southeast so that culture history could be written there as it had elsewhere in the Americas.

The procedure focused on *homologous* similarity, or similarity resulting from shared ancestry. Thus, for example, Frank Setzler's (1934) work at Marksville, Louisiana, resulted in the conclusion that the people who occupied that site were culturally and biologically related to people who deposited artifacts assigned to the Hopewell culture of Ohio (O'Brien and Lyman 1998). By the end of the Birmingham meeting, the Bureau of American Ethnology and the National Museum had successfully diffused to southeastern archaeologists the general idea that typological similarity denoted homologous similarity. The idea made sense from the perspective of Swanton, Linton, Walker, and Collins, all major figures in the discipline at the time, and everyone in attendance adopted it. The take-home message was simple: work from the known to the unknown so that you have (a) a chronological anchor for your temporal sequence of cultures and (b) the most recent evolutionary descendant of a cultural lineage to use as a comparative base for determining historically antecedent cultures. This was not really a new message, but southeastern archaeologists adopted it wholeheartedly and took the direct historical approach to heights unparalleled in Americanist archaeology (O'Brien and Lyman 1999c).

If a picture is worth a thousand words, then it would take about 30 pages of text to explain what the seven figures included at the end of the Birmingham report show very neatly: the state of southeastern archaeology in 1932. After looking at the figures, can there be any doubt that the concept of culture areas was basic to everyone's thinking? Although several presentations in Birmingham, like a few in St. Louis three years earlier, made mention of temporal differences between segments of the archaeological record, the conference as a whole was, as James B. Griffin (1976a:19) later characterized it, "a Culture area approach." Of particular interest from a historical point of view is Figure 7, which is a map showing the distribution of some archaeological complexes in the East. This is what Griffin (1976a:19–20) had to say about the map:

> Stirling refers to it once but in the wrong context. I do not know who made the map. It contains the regions delimited in Stirling's paper but some of them do not follow his boundaries. In addition, there is located on the map Hopewell in Ohio, Illinois and the Upper Mississippi Valley; Fort Ancient and Adena, which were probably put in by Setzler.... In Illinois we see Black Sand, Cahokia, and Illinois Bluff. The latter is probably the Spoon River Mississippi material. This could have been put on by Setzler or Walker.... There are also added the terms Lake Michigan and Upper Mississippi derived from [W. C.] McKern. W. D. Strong had recently joined the Bureau of American Ethnology and undoubtedly helped to add Signal Butte, Mill Creek, Nebraska, Glenwood, and Upper Republican. The map was used in J. R. Swanton's second talk of the conference from the standpoint of attempts to identify the tribal groups responsible. At a later date Kroeber was to commend the map for indicating the presence of Hopewell culture in three different areas.

One might well pose the question, With this heavy reliance on culture-distribution maps and the direct historical approach, wasn't anyone interested in prehistoric chronology? The answer is, yes they were, but they weren't sure how to go about creating a strictly prehistoric chronology. The answer perhaps was beginning to buzz around in the head of one of the youngest attendees at the Birmingham meeting, but that answer was still a few years off. That attendee was Ford, who before the end of the decade would, with Gordon R. Willey, create a prehistoric chronology for the lower Mississippi Valley (Ford 1935a, 1935b, 1936a, 1936b, 1938; Ford and Willey 1940, 1941). But in those crucial years between 1932 and 1935, a group of midwestern archaeologists decided to try a different tack in their relentless pursuit of making sense out of a vexingly complex material record. They decided to ignore temporal differences in the record, at least for the moment, and to concentrate on formal similarities and differences between and among sets of artifacts. Maybe, if assemblages could be categorized into groups that minimized intragroup difference and maximized intergroup difference, this would tell them something important. Efforts to explore the usefulness of this method culminated in the third and final NRC-sponsored conference, this one held at the Marrott Hotel in Indianapolis on December 6–8, 1935.

The Indianapolis Archaeological Conference

Carl Guthe prepared the preface to the mimeographed report that emanated from the Indianapolis conference (NRC 1937), and in the second paragraph he laid out the purpose of the meeting:

> The conference was called for the specific purpose of discussing the technical problems relating to the comparative study of the archaeological cultures in the upper Mississippi Valley and Great Lakes region. Detailed descriptions of the results of the investigation of individual sites were not pertinent to the meeting. The group of delegates was purposely kept small in order to insure the freedom of informal discussion, and was confined to research students who were interested either in the archaeological problems of a restricted part of the area, or in the comparative significance of these problems with relation to similar ones in other areas. (Guthe 1937:v)

The number of attendees, 19, was indeed small, pared down in number from the 40 persons who attended the Conference on Southern Pre-History and well short of the 53 individuals at the St. Louis conference. With two exceptions, amateur archaeologists were not invited to Indianapolis. Three anthropologists from Washington, D.C., attended—Frank M. Setzler of the U.S. National Museum and Frank H. H. Roberts and John Swanton of the Bureau of American Ethnology—but the majority of those at the meeting were from midwestern institutions: Guthe, Emerson F. Greenman, and young archaeologist James B. Griffin (Figure 10) from the University of Michigan; Lloyd A. Wilford from the University of Minnesota; W. C. McKern from the Milwaukee Public Museum; Charles R. Keyes from the State Historical Society of Iowa; Thorne Deuel from the University of Chicago; and Glenn A. Black and Paul Weer from the Indiana Historical Society. Also from Indiana were two nonprofessionals, E. Y. Guernsey and Eli Lilly. Despite the latter's technically nonprofessional status, his contributions to midwestern archaeology—both monetary and in terms of research—were significant (Ruegamer 1980). Cole was absent for health-related reasons but sent a letter that was read to those in attendance.

In his preface Guthe touched on some of the "technical problems," as he put it, related to the comparative study of archaeological cultures in the upper Mississippi Valley and Great Lakes region. The greatest need was for "a uniform methodology and a greater correlation" of the various investigations that had been taking place with increasing frequency over

Figure 10. James B. Griffin (*front center*) of the University of Michigan examining pottery from the Kincaid site, Massac County, Illinois, ca. 1939. Front (*left to right*): Irvin Peithman, Fay-Cooper Cole, Griffin, Charles R. Keyes. Back (*left to right*): W. M. Krogman, Richard Morgan, and Roger Willis. (From Cleland 1976, used by permission of Academic Press; photo courtesy Museum of Anthropology, University of Michigan)

the previous decade (Guthe 1937:v). McKern (1937a:1), who presented the opening paper at the conference, was more specific: "I can't discuss local Wisconsin problems without touching on general problems. These center around an inadequacy of analytical and systematic methods and terminology. Our major problem is determining how to cooperate to mutual advantage with students of cultures similar to those in Wisconsin. We have great difficulty understanding each other because we do not do things in the same way, and lack a systematized terminology. My specific problems relate to cultural manifestations and their place in the classification."

McKern was feeling the effects of a problem that went far beyond the borders of Wisconsin. By the 1930s Americanist archaeologists had come to something of an impasse over the means and terms used to describe and discuss assemblages of artifacts. The term "culture" was ubiquitous in the role of a grouping unit, but it varied tremendously in scope and meaning from one application to the next. McKern (1943:313) later recalled that this "vague and varying use of the word 'culture' to describe manifestations which were so unlike in scope and character, of which some were culturally correlative—but in different degree, while others lay wholly outside the specific field of relationship, led logically and necessarily" to his becoming interested in developing a method of categorizing archaeological phenomena so that they could be discussed and compared systematically.

How to formulate and implement such a method was the key topic addressed at the conference, and in reading through the discussions one gains an appreciation for the complexity of the issues facing archaeologists in the 1930s—not just those working in the Midwest but in all parts of the country. How could archaeologists communicate without a standardized set of terms? How could "cultural manifestations" be classified in terms of time, space, and form if everyone was using a different system? Or, as was beginning to be asked, was it even wise to try and keep track of those three aspects simultaneously? Was it perhaps more practical, given a lack of detailed regional chronologies, to concentrate foremost on form and then bring time and space in as they became known? By 1935 this was a key question in certain quarters, and it was beginning to be answered more and more in the affirmative. The method that grew out of that question and that was formalized in Indianapolis set midwestern archaeology on an interesting course, but one that was to produce little in the way of enduring results, despite statements to the contrary (e.g., Guthe 1952). The chief navigator of that course was McKern.

W. C. McKern and the Midwestern Taxonomic Method

Carl Guthe (1937:vi) noted that the "Indianapolis Conference holds a significant place in the history of the development of Middle Western archaeology," but to understand that significance one needs to backtrack several years to at least 1932 and the first of several unpublished papers of which McKern was the major author. Before arriving at the Milwaukee Public Museum as assistant curator of anthropology in 1925, McKern (Figure 11), who received his undergraduate degree from the University of California, had served research stints at several institutions, including the Bishop Museum in Honolulu and the Bureau of American Ethnology in Washington, D.C. In scanning McKern's early publications based on his research in Wisconsin, one gets the feeling that he was frustrated by the lack of any systematic means of comparing archaeological materials from the state with those from other regions. This impression is corroborated by Alton K. Fisher, who worked with McKern in the late 1920s and early 1930s:

Figure 11. W. C. McKern, longtime archaeologist with the Milwaukee Public Museum and the guiding hand behind the development of the Midwestern Taxonomic Method, ca. 1940.

By the end of the 1929 field season. . . . some cultural distinctions were becoming apparent [across the region]. . . . However, there was no comparative system in general use in the Midwest at that time to facilitate analysis of subtle as well as overt culture traits so as to suggest possible relationships among them. . . . While McKern had every reason to be pleased with the results of his field work between 1925 and the end of 1929, he was not entirely satisfied with his accomplishments. He had not found the means of defining the cultural relationships he felt must exist but which he had not yet been able to demonstrate. (Fisher 1997:118)

Fisher's remembrances of the time he spent with McKern are important because they give us critical insights into not only the problems of the day but also some of the thought processes that went into the formulation of what eventually became known as the Midwestern Taxonomic Method—

a method that was so synonymous with McKern's name that it often was referred to simply as the "McKern classification" (e.g., Griffin 1943). It was discussion of this classificatory method that held the attention of archaeologists at the Indianapolis meeting. As Fisher (1997:119) recalled,

> After the close of the 1929 field season our noon-time discussions began to concentrate on how a cultural classification system could be designed to serve the archaeological needs of the Wisconsin area. It was recognized at the outset that temporal considerations would have to be ignored because no means was available for the relative dating of what had been found. Certain assumptions could have been made about how the prehistoric culture traits had evolved and then one could have arranged the collected data to fit these assumptions. A hypothetical culture sequence could have been created by that approach but that was rejected by both of us as interestingly speculative but not worth the time that would have been required to develop it. What was wanted was a cultural classification system the criteria for which could be agreed to as valid by all who chose to become familiar with it and to use it. When it became unavoidably clear to both of us that temporal and developmental or evolutionary considerations could not be incorporated in the system, it was finally admitted that the system that was needed so urgently would have to be based on morphological or typological considerations alone. A feeling that was more hopeful than optimistic began to grow that when sufficient facts had accumulated, patterns of arrangement could emerge that would not only suggest cultural relationships but perhaps evolutionary sequences as well. Recognition of the restriction imposed upon the search for the needed classification system actually stimulated the search process.

Fisher's recollection underscores the position midwestern archaeology was in during the 1920s and 1930s—a position similar to that occupied by southeastern archaeology during the same period. Although there were hints as to chronological ordering—for example, it was clear what the chronological position of Hopewell and Adena were relative to one another, as it was clear where Fort Ancient fell chronologically—there were few instances of clear stratigraphic orderings, and those that had been found were often idiosyncratic. Missing were repeated orderings at multiple sites—the kind of evidence that ensured that the suspected orderings were not simply fortuitous occurrences. As proposed so forcefully in Birmingham in 1932, one way out of this chronological dilemma was through the use of the direct historical approach, which anchored the chronological ordering in the recent past and allowed the archaeologist to use overlapping traits to extend the sequence backward in time. Seemingly forgotten was the key chronological work of Kroeber (1916a, 1916b) in the Southwest, which demonstrated that sequences could be constructed, through seriation, without turning a single spadeful of dirt.

Swanton's presentation in Indianapolis was on Siouan tribes in the Ohio Valley, but unlike in Birmingham, the whole notion of using ethnohistory and linguistics to sort out the archaeological record received much less attention in Indianapolis. The Midwest had never witnessed the amount of ethnohistorical and ethnological work that the Southeast had, although this had not stopped archaeologists from concocting all manner of schemes to tie their archaeological manifestations to ethnic groups. If anyone doubts either the complexity of the problem or the speculative nature of efforts to tie the midwestern archaeological record to ethnic groups, Griffin's (1943) Appendix A in *The Fort Ancient Aspect* makes convincing reading. After detailing the myriad proposals that had been put forth for the placement and movement of ethnohistorically known groups in the Ohio Valley, Griffin (1943:313) stated that the "confusion of theories mentioned above results from the fact that no one is in position to interpret intelligently the prehistory of the area in terms of tribal migrations." Apparently, at least from Griffin's point of view in 1943, things had not changed significantly in the decade and a half since McKern discarded the "interestingly speculative" notion of constructing any "hypothetical culture sequence" (Fisher 1997:119) for the upper Mississippi Valley and turned instead to a method of classifying archaeological phenomena that relied solely on formal similarities and differences.

Early on it appeared to McKern that to develop a useful classificatory system, time would have to be jettisoned. And if time went out the window, why should space be retained? If it, too, were discarded, then one could concentrate on a comparative examination of empirical units—that is, on artifacts and the attributes they exhibited. Thus form-related units, which anyone could see and measure, would be the building blocks of the classification. Importantly, there could finally be agreement over units; no longer would archaeologists argue about whether shell-tempered pottery was Siouan in origin or grit-tempered pottery Algonquian in origin. *Perhaps,* as Fisher (1997:119) intimated, at some future point "patterns of arrangement" would emerge that would suggest not only "cultural relationships but perhaps evolutionary sequences as well," but for the present archaeologists would have a method of systematizing the artifacts and features encountered in the record—a method that at the very least would facilitate communication and comparison.

In principle the method McKern devised was simple—a branching taxonomy with successively higher levels of inclusiveness—but it was misunderstood from the start. We think part of the misunderstanding stemmed from the fact that in its unadulterated form the method had nothing to do with time and space—two of the three central foci in almost any archaeological endeavor. Prehistorians from the early nineteenth century on were interested in questions of when, where, and what, and to ignore two of the three was viewed in some quarters as foolish if not downright heretical. Thus there was a backlash against the method that continued into the 1940s (e.g., McGregor 1941; see McKern [1944] for a rebuttal), with the most strident criticism coming from Julian Steward, who defended the contributions made to archaeology by the direct historical approach: "[I]t is difficult to see what is gained by scrapping a scheme with historical terms and categories in favor on a non-historical one" (Steward 1942:339; see McKern [1942] for a rebuttal). Although the Midwestern Taxonomic Method was designed to keep time out of the equation, in practice it rarely did. The temporal dimension was too ingrained in Americanist archaeology for it to have been otherwise, despite the best intentions of the method's chief architect.

Perhaps another reason for confusion stemmed from the fact that the method was used almost exclusively in the Midwest and Plains. It was, after all, labeled the *Midwestern* Taxonomic Method (McKern 1939), which made it sound as if it was applicable only in one region of the country. Of course, it wasn't limited to a single area (e.g., McGregor 1941), but the parochialness implied by the name was still an obstacle to overcome. The Southwest had its own classificatory systems and sets of nomenclature, such as the system proposed at the first Pecos Conference in 1927 (Kidder 1927) and the one that emanated from the Globe Conference of 1931 (Gladwin and Gladwin 1934), and in several respects those systems resembled McKern's. There was, however, one major difference, as McKern (1944) well knew: the southwestern schemes admitted time and space, whereas his did not. Southwesternists were not going to give up their classification systems, which were built around all three dimensions of interest—time, space, and form—in favor of one that was built around only form.

As with most methods, the Midwestern Taxonomic Method went through several iterations—and we touch on a few aspects of the different drafts below—but the basic outline of the method remained unchanged from about 1932. The building blocks of the method were called *components*, defined as assemblages of associated artifacts that represented the occupation of a place by a people. Thus a component was not viewed as being equivalent to a site unless a place had experienced *only* a single occupation (McKern 1939:308, 1944:445)—a key point missed by some archaeologists (e.g., Setzler 1940). Artifact trait lists were used to create higher-level groups. An archaeologist polled available components and identified those traits that linked—were shared by—various components, which were then placed together in a group. Simultaneously, one used those same trait lists to identify traits that could be used to isolate one group of components from another group. Five levels of groups were eventually recognized. From least to most inclusive, these were *focus, aspect, phase, pattern,* and *base*. Three kinds of traits were distinguished: *linked traits*, which were common to more than one unit; *diagnostic traits*, which were limited to a single unit; and *determinants*, which were traits that occurred in all members of a unit but in no other unit. If this sounds rather confusing, note that even those who worked alongside McKern in refining the method were confused on occasion, not only over the different kinds of traits but over how they were to be identified. Figure 12 is our effort to slice through the confusion and show the difference among linked traits, diagnostic traits, and determinants.

According to Fisher (1997:119), it was he, not McKern, who first proposed the method of classification that would become synonymous with his supervisor's name:

> About that time I recalled my relatively recent studies in biology during which I had become quite familiar with the taxonomic system of Linnaeus. It was based primarily on relationship of form, originally applied to the classification of plants but later extended to animate creatures with equal success. If that classification system could show morphological relationships between animal forms as diverse as mastodons and earthworms, might it not be possible to show some relationship between the creations of man as demonstrated by form or structure alone?

This insight was significant in that it eschewed any question of equating archaeological remains with ethnic groups and instead *ultimately* sought evolutionary, or phylogenetic, relations among sets of artifacts (Lyman et al. 1997; O'Brien and Lyman 2000). But for the initial sorting, time was ignored in favor of morphological similarity. Fisher was correct:

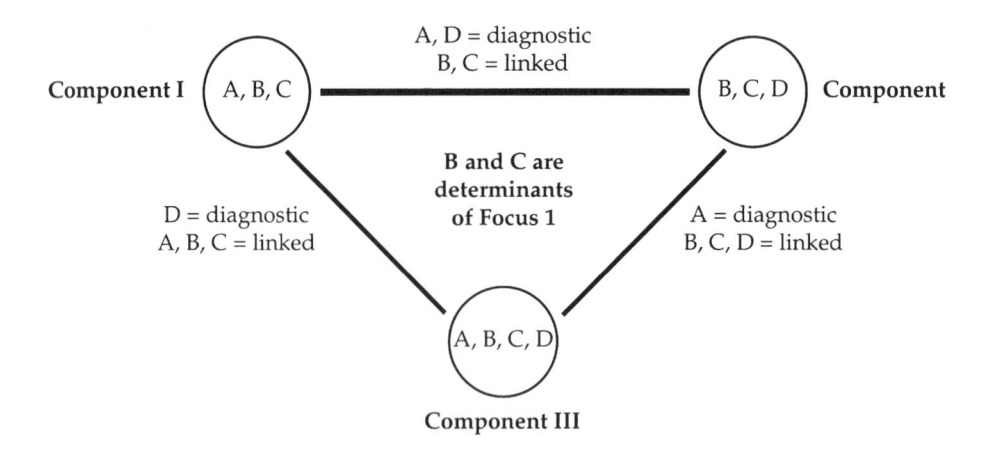

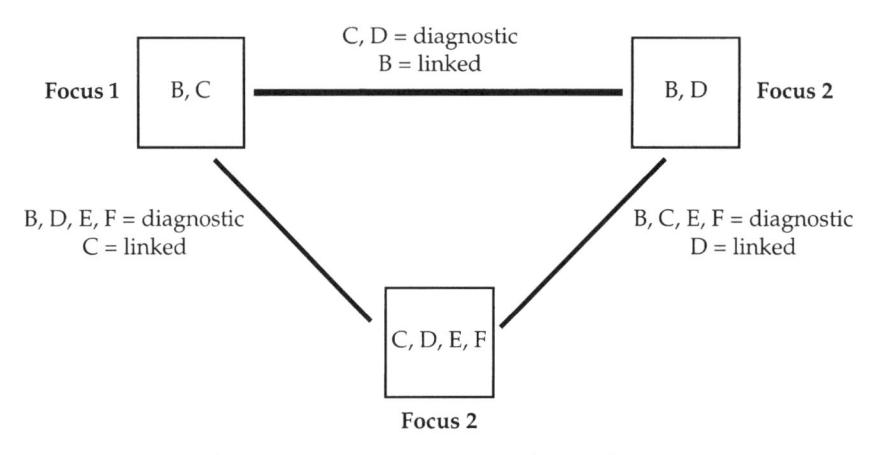

Figure 12. The analytical relations among traits, components, and foci in the Midwestern Taxonomic Method (after Lyman et al. 1997).

in basic principle what he proposed to McKern *was* similar to the way in which Carolus Linnaeus approached the taxonomic classification of organisms in the late eighteenth century. Both methods produce nested categories, and one could make a rough correspondence between components and populations, foci and species, aspects and genera, and so on. Strictly speaking, however, the Linnaean taxonomic system is not an evolutionary scheme; certainly Linnaeus had no evolutionistic pretensions when he first developed the method of classification. That the classification was later shown to have phylogenetic implications had nothing to do with how and why it was created. Similarly, McKern's taxonomic method was not devised to show evolutionary relationships, although it was admitted from the start that various formal relationships that it revealed *might* be phylogenetic.

Alice Kehoe (1990:34) noted that there "is an interesting parallel to McKern's method in the currently controversial method of cladistics in biology." Such a statement is based on an ill-informed view of what cladistics is and is not. Although both are based on the identification of varying degrees of morphological similarity in character states, McKern's method was an application of numerical phenetics, or numerical taxonomy (Lyman et al. 1997; O'Brien and Lyman 2000), in most cases without recourse to the actual quantitative measurement of similarity. Indeed, some of McKern's contemporaries (e.g., Kroeber 1940b, 1942) were quick to point out that the failure to quantitatively measure similarity was a major flaw of the method. In unrelated fashion, cladistics is based solely on the ability to differentiate between not only analogous and homologous traits but, with respect to the latter, shared derived traits and shared ancestral traits. The Midwestern Taxonomic Method made no attempt to separate analogous traits from homologous traits. This was no deficiency of the method; McKern never intended it to do so.

Fisher (1997:120) indicated that at first McKern was skeptical of the method, but after

> considerable discussion and thought on the matter . . . he began to test the idea with data he had collected, and he was pleased to find that it often was successful. . . . When it became evident that there might be a reasonable prospect of success at designing different levels or degrees of relationship between lithic and bone artifacts, pottery, earthworks, and burials and between complexes of such cultural manifestations, the need to become specific in defining the various proposed categories of relationship claimed [McKern's] attention.

McKern's pilot run at introducing the method formally was to be at the annual meeting of the Central Section of the American Anthropological Association, which was held in Ann Arbor, Michigan, in April 1932, but illness precluded his attendance (Griffin 1943:327). Instead, the first presentation was made at a meeting of the Illinois State Academy of Science held at the University of Chicago the following month. McKern revised his paper in light of suggestions he received, and Guthe circulated it to interested parties. The paper was titled "A Suggested Classification of Cultures." McKern revised the paper again, incorporating more suggestions, and on December 10, 1932, he and a small group of archaeologists—Samuel A. Barrett, a former Kroeber student at the University of California and director of the Milwaukee Public Museum (and the man whom McKern had replaced as curator of anthropology); A. R. Kelly of the National Park Service, who after receiving his doctorate at Harvard in 1929 had worked at the University of Illinois until 1933; and Cole, Deuel, Griffin, and Guthe—met at the University of Chicago to discuss the paper. McKern revised it yet again, and on April 4, 1933, the paper was sent out under the authorship of McKern, Deuel, and Guthe (McKern et al. 1933).

McKern revised the paper once more, changing the title to "Certain Culture Classification Problems in Middle Western Archaeology." He presented it the following year as his presidential address to the Central Section, and the CSAS issued the paper through its *Circular* series (McKern 1934). It was the content of that paper that formed the major points of discussion at the Indianapolis meeting in December 1935 (McKern 1937b). The paper assumed such a key role at the meeting that it was appended, without modification, to the published report on the conference. Guthe (1937:vi) had this to say about McKern's paper in his preface to the proceedings: "This paper constitutes the first concise statement of the principles upon which this classification is based, and the detailed methods by which it may be applied. It is included here because the discussions at the conference assumed that the delegates had a knowledge of its contents."

The delegates indeed had a knowledge of its content—several of them had made significant contributions to the paper—just as they had a knowledge of both McKern's deep commitment to the method and how he had defended it:

> I have received such questions as this: Why call the cultural manifestation of the pre-literate Iroquois, Upper Mississippi, or any name other than Iroquois? In some instances we may have sufficient data to verify identification with a known historic group, such as the Iroquois. However, in most instances, we cannot immediately bridge the gap . . . and in many instances we cannot hope ever to be able to do so. . . . The only taxonomic basis for dealing with all cultural manifestations . . . is that of culture type as illustrated by trait-indicative materials and features encountered at former habitation sites. If in the future it becomes possible to name the historic ethnic group of which the pre-literate group is the progenitor, no

confusion should result from the statement that, for example, Upper Mississippi Oneota is Ioway Sioux; no more so than from the statement that *Elephas primigenius* is the mammoth. (McKern 1937b:70)

McKern chose a poor analog for his last point. Mammoth and *E. primigenius* are simply different names for the same creature; one does not have to show any kind of a connection between the two names to use them interchangeably. This is decidedly not the case with Upper Mississippi Oneota, an archaeological manifestation, and Ioway Sioux, an ethnic and linguistic unit. Here it must be demonstrated that two very different kinds of units have an equivalence. McKern's mammoth example would have been better had he said something like, "No confusion should result from saying that a particular set of fossils represents *Elephas primigenius*," because this would have underscored the necessity of definitive criteria for distinguishing between the fossils of mammoth and those of some other large quadruped. His comments are strong evidence that he viewed his archaeological units as equivalents of ethnic units; he just didn't know which archaeological unit went with which ethnic unit, and until he did, he didn't want to guess.

McKern was determined to leave critics of his proposed method defenseless. He attacked two of the prized possessions in the archaeological tool kit of the early twentieth century: the direct historical approach and the culture-area concept. In attacking the former, he stated,

> Aside from the inadequacy of the direct-historical method in supplying the archaeologist with a means of attachment to the ethnological classification, the latter, even if applicable, would not ideally answer the needs of the archaeologist. One ethnological classification divides the aborigines into linguistic stocks which are first subdivided into more specific linguistic groups and, finally, into socio-political groups. The criteria for classification are social, primarily linguistic. The major portion of the data available to the archaeologist relates to material culture, and in no instance includes linguistic data. Consequently this ethnological classification does not satisfy archaeological requirements. (McKern 1937b:71)

McKern then went after the culture-area concept:

> It may be said that we have the ethnologically conceived culture areas to supply a basis for archaeological classification. However, these so-called culture areas involve two factors which the archaeologist must disregard in devising his culture classification if he is to avoid hopeless confusion;

these are the spatial and temporal factors. First, the culture area attempts to define, or at least limit, geographical distribution. Unfortunately, the American aborigines did not always succeed in confining their cultural divisions within a continuous area, or in keeping culturally pure an area of any important size. Second, the archaeologist considers the American Indians from the standpoint of all time, and certainly, there can be no cultural areas devised which can include an unlimited temporal factor. (McKern 1937b:71)

Applying the Midwestern Taxonomic Method

Armed with these caveats, participants at the Indianapolis conference got down to the business of using McKern's method to sort out the archaeological record of the Midwest. From our perspective the reports from the Saturday sessions are the most interesting because they chronicle the difficulties that archaeologists encountered in actually trying to use the method. Up to that point few attempts had been made to do so. The ones with which we are familiar are McKern's discussion of data from Wisconsin that appeared in the 1934 draft of his paper on the Midwestern Taxonomic Method and four treatments that appeared the following year: Griffin's (1935) preliminary analysis of the Fort Ancient Aspect; Strong's (1935) and Wedel's (1935) treatments of Plains data; and Deuel's (1935) handling of data from the upper Mississippi Valley, which was roundly criticized by several of his colleagues (e.g., Griffin 1943; Guthe 1936; McKern 1938). Deuel's paper and the subsequent treatment of Illinois data by Cole and Deuel (1937) highlight the conceptual difficulties that archaeologists had in actually applying the Midwestern Taxonomic Method to a set of data. In the work of Strong and Wedel we do not see a pure application of the method but rather something of a hybrid of the Midwestern Taxonomic Method and the direct historical approach.

At the time of the Indianapolis conference there were four levels in the classificatory system of the method—basic culture, phase, aspect, and focus—the same four that were in the 1933 draft (McKern et al. 1933). At one of the Saturday sessions Guthe suggested dropping the term "basic culture" because of the confusion surrounding the term, and in its place he suggested "base," with a new level, "pattern," to be inserted just below it. Pattern was a term that had been discussed for some time, and its eventual insertion created the five-tier system that appeared in the published version of the paper (McKern 1939). It was a common misconception among those who were not part of the group that devised the Midwestern

Taxonomic Method that "component" was the sixth, and lowest, tier in the system. McKern and others consistently warned that this was not the case. Rather, a component is "the manifestation of a given culture at a single site" (McKern et al. 1933:4), or "the manifestation of any given focus at a specific site" (McKern 1937b:73). This unit "serves to distinguish between a site which may bear evidence of several cultural occupations, each foreign to the other, and a single, specified manifestation at a site" (McKern 1937b:73–74).

Conference participants had a difficult time deciding whether known cultural manifestations should be labeled as aspects or foci, and some were irritated that their favorite manifestation might lose its primacy. Take, for example, the following exchange:

> McKern: It seems to me that the majority of Hopewell traits are un-Woodland.
>
> Deuel: Outside of Ohio, our Central Basin largely consists of Woodland characteristics, and there are a number of sites called Hopewell that have traits like Marksville [Avoyelles Parish, Louisiana] and others which cannot be placed.
>
> Roberts: Would you say that in your Central Basin, except for Ohio, you have about an equal division of Woodland and Mississippi traits? It seems to me that your separate Pattern here is Hopewell. You may find out that it is a northern extension of your southern pattern. Why not make the Pattern Hopewell?
>
> Guthe: As a matter of convenience, what is there wrong in thinking in terms of Aspects and Phases? Include a Hopewell Phase under the Central Basin Pattern.
>
> McKern: Why can't we say an unnamed Pattern under which we get Hopewell?
>
> Setzler: Why not use Hopewellian Phase instead of Hopewell?
>
> McKern: Hopewell is also a Component in itself. Use the Scioto Valley [Ohio] as Focus. (NRC 1937:61)

There is also clear evidence that try as they might, many of the participants couldn't shake their tendencies to hold to subjective impressions of evolutionary relationships between various units. For example, Setzler stated, "I want a single Pattern called Mississippi, with all pottery-agriculture divisions listed under it." He then asked, "Can't you make your divisions under Phases instead of the Pattern?" (NRC 1937:60). Deuel realized what Setzler was getting at: "It seems to me what is bothering

Setzler is the fact that he sees a genetic relationship between the Gulf cultures and the Mississippi cultures, which should be if the two are classified on the basis of their inherent traits" (NRC 1937:60).

It was the identification of these "inherent traits" that was the final undoing of the Midwestern Taxonomic Method. That and deciding not only what a trait was but whether any particular trait was a linked trait, a diagnostic trait, or a determinant, any of which could be "inherent" to a given unit regardless of whether that unit was a focus, an aspect, or a phase. This problem apparently became so acute that in what was a well-thought-out application—Griffin's (1943) *The Fort Ancient Aspect*—the author took an entirely different tack: "The concepts 'determinant,' 'determinant trait,' 'determinant complex,' 'diagnostic,' 'diagnostic trait,' 'diagnostic complex,' and 'link traits' have not been seriously employed in this paper, partly because of the confusion and contradiction in the present use of such jargon and partly because there was no apparent need for such terms" (Griffin 1943:335). Although Griffin's monograph was published in 1943, the analysis was completed in 1939, three years after he finished his doctoral dissertation at the University of Michigan and while he was assistant curator of archaeology at the university's Museum of Anthropology. Even as early as 1939 Griffin must have seen that applications of the Midwestern Taxonomic Method were hopelessly confused and tautological—Cole and Deuel's (1937) *Rediscovering Illinois* being a case in point. Although Griffin (1943:338) did not refer to that work by name, he obviously had it in mind when he commented on how some archaeologists working in the Mississippi Valley chose determinants: "A few 'determinants' are chosen from a small number of sites, and these same sites are then used to illustrate that the selected list recurs at these same sites." The alternative Griffin selected—establishing a complex of traits and ignoring determinants—became the cornerstone of archaeology throughout the 1940s and 1950s, culminating in the formulations of Philip Phillips and Gordon R. Willey (1953; Willey and Phillips 1955, 1958) that emphasized the temporal and spatial dimensions of archaeological phenomena. The phase unit they proposed came to dominate Americanist archaeology in the 1960s and is, in many respects, simply the result of jettisoning the higher-level units of the Midwestern Taxonomic Method and of modifying the focus to include explicit temporal and spatial parameters (as suggested by Harold S. Gladwin [1936] and Harold S. Colton [1939]).

Epilogue

At the last session of the Indianapolis conference, McKern noted, "It seems to me before we depart that we have gotten a great deal out of this meeting. It seems advisable that we should have such meetings at least once a year" (NRC 1937:69). Guthe, however, announced that "We are confronted with several problems regarding further meetings of this sort. The National Research Council is trying to withdraw from projects it has supported for a long time. According to present plans, the CSAS will go out of existence in June or July 1937, which means that . . . [the] machinery will not exist so that we can get money from a central organization" (NRC 1937:69).

Times and interests change, as do federal funding priorities, and by late 1935 the NRC felt it had supported archaeology long enough. Besides, other branches of the government had become heavily involved in archaeology—the Federal Emergency Relief Administration was created in May 1933, the Civil Works Administration later that year, and the Works Progress Administration in 1935—primarily in an effort to stabilize the economy and get people back to work. Archaeology, being a labor-intensive endeavor, was the perfect vehicle for employing large numbers of people. Ironically, the CSAS disbanded shortly after these programs began and just as millions of federal dollars were starting to pour into local and state coffers to fund archaeological projects. In some quarters the work that resulted from relief efforts was highly innovative (Haag 1985; Lyon 1996; Setzler and Strong 1936), but in others it was less than spectacular (Johnson 1947, 1966). At the point where a strong central body such as the CSAS could perhaps have done the most good in terms of quality control, it was dissolved. But not for long, for early in 1939 the Works Progress Administration asked the National Research Council to create a committee to examine the state of archaeology in the United States and to determine whether federal relief archaeology was producing the kind of results it should. Out of this request grew the Committee on Basic Needs in American Archaeology. And to whom did the NRC turn for assistance in organizing the committee? None other than the tireless Carl Guthe.

When the CSAS was dissolved, no one thought that the immediate problems facing archaeology in the East had been solved. In fact, the majority of sentiment ran in the opposite direction. Midwestern archaeologists had a new method for classifying archaeological manifestations, and southeastern archaeologists had the direct historical approach to help solve their chronological problems, but some of the problems on which the committee had focused from the beginning were as bad or worse in 1935 than they had been 15 years earlier. One of these was the destruction of archaeological sites, which if anything had accelerated in the 1930s despite the best efforts of Guthe and his colleagues. This is how Setzler and Strong (1936:308–309) saw the problem in their mid-1930s assessment of federal relief efforts:

> The present actual status of archaeological conservation in the United States ... is deplorable. ... The Antiquities Act of 1906 forbids unauthorized archaeological excavation on public lands, but the law is difficult to enforce and, so long as archaeological specimens can be sold on the open market, can have at best a very limited effect. ... It is a sad paradox that at this time, when trained men are becoming available and new techniques for determining archaeological history are reaching a high pitch of development, the materials themselves should be vanishing like snow before the sun.

One bright spot in the mid-1930s was the creation of yet another organization, which in many respects acted in the same capacity as the Committee for State Archaeological Surveys had since its inception. However, the new organization differed in structure in that it was a national body and was composed of nonprofessional as well as professional archaeologists. The genesis of the organization was a query posed to the committee in 1933 as to why there was no national society dedicated solely to archaeology in the Americas (Guthe 1967). The committee agreed to look into forming such an organization, and in April 1934 a prospectus was mailed to 192 persons with whom the committee corresponded (Griffin 1985). These included nonprofessionals as well as professionals because, as Griffin (1985:265) later pointed out, if only the latter had been included, their dues would have been prohibitively high in order to fund publication of the journal that the organization proposed to publish. Given the mix of the membership, what should the society be called? After toying with several names, the committee decided on the Society for American Archaeology, the organizational meeting of which took place on December 28, 1934, following the annual dinner of Section H (Anthropology) of the American Association for the Advancement of Science, held that year at the Hotel Roosevelt in Pittsburgh (Guthe 1935).

Despite worries on the part of some of the founders that the new organization would be viewed by some as a vehicle for moving archaeology away from the more traditional societies such as the American Anthropological Association (Guthe 1935), this was not the intent: "the Society was not the expression of a separatist movement, but an attempt to bring anthropologists using the archaeological method into closer contact with the public, and to establish a wider appreciation of the methods and principles of scientific study" (Guthe 1967:438). Further, it was felt that under the conditions, "the original objectives of the Committee [on State Archaeological Surveys] would have a better chance of attainment through such a national membership organization" (Guthe 1967:438). We assume that by late 1934, a year before the Indianapolis conference, Guthe could see the handwriting on the wall: the NRC was going to shift its support away from the committee, and there would have to be an organization capable of assuming its duties. To that end, he approached the Carnegie Corporation, which had helped fund the activities of the CSAS since 1929,[26] asking if the last round of funding could be shifted to the Society for American Archaeology. The corporation agreed,[27] and the newly created organization assumed the duties that had previously been the charge of the Wissler-Kidder-Guthe committee. That committee was discharged at the end of June 1937, having been in existence for 17 years.

In assessing the accomplishments of the CSAS, especially as those are reflected in the three regional conferences the committee sponsored, we are struck by the parallels between Americanist archaeology in the 1920s and early 1930s and Americanist archaeology today. The destruction of archaeological sites did not abate after the Society for American Archaeology took over the functions of the CSAS in 1937, and those in the discipline today are as concerned with the problem as their forebears were. Similarly, chronology is as important today as it was during that earlier period, and although modern archaeologists have access to a battery of methods that earlier generations of archaeologists could not have imagined, some local chronological sequences in the Midwest and Southeast are only slightly more developed than those of the mid-1930s. Today's archaeologists are also as interested in classification as McKern, Guthe, and Griffin were when they were debating the finer points of the Midwestern Taxonomic Method. The descendant of that method—the phase-centered approach to categorizing archaeological manifestations—has, since the early 1940s, been integral to archaeological systematics as used over much of North America. Discussions at the Birmingham conference showed there was considerable need for a systematic method of categorizing archaeological remains. Subsequent discussions at the Indianapolis conference demonstrated what one such method might look like, but it also demonstrated the incredible complexity of the archaeological record and the difficulties involved in fitting it into a taxonomy.

As we peruse the discussions that took place at the various conferences, we often catch a germ of an idea that would later become a central focus in Americanist archaeology. Or maybe it was a simple statement or suggestion that foreshadowed events to come—events that became milestones in terms of how they moved the discipline forward either methodologically or in terms of its knowledge base. For example, considerable debate at the Indianapolis conference revolved around the concept of Middle Mississippi—both how to recognize it and how to classify it. During the discussions, Swanton asked, "How are you going to get anywhere with Middle Mississippi until you investigate the Arkansas–west Tennessee district?" (NRC 1937:64). At least one person in the room must have thought about that question, because within a few years Griffin, along with Philip Phillips and James A. Ford, would begin a decade-long project in the Arkansas-Mississippi-Tennessee portion of the Mississippi Valley that resulted in a monograph (Phillips et al. 1951) that in our opinion is one of the most important works ever written in Americanist archaeology.

Taken in the aggregate, the three volumes that emanated from the conferences sponsored by the CSAS contain an extensive array of information on how archaeologists working in the eastern United States during the 1920s and 1930s organized their study of the past and how they arrived at some of their conclusions about the past. Some of that information is contained elsewhere, either in monographs written during that period or in the reminiscences of those who worked during those times, but it is not the same as reading the actual exchanges that took place at meetings and hearing the way in which ideas were shaped through discussion and debate. In closing, we note that our sentiments are identical to those of Griffin (1976a:171): "If historians of American Archaeology really want to know what a significant number of American archaeologists were working on [between 1929 and 1935] and their views of the then current knowledge of the participants, these reports need to be read."

Notes

1. Clark Wissler to C. E. Seashore, letter, October 14, 1921. NRC Archives, CSAS, Washington, D.C.

2. Franz Boas is often portrayed as the leading figure in American anthropology during the period 1900–1920, but in our opinion this is based in large part on his flamboyant personality and the quality of students he produced at Columbia. Clark Wissler, who claimed fewer students and whose manner was much more reserved, produced work that would endure far longer than Boas's. For a readable account of Wissler's professional life, see Freed and Freed (1983).

3. Clark Wissler to W. C. Mills of the Ohio State Archaeological and Historical Society, letter, November 4, 1921. NRC, CSAS, Washington, D.C.

4. Reprinted as number 97 in the *Reprint and Circular Series of the National Research Council,* Washington, D.C. (1930).

5. The *Proceedings* were published as part of the *Transactions of the Academy of Science of St. Louis.* Missouri Historical Society minutes for June 17, 1880, p. 2. Missouri Historical Society Archives, St. Louis.

6. Archaeological Institute of America Archives, vol. 1:23, Missouri Historical Society Archives, St. Louis.

7. Archaeological Institute of America Archives, vol. 1:23, Missouri Historical Society Archives, St. Louis.

8. Wissler to C. E. Seashore, letter, March 20, 1922. NRC Archives, CSAS, Washington, D.C.

9. Wissler to Charles H. Danforth, letter, October 25, 1921. NRC Archives, CSAS, Washington, D.C.

10. Wissler to the CSAS, memorandum, November 7, 1921, reporting on a joint letter from R. J. Terry and C. H. Danforth of the Anthropological Society of St. Louis. NRC Archives, CSAS, Washington, D.C.

11. "Report of the Chairman on a Trip through the Mississippi Valley, September, 1928," unsigned but written by Carl E. Guthe. NRC Archives, CSAS, Washington, D.C.

12. The complete list of summaries appearing in *American Anthropologist* is as follows (titles and volume numbers can be found in the reference list): 1921 (Wissler 1922), 1922 (Wissler 1923a), 1924 (Kidder 1925), 1925 (Kidder 1926), 1926 (Kidder 1927), 1927 (Guthe 1928), 1928 (Guthe 1929), 1929 (Guthe 1930a), 1930 (Guthe 1931), 1931 (Guthe 1932a), 1932 (Guthe 1933), 1933 (Guthe 1934). There apparently was no summary for 1923.

13. Wissler to Seashore, letter, March 20, 1922. NRC Archives, CSAS, Washington, D.C.

14. Wissler to Albert E. Jenks, director of the Division of Anthropology and Psychology, letter, January 2, 1924; Jenks to Wissler, letter, January 4, 1924. NRC Archives, CSAS, Washington, D.C.

15. Kidder served simultaneously as chairman of the CSAS and chairman of

the Division of Anthropology and Psychology for the period July 1, 1926–June 30, 1927. In 1927 he said of his tenure, "I believe that all chairmen go through four periods: (1) bewilderment, (2) a great burst of energy, (3) discouragement, and (4) a return to normalcy. The greatest problem of the chairman is that he is given a large handsome machine and no gas to run it" (Stevens 1952:123).

16. A. V. Kidder to Vernon Kellogg, memorandum, June 14, 1927; Kellogg to Kidder, memorandum, June 29, 1927. NRC Archives, CSAS, Washington, D.C.

17. Undated manuscript (probably late 1927) by Guthe titled "The Ceramic Repository for the Eastern United States, at the University of Michigan, under the Auspices of the National Research Council." NRC Archives, CSAS, Washington, D.C.

18. "Report of the Chairman on a Trip through the Mississippi Valley, September, 1928," unsigned but written by Guthe. NRC Archives, CSAS, Washington, D.C.

19. Knight Dunlap to Edmund Day, director of the Laura Spelman Rockefeller Memorial, letter, April 11, 1928. NRC Archives, CSAS, Washington, D.C.

20. In the official list of attendees, Futrall is listed as the president of the University of Arkansas at Batesville. This is incorrect; there was no branch of the university at Batesville. Futrall was president of the University of Arkansas at Fayetteville from 1913 until his death in 1939. In addition to being a classicist and an avocational archaeologist, he founded the university's football program, serving as coach for its first three seasons. He is also credited with helping form the Southwest Conference for intercollegiate athletics.

21. A number of anthropologists mentioned in this essay—more than just archaeologists working in the East—were influenced early in their careers by Putnam. For example, Berthold Laufer, Gerard Fowke, Roland Dixon, A. L. Kroeber, and John Swanton were at various times all members of the Jesup North Pacific Expedition sponsored by the American Museum of Natural History. Franz Boas, who at the time was assistant curator at the American Museum, more than anyone set the scientific direction for the expedition, but Putnam certainly had a hand in the project's formulation. Further, it was Putnam who brought Boas to the museum in the first place.

22. There actually was a meeting that took place between the Conference on Midwestern Archaeology and the Conference on Southern Pre-History, but technically it was not sponsored by the CSAS. We say "technically," because although the committee did not publicize or fund it, many of the same archaeologists who participated in the sponsored conferences attended the meeting held in Vermillion, South Dakota, on August 31 and September 1, 1931. A two-page summary

was published in 1931 as number 9 in the committee's *Circular* series. The meeting is of historical interest because even in the short summary statement one sees how archaeologists working in the upper Plains and Midwest were beginning to wrestle with the problem of cultural classification—the single issue that led to the third NRC-sponsored archaeological conference, which was convened in Indianapolis in 1935.

23. The report must have been printed in 1933, but it carries no date other than that of the meeting. We cite the papers in the report as 1932.

24. Brannon later served as the director of the Alabama State Department of Archives and History in Montgomery. He was a prolific author, publishing numerous articles in the *Alabama Historical Quarterly* between 1930 and 1962. His most widely cited publication is *The Organization of the Confederate Post Office Department at Montgomery* (1960; published privately).

25. Albert L. Barrows to John R. Swanton, letter, July 13, 1932. NRC Archives, CSAS, Washington, D.C.

26. Report made to the Carnegie Corporation by the CSAS covering the period 1929–1934. NRC Archives, CSAS, Washington, D.C.

27. Report made to the Carnegie Corporation by the CSAS covering the period July 1, 1935–June 30, 1936. NRC Archives, CSAS, Washington, D.C.

References

Archaeological Institute of America
    1880   *Archaeological Institute of America, Annual Report* (1879) 1.

Blake, L., and J. Houser
    1978   The Whelpley Collection of Indian Artifacts. *Academy of Science of St. Louis, Transactions* 32:1–64.

Brannon, P. A.
    1960   *The Organization of the Confederate Post Office Department at Montgomery.* Privately printed.

Broadhead, G. C.
    1880   Prehistoric Evidences in Missouri. *Smithsonian Institution, Annual Report* (1879):350–359.

Browman, D. L.
    1978   The "Knockers": St. Louis Archaeologists from 1904–1921. *Missouri Archaeological Society Newsletter* 319:1–6.
    2000   The Nineteenth-Century Peabody Museum Stratigraphic Excavation Method. Paper presented at the 65th Annual Meeting of the Society for American Archaeology, Philadelphia.

Browman, D. L., and D. R. Givens
- 1996 Stratigraphic Excavation: The First "New Archaeology." *American Anthropologist* 98:80–95.

Bushnell, D. I., Jr.
- 1904 The Cahokia and Surrounding Mound Groups. *Peabody Museum of American Archaeology and Ethnology, Papers* 3.

Cleland, C. E.
- 1976 Cultural Change and Continuity: Essays in Honor of James Bennett Griffin. Academic Press, New York.

Cochrane, R. C.
- 1978 *The National Academy of Sciences: The First Hundred Years, 1863–1963.* National Academy of Sciences, Washington, D.C.

Cole, F.-C.
- 1929 [Untitled Remarks]. In Report of the Conference on Midwestern Archaeology, Held in St. Louis, Missouri, May 18, 1929. *National Research Council, Bulletin* 74:112.
- 1932 Exploration and Excavation. In *Conference on Southern Pre-History*, pp. 74–78. NRC, Washington, D.C.

Cole, F.-C., and T. Deuel
- 1937 *Rediscovering Illinois: Archaeological Explorations in and around Fulton County.* University of Chicago Press, Chicago.

Collins, H. B., Jr.
- 1927 Potsherds from Choctaw Village Sites in Mississippi. *Washington Academy of Sciences Journal* 17:259–263.
- 1932 Archaeology of Mississippi. In *Conference on Southern Pre-History*, pp. 37–42. NRC, Washington, D.C.

Colton, H. S.
- 1939 Prehistoric Culture Units and Their Relationships in Northern Arizona. *Museum of Northern Arizona, Bulletin* 17.

Committee on State Archaeological Surveys (CSAS)
- 1930 Guide Leaflet for Amateur Archaeologists. *National Research Council Reprint and Circular Series* No. 93. Washington, D.C.

Deuel, T.
- 1935 Basic Cultures of the Mississippi Valley. *American Anthropologist* 37:429–446.

Coon, Carleton S., and James M. Andrews IV
- 1943 Studies in the Anthropology of Oceania and Asia, *Papers of the Peabody Museum of Archaeology and Ethnology* 20.

Dexter, R. W.
- 1966 Putnam's Problems Popularizing Anthropology. *American Scientist* 54:315–332.

Dixon, R. B.
- 1913 Some Aspects of North American Archeology. *American Anthropologist* 15:549–577.

Dorsey, J. O., and J. R. Swanton
- 1912 A Dictionary of the Biloxi and Ofo Languages, Accompanied with Thirty-One Biloxi Texts and Numerous Biloxi Phrases. *Bureau of American Ethnology, Bulletin* 47.

Dunnell, R. C.
- 1971 *Systematics in Prehistory.* Free Press, New York.
- 1986 Methodological Issues in Americanist Artifact Classification. *Advances in Archaeological Method and Theory* 9:149–207.

Fisher, A. K.
- 1997 Origins of the Midwestern Taxonomic Method. *Midcontinental Journal of Archaeology* 22:117–122.

Ford, J. A.
- 1935a Outline of Louisiana and Mississippi Pottery Horizons. *Louisiana Conservation Review* 4(6):33–38.
- 1935b Ceramic Decoration Sequence at an Old Indian Village Site near Sicily Island, Louisiana. *Louisiana Department of Conservation, Anthropological Study* No. 1.
- 1936a Analysis of Indian Village Site Collections from Louisiana and Mississippi. *Louisiana Department of Conservation, Anthropological Study* No. 2.
- 1936b Archaeological Methods Applicable to Louisiana. *Louisiana Academy of Sciences, Proceedings* 3:102–105.
- 1938 A Chronological Method Applicable to the Southeast. *American Antiquity* 3:260–264.
- 1954a Comment on A. C. Spaulding, "Statistical Technique for the Discovery of Artifact Types." *American Antiquity* 19:390–391.
- 1954b Spaulding's Review of Ford. *American Anthropologist* 56:109–112.
- 1954c On the Concept of Types: The Type Concept Revisited. *American Anthropologist* 56:42–57.
- 1962 A Quantitative Method for Deriving Cultural Chronology. *Pan American Union, Technical Manual* 1.

Ford, J. A., P. Phillips, and W. G. Haag
- 1955 The Jaketown Site in West-Central Mississippi. *American Museum of Natural History, Anthropological Papers* 46(1).

Ford, J. A., and C. H. Webb
- 1956 Poverty Point, a Late Archaic Site in Louisiana. *American Museum of Natural History, Anthropological Papers* 46(1).

Ford, J. A., and G. R. Willey
- 1940 Crooks Site, a Marksville Period Burial Mound in La Salle Parish, Louisiana. *Louisiana Department of Conservation, Anthropological Study* No. 3.
- 1941 An Interpretation of the Prehistory of the Eastern United States. *American Anthropologist* 43:325–363.

Fowke, G.
- 1910 Antiquities of Central and Southeastern Missouri. *Bureau of American Ethnology, Bulletin* 37.
- 1928 Archaeological Investigations—II. *Bureau of American Ethnology, Annual Report* 44:399–540.

Freed, S. A., and R. S. Freed
- 1983 Clark Wissler and the Development of Anthropology in the United States. *American Anthropologist* 85:800–825.

Gatschet, A. S., and J. R. Swanton
- 1932 A Dictionary of the Atakapa Language Accompanied by Text Material. *Bureau of American Ethnology, Bulletin* 108.

Gibson, J. L.
- 1982 *Archeology and Ethnology on the Edges of the Atchafalaya Basin, South Central Louisiana.* Report submitted to the U.S. Army Corps of Engineers, New Orleans.

Gladwin, H. S.
- 1936 Editorials: Methodology in the Southwest. *American Antiquity* 1:256–259.

Gladwin, W., and H. S. Gladwin
- 1934 A Method for the Designation of Cultures and Their Variations. *Medallion Papers* 15.

Greenman, E. F.
- 1929 A Form for Collection Inventories. In Report of the Conference on Midwestern Archaeology, Held in St. Louis, Missouri, May 18, 1929. *National Research Council, Bulletin* 74:82–86.

Griffin, J. B.
- 1935 An Analysis of the Fort Ancient Culture. Notes on file, Museum of Anthropology, University of Michigan.
- 1943 The Fort Ancient Aspect: Its Cultural and Chronological Position in

	Mississippi Valley Archaeology. *University of Michigan Museum of Anthropology, Anthropological Papers* No. 28.
1974	Foreword. In *The Adena people,* by W. S. Webb and C. E. Snow, v–xix. University of Tennessee Press, Knoxville.
1976a	A Commentary on Some Archaeological Activities in the Mid-Continent 1925–1975. *Midcontinental Journal of Archaeology* 1:5–38.
1976b	Carl Eugen Guthe, 1893–1974. *American Antiquity* 41:168–177.
1985	The Formation of the Society for American Archaeology. *American Antiquity* 50:261–271.

Guthe, C. E.

1928	Archaeological Field Work in North America during 1927. *American Anthropologist* 30:501–524.
1929	Archaeological Field Work in North America during 1928. *American Anthropologist* 31:332–360.
1930a	Archaeological Field Work in North America during 1929. *American Anthropologist* 32:342–374.
1930b	The Committee on State Archaeological Surveys of the Division of Anthropology and Psychology, National Research Council. *International Congress of Americanists, Proceedings* 23:52–59.
1931	Archaeological Field Work in North America during 1930. *American Anthropologist* 33:459–486.
1932a	Archaeological Field Work in North America during 1931. *American Anthropologist* 34:476–509.
1932b	[Untitled Remarks] In *Conference on Southern Pre-History,* p. 1. NRC, Washington, D.C.
1933	Archaeological Field Work in North America during 1932. *American Anthropologist* 35:483–511.
1934	Archaeological Field Work in North America during 1933. *American Anthropologist* 36:595–598.
1935	The Society for American Archaeology Organizational Meeting. *American Antiquity* 1:141–151. [unsigned]
1936	Review of Thorne Deuel's "Basic Cultures of the Mississippi Valley." *American Antiquity* 1:249–250.
1937	Preface. In *The Indianapolis Archaeological Conference,* pp. v-vii. NRC, Washington, D.C.
1940	(editor) *International Directory of Anthropologists,* 2nd ed. NRC, Washington, D.C.
1952	Twenty-Five Years of Archeology in the Eastern United States. In *Ar-*

cheology of Eastern United States, edited by J. B. Griffin, pp. 1–12. University of Chicago Press, Chicago.

1967   Reflections on the Founding of the Society for American Archaeology. *American Antiquity* 32:433–440.

Haag, W. G.

1985   Federal Aid to Archaeology in the Southeast, 1933–1942. *American Antiquity* 50:272–280.

Hale, G. E.

1916   The National Research Council. *Science* 44:264–266.

1919   The National Importance of Science and Industrial Research: The Purpose of the National Research Council. *National Research Council, Bulletin* 1:1–7.

Harn, A.D.

1980   The Prehistory of Dickson Mounds: The Dickson Excavation, 2nd ed. *Illinois State Museum Reports of Investigations* 35.

Hodge, F. W.

1929   The Importance of Systematic and Accurate Methods in Archaeological Investigation. In Report of the Conference on Midwestern Archaeology, Held in St. Louis, Missouri, May 18, 1929. *National Research Council, Bulletin* 74:19–24.

Holmes, W. H.

1886   Ancient Pottery of the Mississippi Valley. *Bureau of Ethnology, Annual Report* 4:361–436.

1892   Modern Quarry Refuse and the Paleolithic Theory. *Science* 20:295–297.

1893a   Are There Traces of Man in the Trenton Gravels? *Journal of Geology* 1:15–37.

1893b   Traces of Glacial Man in Ohio. *Journal of Geology* 1:147–163.

1897   Stone Implements of the Potomac-Chesapeake Tidewater Province. *Bureau of American Ethnology, Annual Report* 15:13–152.

1903   Aboriginal Pottery of the Eastern United States. *Bureau of American Ethnology, Annual Report* 20:1–201.

Hrdlička, A.

1907   Skeletal Remains Suggesting or Attributed to Early Man in North America. *Bureau of American Ethnology, Bulletin* 33.

1918   Recent Discoveries Attributed to Early Man in North America. *Bureau of American Ethnology, Bulletin* 66.

Indiana Academy of Science

1921   *Indiana Academy of Science, Proceedings, 1920.* Indianapolis.

Jennings, J. D.
- 1974 *Prehistory of North America,* 2nd ed. McGraw-Hill, New York.

Johnson, F.
- 1947 The Work of the Committee for the Recovery of Archaeological Remains: Aims, History, and Activities to Date. *American Antiquity* 12:212–215.
- 1966 Archaeology in an Emergency. *Science* 152:1592–1597.

Judd, N. M.
- 1929 The Present Status of Archaeology in the United States. *American Anthropologist* 31:401–418.

Kehoe, A. B.
- 1990 The Monumental Midwestern Taxonomic Method. In The Woodland Tradition in the Western Great Lakes: Papers Presented to Elden Johnson, edited by G. E. Gibbon, pp. 31–36. *University of Minnesota, Publications in Anthropology* No. 4.

Kelly, J. E.
- 2000 Introduction. In *The Cahokia Mounds,* by W. K. Moorehead. University of Alabama Press, Tuscaloosa. [reprint of original]

Kidder, A. V.
- 1916 Chronology of the Tano Ruins, New Mexico. *American Anthropologist* 18:159–180.
- 1925 Notes on State Archaeological Surveys during 1924. *American Anthropologist* 27:581–587.
- 1926 Notes on State Archaeological Surveys in 1925. *American Anthropologist* 28:679–694.
- 1927 Southwestern Archaeological Conference. *Science* 66:489–491.

Kidder, M. A., and A. V. Kidder
- 1917 Notes on the Pottery of Pecos. *American Anthropologist* 19:325–360.

Kniffen, F. B.
- 1936 A Preliminary Report on the Indian Mounds of Plaquemines and St. Bernard Parishes. In Reports on the Geology of Plaquemines and St. Bernard Parishes, by R. J. Russell, pp. 407–422. *Department of Conservation, Louisiana Geological Survey, Geological Bulletin* No. 8.
- 1938 Indian Mounds of Iberville and Ascension Parishes. In Reports on the Geology of Iberville and Ascension Parishes, by R. J. Russell, pp. 189–207. *Department of Conservation, Louisiana Geological Survey, Geological Bulletin* No. 13.

Kroeber, A. L.
1916a   Zuñi Culture Sequences. *National Academy of Sciences, Proceedings* 2:42–45.
1916b   Zuñi Potsherds. *American Museum of Natural History, Anthropological Papers* 18(1):1–37.
1940a   The Work of John R. Swanton. In Essays in Historical Anthropology of North America: Publications in Honor of J. R. Swanton. *Smithsonian Miscellaneous Collections* 100:1–9.
1940b   Statistical Classification. *American Antiquity* 6:29–44.
1942   Tapajó Pottery. *American Antiquity* 7:403–405.

Lemley, H. J.
1936   Discoveries Indicating a Pre-Caddo Culture on Red River in Arkansas. *Texas Archeological and Paleontological Society, Bulletin* 8:25–55.

Lemley, H. J., and S. C. Dickinson
1937   Archaeological Investigations on Bayou Macon in Arkansas. *Texas Archeological and Paleontological Society, Bulletin* 9:11–47.

Lewis, T. H.
1883   Mounds of the Mississippi Basin. *Magazine of American History* 9(3):177–182.

Linton, R.
1932   The Interest of Scientific Men in Pre-History. In *Conference on Southern Pre-History*, pp. 3–4. NRC, Washington, D.C.

Lonergan, D.
1991   Hodge, Frederick Webb. In *International Dictionary of Anthropologists*, edited by C. Winters, pp. 293–295. Garland, New York.

Lyman, R. L., and M. J. O'Brien
1997   The Concept of Evolution in Early Twentieth-Century Americanist Archeology. In Rediscovering Darwin: Evolutionary Theory and Archeological Explanation, edited by C. M. Barton and G. A. Clark, pp. 21–48. *American Anthropological Association, Archeological Papers* No. 7.
1999   Americanist Stratigraphic Excavation and the Measurement of Culture Change. *Journal of Archaeological Method and Theory* 6:55–108.
2000   Chronometers and Units in Early Archaeology and Paleontology. *American Antiquity* 65:691–707.

Lyman, R. L., M. J. O'Brien, and R. C. Dunnell
1997   *The Rise and Fall of Culture History*. Plenum Press, New York.

Lyon, E. A.
1996   *A New Deal for Southeastern Archaeology*. University of Alabama Press, Tuscaloosa.

Mark, J.
- 1980 *Four Anthropologists: An American Science in Its Early Years.* Science History Publications, New York.

Mason, O. T.
- 1896 Influence of Environment upon Human Industries or Arts. *Smithsonian Institution, Annual Report* (1894), pp. 639–665.
- 1905 Environment. In Handbook of American Indians, edited by F. W. Hodge, pp. 427–430. *Bureau of American Ethnology, Bulletin* 30.

McGregor, J. C.
- 1941 *Southwestern Archaeology.* Wiley, New York.

McKern, W. C.
- 1934 Certain Culture Classification Problems in Middle Western Archaeology. *National Research Council, Committee on State Archaeological Surveys, Circular* No. 17.
- 1937a Wisconsin. In *The Indianapolis Archaeological Conference,* pp. 1–2. NRC, Washington, D.C.
- 1937b Certain Culture Classification Problems in Middle Western Archaeology. In *The Indianapolis Archaeological Conference,* pp. 70–82. NRC, Washington, D.C.
- 1938 Review of "Rediscovering Illinois" by F.-C. Cole and T. Deuel. *American Antiquity* 3:368–374.
- 1939 The Midwestern Taxonomic Method as an Aid to Archaeological Culture Study. *American Antiquity* 4:301–313.
- 1942 Taxonomy and the Direct Historical Approach. *American Antiquity* 8:170–172.
- 1943 Regarding Midwestern Archaeological Taxonomy. *American Anthropologist* 45:313–315.
- 1944 An Inaccurate Description of Midwestern Taxonomy. *American Antiquity* 9:445–556.

McKern, W. C., T. Deuel, and C. E. Guthe
- 1933 Paper on the problem of culture classification. NRC Archives, Washington, D.C.

Meltzer, D. J.
- 1983 The Antiquity of Man and the Development of American Archaeology. *Advances in Archaeological Method and Theory* 6:1–51.
- 1985 North American Archaeology and Archaeologists 1879–1934. *American Antiquity* 50:249–260.

Meltzer, D. J., and R. C. Dunnell
- 1992 Introduction. In *The Archaeology of William Henry Holmes,* edited by

D. J. Meltzer and R. C. Dunnell, pp. vi-l. Smithsonian Institution Press, Washington, D.C.

Mercer, H. C.
- 1897 Researches upon the Antiquity of Man in the Delaware Valley and the Eastern United States. *University of Pennsylvania Series in Philology, Literature, and Archaeology* 6.

Mills, W. C.
- 1906 Baum Prehistoric Village. *Ohio State Archaeological and Historical Society Quarterly* 15:45–136.
- 1907 Explorations of the Edwin Harness Mound. *Ohio State Archaeological and Historical Society Quarterly* 16:113–193.

Moore, C. B.
- 1892 Certain Shell Heaps of the St. John's River, Florida, Hitherto Unexplored. *American Naturalist* 26:912–922.
- 1894 Certain Sand Mounds of the St. John's River, Florida, Pts. 1–2. *Academy of Natural Sciences of Philadelphia, Journal* 10:5–103; 129–246.
- 1896 Certain River Mounds of Duval County, Florida. *Academy of Natural Sciences of Philadelphia, Journal* 10:448–502.
- 1902 Certain Aboriginal Remains of the Northwest Florida Coast, Pt. 2. *Academy of Natural Sciences of Philadelphia, Journal* 12:127–355.
- 1905 Certain Aboriginal Remains of the Black Warrior River. *Academy of Natural Sciences of Philadelphia, Journal* 13:125–244.
- 1907 Moundville Revisited. *Academy of Natural Sciences of Philadelphia, Journal* 13:337–405.
- 1908 Certain Mounds of Arkansas and of Mississippi—Part II: Mounds of the Lower Yazoo and Lower Sunflower Rivers, Mississippi. *Academy of Natural Sciences of Philadelphia, Journal* 13:564–592.
- 1909 Antiquities of the Ouachita Valley. *Academy of Natural Sciences of Philadelphia, Journal* 14:7–170.
- 1910 Antiquities of the St. Francis, White and Black Rivers, Arkansas. *Academy of Natural Sciences of Philadelphia, Journal* 14:255–364.
- 1911 Some Aboriginal Sites on Mississippi River. *Academy of Natural Sciences of Philadelphia, Journal* 14:367–480.
- 1912 Some Aboriginal Sites on Red River. *Academy of Natural Sciences of Philadelphia, Journal* 14:482–644.
- 1913 Some Aboriginal Sites in Louisiana and Arkansas. *Academy of Natural Sciences of Philadelphia, Journal* 16:7–99.

Moorehead, W. K.
- 1892a *Primitive Man in Ohio.* Putnam, New York.

1892b   Recent Archaeological Discoveries in Ohio. *Scientific American Supplement* 34:13,886–13,890.

1897   Report of Field Work Carried out in the Muskingum, Scioto, and Ohio Valley during the Season of 1896. *Ohio State Archaeological and Historical Society Quarterly* 5:165–274.

1899   Report of Field Work Carried out in Various Portions of Ohio. *Ohio State Archaeological and Historical Society Quarterly* 7:110–203.

1922   The Hopewell Mound Group of Ohio. *Field Museum of Natural History Anthropological Series* 6:73–184.

1927   A Report of Progress on the Exploration of the Cahokia Group. In The Cahokia Mounds (part I), edited by F. C. Baker, pp. 9–56. *University of Illinois, Museum of Natural History, Contribution* No. 28.

1929a   The Cahokia Mounds. *University of Illinois Bulletin* 26(4).

1929b   Mound Areas in the Mississippi Valley and the South. In Report of the Conference on Midwestern Archaeology, Held in St. Louis, Missouri, May 18, 1929. *National Research Council, Bulletin* 74:74–78.

Morgan, L. H.

1877   *Ancient Society.* Holt, New York.

National Research Council (NRC)

1921   *National Research Council, Fifth Annual Report.* Washington, D.C.

1937   *The Indianapolis Archaeological Conference.* Washington, D.C.

Nelson, N. C.

1916   Chronology of the Tano Ruins, New Mexico. *American Anthropologist* 18:159–180.

1928   Review of "The Antiquity of the Deposits of Jacob's Cavern," by V. C. Allison. *American Anthropologist* 30:329–335.

O'Brien, M. J.

1996   *Paradigms of the Past: The Story of Missouri Archaeology.* University of Missouri Press, Columbia.

O'Brien, M. J., and R. L. Lyman

1998   *James A. Ford and the Growth of Americanist Archaeology.* University of Missouri Press, Columbia.

1999a   *Seriation, Stratigraphy, and Index Fossils: The Backbone of Archaeological Dating.* Kluwer Academic/Plenum, New York.

1999b   *Measuring the Flow of Time: The Works of James A. Ford, 1935–1941.* University of Alabama Press, Tuscaloosa.

1999c   The Bureau of American Archaeology and Its Legacy to Southeastern Archaeology. *Journal of the Southwest* 41:407–440.

2000 *Applying Evolutionary Archaeology: A Systematic Approach.* Kluwer Academic/Plenum, New York.

O'Brien, M. J., R. L. Lyman, and J. Darwent
2000 Time, Space, and Marker Types: Ford's 1936 Chronology for the Lower Mississippi Valley. *Southeastern Archaeology* 19:46–62.

Parker, A. C.
1929 The Value to the State of Archaeological Surveys. In Report of the Conference on Midwestern Archaeology, Held in St. Louis, Missouri, May 18, 1929. *National Research Council, Bulletin* 74:31–41.

Phillips, P.
1973 Introduction. In *The Archaeological Reports of Frederic Ward Putnam*, pp. ix–xii. AMS Press, New York.

Phillips, P., J. A. Ford, and J. B. Griffin
1951 Archaeological Survey in the Lower Mississippi Alluvial Valley, 1940–1947. *Harvard University, Peabody Museum of Archaeology and Ethnology, Papers* 25.

Phillips, P., and G. R. Willey
1953 Method and Theory in American Archaeology: An Operational Basis for Culture-Historical Integration. *American Anthropologist* 55:615–633.

Pool, K. J.
1989 A History of Amateur Archaeology in the St. Louis Area. *The Missouri Archaeologist* 50.

Putnam, F. W.
1887 The Proper Method of Exploring an Earthwork. *Ohio Archaeological and Historical Quarterly* 1:60–62.

Rau, C.
1876 The Archaeological Collections of the United States National Museum in Charge of the Smithsonian. *Smithsonian Contributions to Knowledge* 22(4).

Ruegamer, L.
1980 *A History of the Indiana Historical Society 1830–1980.* Indiana Historical Society, Indianapolis.

Setzler, F. M.
1933a Hopewell Type Pottery from Louisiana. *Washington Academy of Sciences, Journal* 23:149–153.
1933b Pottery of the Hopewell Type from Louisiana. *United States National Museum, Proceedings* 82(22):1–21.
1934 A Phase of Hopewell Mound Builders in Louisiana. *Explorations and*

    Field-Work of the Smithsonian Institution in 1933, pp. 38–40. Smithsonian Institution, Washington, D.C.
1940 Archaeological Perspectives in the Northern Mississippi Valley. *Smithsonian Miscellaneous Collection* 100:253–291.

Setzler, F. M., and W. D. Strong
1936 Archaeology and Relief. *American Antiquity* 1:301–309.

Shetrone, H. C.
1920 The Culture Problem in Ohio Archaeology. *American Anthropologist* 22:144–172.

Smith, B. D.
1990 The Division of Mound Exploration of the Bureau of Ethnology and the Birth of Modern American Archeology. In *Edward Palmer's Arkansaw Mounds,* edited by M. D. Jeter, pp. 27–37. University of Arkansas Press, Fayetteville.

Spaulding, A. C.
1953 Statistical Techniques for the Discovery of Artifact Types. *American Antiquity* 18:305–313.
1954a Reply to Ford. *American Anthropologist* 56:112–114.
1954b Reply to Ford. *American Antiquity* 19:391–393.

Spier, L.
1917 An Outline for a Chronology of Zuñi Ruins. *American Museum of Natural History, Anthropological Papers* 18(3):207–331.

Squier, E., and E. H. Davis
1848 Ancient Monuments of the Mississippi Valley. *Smithsonian Contributions to Knowledge* 1.

Stevens, S. S.
1952 The NAS-NRC and Psychology. *American Psychologist* 7:119–124.

Steward, J. H.
1942 The Direct Historical Approach to Archaeology. *American Antiquity* 7:337–343.

Stirling, M. W.
1929a Comments on Mr. Hodge's Paper. In Report of the Conference on Midwestern Archaeology, Held in St. Louis, Missouri, May 18, 1929. *National Research Council, Bulletin* 74:4–28.
1929b [Untitled Remarks]. In Report of the Conference on Midwestern Archaeology, Held in St. Louis, Missouri, May 18, 1929. *National Research Council, Bulletin* 74:109–112.
1932 The Pre-Historic Southern Indians. In *Conference on Southern Pre-History,* pp. 20–31. NRC, Washington, D.C.

Strong, W. D.
- 1935 An Introduction to Nebraska Archeology. *Smithsonian Miscellaneous Collections* 93(10).
- 1936 Anthropological Theory and Archaeological Fact. In *Essays in Anthropology: Presented to A. L. Kroeber in Celebration of His Sixtieth Birthday, June 11, 1936,* edited by R. H. Lowie, pp. 359–368. University of California Press, Berkeley.

Swanton, J. R.
- 1919 A Structural and Lexical Comparison of the Tunica, Chitimacha, and Atakapa Languages. *Bureau of American Ethnology, Bulletin* 68.
- 1940 Linguistic Material from the Tribes of Southern Texas and Northeastern Mexico. *Bureau of American Ethnology, Bulletin* 127.

Swanton, J. R., and H. S. Halbert (editors)
- 1915 A Dictionary of the Choctaw Language, by C. Byington. *Bureau of American Ethnology, Bulletin* 46.

Taylor, J. L. B.
- 1921a Discovery of a Prehistoric Engraving Representing a Mastodon. *Science* 54:357–358.
- 1921b Did the Indian Know the Mastodon? *Natural History* 21:591–597.

Thomas, C.
- 1884 Who Were the Mound Builders? *American Antiquarian* 6:90–99.
- 1894 Report on the Mound Explorations of the Bureau of Ethnology. *Bureau of Ethnology, Annual Report* 12:3–742.

Thomas, C., and J. R. Swanton
- 1911 Indian Languages of Mexico and Central America and Their Geographical Distribution. *Bureau of American Ethnology, Bulletin* 44.

Trubowitz, N. L.
- 1993 The Overlooked Ancestors: Rediscovering the Pioneer Archaeology Collections at the Missouri Historical Society. Paper presented at the 58th Annual Meeting of the Society for American Archaeology, St. Louis.

Tylor, E. B.
- 1871 *Primitive Culture.* Murray, London.

Walker, W. M.
- 1932 Pre-Historic Cultures of Louisiana. In *Conference on Southern Pre-History,* pp. 42–48. NRC, Washington, D.C.

Wedel, W. R.
- 1935 Reports of Field Work by the Archaeological Survey of the Nebraska State Historical Society. *Nebraska History Magazine* 15(3).

1938   The Direct-Historical Approach in Pawnee Archaeology. *Smithsonian Miscellaneous Collections* 97(7).

Willey, G. R.
1936   A Survey of Methods and Problems in Archaeological Excavation, with Special Reference to the Southwest. Unpublished M.A. thesis, University of Arizona, Tucson.

Willey, G. R., and P. Phillips
1955   Method and Theory in American Archaeology, II: Historical-Developmental Interpretation. *American Anthropologist* 57:723–819.
1958   *Method and Theory in American Archaeology.* University of Chicago Press, Chicago.

Willey, G. R., and J. A. Sabloff
1993   *A History of American Archaeology,* 3rd ed. Freeman, New York.

Wilson, T.
1899   Arrowheads, Spearheads, and Knives of Prehistoric Times. *United States National Museum, Annual Report* (1897), Part I, pp. 811–988.

Wissler, C.
1914   Material Cultures of the North American Indians. *American Anthropologist* 16:447–505.
1916   Correlations between Archeological and Culture Areas in the American Continents. In *Holmes Anniversary Volume: Anthropological Essays,* edited by F. W. Hodge, pp. 481–490, Washington, D.C.
1917   *The American Indian.* McMurtrie, New York.
1922   Notes on State Archaeological Surveys [for 1921]. *American Anthropologist* 24:233–242.
1923a   Notes on State Archaeological Surveys [for 1922]. *American Anthropologist* 25:110–116.
1923b   *Man and Culture.* Crowell, New York.
1924   The Relation of Nature to Man as Illustrated by the North American Indian. *Ecology* 5:311–318.

Wissler, C., A. W. Butler, R. B. Dixon, F. W. Hodge, and B. Laufer
1923   *State Archaeological Surveys: Suggestions in Method and Technique.* NRC, Washington, D.C.

Woodbury, R. B.
1973   *Alfred V. Kidder.* Columbia University Press, New York.

Wright, J. H., J. D. McGuire, F. W. Hodge, W. K. Moorehead, and C. Peabody
1909   Report of the Committee on Archeological Nomenclature. *American Anthropologist* 11:114–119.

Bulletin of the National Research Council
December 1929
Number 74

Report of the Conference on Midwestern
Archaeology, Held in St. Louis, Missouri
May 18, 1929

Including a Report of an Open Meeting of
the Committee, Held May 17, 1929

Issued under the Auspices of the Committee on
State Archaeological Surveys of the Division of
Anthropology and Psychology of the National
Research Council

[Frontispiece of original report] The Serpent Mound, Adams County, Ohio. The largest known "effigy mound," 1330 feet in length. It probably had a religious significance to its builders. The location is now a state park belonging to the Ohio Archaeological and Historical Society.

PREFACE

The Committee on State Archæological Surveys of the National Research Council has been active for several years in promoting the conservation and scientific study of the prehistoric Indian sites scattered through the states bordering on the Mississippi River and its tributaries. The appalling destruction of these sites by individuals ignorant or careless of their value, and by commercial exploiters, has extinguished much valuable historical evidence. Inefficient excavation by untrained " archæologists " has contributed its quota of ruin. On the other hand, the benefits derived through conserving of the mounds and making them available to the public as parks, and through fostering of their scientific study, have been signally demonstrated by several of the states.

As a detail in its efforts to promote the conservation, study, and public use of the sites, the Committee decided to hold a Conference on Midwestern Archæology, at which archæologists familiar with the region, representatives of the states, and other public spirited citizens might discuss the problems; and through which the facts might effectively be laid before the public. St. Louis, the " Mound City," was selected as the place for the Conference, both because of the wealth of archæological material in the State of Missouri and because of the generous cooperation of the Governor and citizens of that state. The Conference was held on Saturday, May 18, 1929, preceded on Friday, May 17, by an open meeting of the Committee for the presentation of technical reports in the morning; an expedition to the Cahokia mounds (see Figures 1 and 2) under the direction of Professor Moorehead in the afternoon; and an illustrated public lecture by Director H. C. Shetrone of the Ohio State Museum in the evening.

By means of many lantern slides, Mr. Shetrone introduced the audience to the great variety of earthworks scattered through the " General Mound Area " including the " effigy mounds " of Wisconsin, representing birds and animals; the Great Serpent Mound (see frontispiece), the great conical mounds, and the geometrical mound of Ohio; and the truncated pyramid flood-refuges of the lower Mississippi valley. The lecturer portrayed vividly the excavation work of the archæologist, by which he reveals the physical and social characteristics and the arts and industries of the mound-builders. Copious illustrations of the artistic pottery, and the implements and ornaments of stone, shell, bone, copper, silver and pearl, of these peoples, were presented.

The Conference concluded Saturday evening with a dinner, from which addresses were radioed through Station KMOX. On Friday and Saturday, radio talks relevant to the conference purposes were given over Stations KMOX and KWK.

The principal addresses at the Conference and the preliminary sessions, with the additional radio speeches, are included in this bulletin, together with extracts from the discussions. These statements, by men thoroughly familiar with the matters whereof they speak, present the situation and the needed remedies more effectively than would be possible in a single expository account.

The National Research Council expresses its grateful appreciation to the Committee on State Archæological Surveys, to the members of the Conference, to Governor Caulfield of Missouri, and to many citizens of the state who gave their time and energy towards making a success of the Conference.

                        KNIGHT DUNLAP,
                            Chairman, Division of
                            Anthropology and Psychology,
                            National Research Council.

CONTENTS

FRONTISPIECE—The Serpent Mound   PAGE

PREFACE 3

PART I. PAPERS AND DISCUSSIONS BEFORE THE CONFERENCE

    Address of welcome. Governor Henry S. Caulfield 7
    Reply. Knight Dunlap 10
    The conservation of public sites. Fay-Cooper Cole 11
    Discussion of Mr. Cole's paper. Rufus Dawes 16
    The importance of systematic and accurate methods in archæological
        investigation. F. W. Hodge 19
    Discussion of Mr. Hodge's paper. M. W. Stirling.................... 24
    General discussion 28
    The value to the state of archæological surveys. A. C. Parker 31
    Discussion of Mr. Parker's paper. G. R. Throop 42
    Archæology as a human interest. Clark Wissler 44
    Discussion of Mr. Wissler's paper. Wm. John Cooper 48
    General Discussion 50

PART II. PAPERS AND DISCUSSIONS PRESENTED AT THE OPEN MEETING OF THE
    COMMITTEE ON STATE ARCHÆOLOGICAL SURVEYS

    Excavation of the Nicholls Mound of Wisconsin. W. C. McKern 57
    Discussion of Mr. McKern's paper 60
    The Williams and Glover Sites in Christian County, Kentucky. William
        S. Webb 61
    Discussion of Mr. Webb's paper 62
    Some recent notable finds of urn burials in Alabama. P. A. Brannon 63
    Discussion of Mr. Brannon's paper 67
    Trailing DeSoto. John R. Fordyce 67
    Discussion of Mr. Fordyce's paper 73
    Mound areas in the Mississippi Valley and the South. W. K. Moorehead 74
    Discussion of Mr. Moorehead's paper 78
    A form for recording data of field surveys. S. A. Barrett 79
    Discussion of Mr. Barrett's paper 80
    A form for collection inventories. E. F. Greenman 82
    Discussion of Mr. Greenman's paper 86

PART III. ADDRESSES AT THE DINNER, AND ADDITIONAL RADIO ADDRESSES

    S. A. Barrett 89
    C. E. Guthe 92
    R. B. Dixon 95
    W. K. Moorehead 97
    P. E. Cox 98
    Thomas S. Knapp 100
    G. R. Throop 102
    F.-C. Cole 103
    S. A. Barrett 104
    C. E. Guthe 105
    Rufus Dawes 107
    M. W. Stirling 109

	PAGE
Resolutions and Appreciations	112
Appendix A. List of Members of the Committee on State Archæological Surveys	113
Appendix B. List of Members of the Conference on Midwestern Archæology	114
Appendix C. Plates	
Figure 1—Monk's Mound, Cahokia group	117
Figure 2—The " Red " Mound, Cahokia group	117
Figure 3—Pottery vessels	117
Figure 4—An example of intelligent methods of excavation	118
Figure 5—Post-molds, under a mound	119
Figure 6—A fine specimen of Hopewell pottery	119
Figure 7—Examples of early Indian handiwork	120

PART I
PAPERS AND DISCUSSIONS AT THE CONFERENCE

ADDRESS OF WELCOME

Governor Henry S. Caulfield

Mr. President, Ladies and Gentlemen: I feel very humble coming before this august body of learned and eminent men, but there is one thing which strengthens me, and which is illustrated by the story of the little Massachusetts constable who said to the big bully, "You can shake me, but when you shake me you have got to shake the whole State of Massachusetts." And so I am strengthened by the thought that I am not here as an individual, but as the representative of the great State of Missouri.

I am proud of being a Missourian, and so proud of being its Governor that I am very glad and happy to be here with you. I am also strengthened by the presence here of Mr. W. E. Freeland, who is the Republican floor leader of the House of Representatives in Missouri.

I am very glad to welcome you here to the State of Missouri, and I think the state should be very proud to have such a body as this assembled within her borders. And may I say to you that I think you are to be congratulated upon your choice of a place of meeting. I am not in charge of the convention bureau or anything of that sort; but I think the head of the convention bureau would agree with me that St. Louis is an ideal place for any convention or conference, because of its central location, because of its being a railroad center, and because of the many other conveniences it offers.

But there is a special reason why you should gather here, and that reason you know a great deal better than I do. It is because this place is full of interest on account of the very matters which you are assembled here to discuss. St. Louis was formerly called the Mound City, and of course that was because of the large number of mounds that were here. I think that quite a group, of nine or ten great mounds, were on the original site of St. Louis; and unless they have been recently obliterated there are some here yet, I am told. There were some of them in Forest Park. There is going to be an assemblage this afternoon of the Colonial Dames, and others who dare face this weather, to put a tablet where there was formerly the sepulchral mound, called the Big Mound, and you know that was at the foot of Broadway and Mound Street. That is one reason, I suppose, that they called it Mound Street. I lived on Mound Street when I was a little boy, and so when I read the other day concerning Big Mound that it was destroyed in 1869, I did not believe it. It is either my imagination, or the author was wrong, because I have a very distinct recollection of seeing at least parts of that mound when I was a youngster playing there, and I know I was not born until four years after 1869.

But that mound is gone now, and so are almost all other mounds here in St. Louis, except some small ones that have attracted no great interest. There has been a great deal of destruction of these mounds in the State of Missouri. But even so, Missouri still is, I believe, a wonderful place for research and study for archæologists. We not only have some mounds remaining which should be preserved, but we have an abundance of caves. The Ozarks abound with wonderful caves; and I am told that that is a place where the archæologist may find much matter for study.

I was asking a while ago whether there was any money in this mound-preserving business. You know nowadays we, of course, have a great many people who are interested in things from a cultural standpoint, and I believe that there are enough of those to preserve the mounds. But if you want to insure their preservation, just show how there is going to be some money in it for somebody, and I will guarantee that they will be preserved. And I am very glad to be told that there is money in it. Here and there where they have been preserved, the sight-seers, coming with this spendid road system we now have everywhere, actually pay, in admission fees alone, enough to justify the preservation of the mounds. And in making this land more interesting we bring innumerable tourists, and consequently can figure it as a great financial benefit to the city and the state.

But lest you get a false idea of Missouri, I want to say that Missouri has not proceeded solely along the lines of money prosperity. We have a most beautiful capitol. Maybe somebody will dispute this, but I think I have a right to say as the Governor of Missouri that we have the most beautiful capitol building in the United States. I am told the mural decorations in the Missouri capitol are more beautiful than any in the United States except those in the Congressional Library and in the Boston Art Institute. I don't know whether this is true or not, but I will say it anyway.

Still, I do wish to impress on you that Missouri does things for other purposes than making money. We have now established a park system of some 28,000 acres in the beautiful Ozarks. That is another proof that we do not commercialize everything, because those parks are not commercialized at all. If you can show the people of Missouri that there is a reason for preserving the mounds, it does not have to be a purely commercial or financial reason. If you can show the people of Missouri that there is an artistic reason, or a romantic reason, for maintaining these mounds, I believe that, through their government, they will preserve them. And I may say to you that if your meeting here can only start the thought in that direction, you will have accomplished a great thing for the people of Missouri.

Of course, I do not overlook the fact that all around us, in all these states, they have rich treasures in the way of mounds and caves and places where the archæologist can delve and enjoy himself, but I am particularly interested in Missouri.

I think there is another reason, perhaps, why your study helps the people of the state. It creates in the mind of the people a veneration for the soil. You know it makes us like a place better if there is a mystery or something wonderful about it; and you know all of us like to think the Indians used to be here. I remember when I was a little boy up in Louisiana, Missouri, we used to go up on the hill (a big mountain, we called it) and look for arrowheads. I don't know of anything that appeals more to the imagination of a boy than the thought that there was a race here before us that has vanished; and that sort of thought is good for people. It upbuilds them spiritually, and that is a thing we need very much to create and foster—the spiritual, the imaginative in our people.

I might tell you a little thing that occurred some time ago that interested me, concerning the usefulness of the finds in these tombs and mounds. I was talking to the manager of the American Car and Foundry Company, and we were especially interested in a rivet-heating furnace, and its efficiency. He told me that he undertook to construct a furnace some time ago, and got a plan of one that had been found in one of the tombs along the Nile. I suppose it was about two thousand years old. He constructed his furnace in exact accordance with that ancient plan, and he said it was working efficiently. It is very interesting to me to think that the same kind of furnace that worked efficiently two or three thousand, or I don't know how many years ago, is right here in St. Louis operating with efficiency, and is just as good as any modern furnace they have.

. . . . Now, I have talked longer than I intended. I came here for no other purpose than to greet you and to welcome you to Missouri. And when I say welcome, I mean welcome. I really think it is a very wonderful thing for our people to have such men as are assembled here today come among us and hold this conference. It is a tremendous compliment to St. Louis and to Missouri, and on behalf of the people of Missouri I desire to stress our very great appreciation that you are here, and to welcome you, and hope that you will enjoy yourselves.

REPLY BY THE CHAIRMAN
KNIGHT DUNLAP

I wish to back up what the Governor says about the beauty of the capitol at Jefferson City, and its beautiful situation. I wondered, when I was out there, how he kept his mind on his work with that beautiful expanse before him, but I found he did it most completely.

On behalf of the National Research Council, I wish to express my deep appreciation, not only of the cordial welcome which Governor Caulfield has just extended, but also of the cooperation which we have received from those citizens of this state and this city, who have made this conference possible.

The organization of a conference of this scope, and at this distance from Washington, would, under the best of circumstances, be a difficult matter. There have arisen, moreover, extraordinary difficulties which we could not foresee when the original plans for the conference were made last year. Abnormal conditions of governmental activity affecting many persons, not only in Washington but throughout the Middle West, have rendered impossible the active cooperation of many who would otherwise have been with us. Thanks to the labor of the Committee on State Archæological Surveys, and of many local organizations and individuals, we have gathered here today a distinguished company to consider a momentous problem.

The National Research Council may very properly concern itself with this matter. It has been actively promoting the preservation of the precious archæological material scattered through the regions from the Alleghenies westward to the Rockies, in the interests of the scientific study of these materials, and of the profit of the people of this great area through such preservation and study. The Council has neither purposes nor means of dictation or control. Its function in the matter is to encourage state and local interest and pride, to aid in organization, to promote discussions and cooperation, and to assist in dissemination of information. With these purposes in view this conference has been organized. The speakers have been requested to present their statements in such manner and detail as shall be informative and useful not only to the representatives assembled here today, but to the people of the United States during the next few years.

[10]

THE CONSERVATION OF PUBLIC SITES

Fay-Cooper Cole

Everyone is interested in the early inhabitants of America. Every boy has played Indian and at some time has longed for the traditional carefree life of our predecessors. In each village and city of the Mississippi valley are men who have tramped the fields gathering evidences of this early life, and many are the collections scattered over our states. Museums exhibit cases of Indian remains, while historical and other learned societies publish books dealing with the Red Man. Undoubtedly there is an interest; and yet in nearly every state and township, we are rapidly destroying every vestige of Indian life.

Imagine a valuable, illustrated historic book of 655 pages placed in your county courthouse. Person after person comes in and looks it over. One rips out a leaf and stuffs it in his pocket. Another, somewhat more careful, takes his penknife and removes an illustration, but in so doing destroys the reading matter on the opposite side of the page. Someone mildly protests, saying that the pictures and pages will soon be scattered and of no value to anyone; but he is met with the reply that the book belongs to all the people and if one does not get his share now, another will. And so the destruction goes on until only fifty pages of the book are left. Then suddenly the people of the county come to realize that they have allowed the destruction of a priceless historic document, a volume which would have brought thousands of visitors to the county and thus added to its fame and to its revenue.

The destruction of this book is not fancy but fact. In one county in Illinois there have been located 655 Indian mounds varying from simple burial plots to pyramids of considerable size; from earth works to effigy mounds. And all but fifty of these have been dug into and for the most part looted, and their historical significance lost. Three years ago when the Archæological Survey of Illinois attempted to recover the prehistory of that county it found many collections, each with a few specimens, but in less than a half dozen cases was it possible to get accurate information as to the place from which the materials had come and the conditions of the finds. Six hundred pages had been torn from the book of that county's prehistory—pages that can never be rewritten. The looted mounds bear eloquent testimony to the fact that the volume was beautifully illustrated, but today the county possesses only a few battered remains.

This case is not the exception but the rule. The State of Michigan had hundreds of mounds, yet almost without exception they have been disturbed and most of them looted. A similar record is to be found in nearly every state in the Mississippi valley. Yet it is not primarily the work of vandals. In nearly all cases it is due to lack of knowledge. The mounds, or at least the records contained in them, might have been preserved had the owners realized their importance.

Last year a town in central Illinois planned a picnic and it was advertized that as a part of the celebration an Indian mound would be opened. An associate of the speaker investigated, fearing that the promoters might not be properly equipped to carry out the work. He suspected they might be planning to use horses and scrapers, or perhaps even a tractor. But this was not their intention. All their plans were laid to blow it up with a stick of dynamite to see what it contained. When the value of the mound was explained to them, they not only gave up the idea of the explosion, but left the site entirely untouched.

Many untrained men open mounds out of curiosity or in search of relics. Most of them have learned from experience that our mounds are not as a rule very productive in specimens suitable for a collection. The work is hard, the finds few, and of little monetary value, so that most of the mounds are only pitted. A hole is dug at the top near the center, and when the digging proves unprofitable the site is abandoned. Dampness slowly penetrates to the lower levels, ruining in a few years material which has survived generations.

Most of the destruction of the mounds has occurred in this manner. Material of great historic importance has been destroyed, while little has been acquired. The occasional find of a really beautiful specimen has often set the whole countryside to digging, although in all probability the piece in question was found on or near the surface. The seasoned collector of Indian relics has learned that old camp sites are so much more profitable than the mounds that he seldom indulges in digging the latter, but a new crop of the curious and untrained carries on the work of destruction each year. In a few instances professional dealers in Indian relics have worked great havoc. Having secured permission from the owners, they riddle the mounds in the hope of finding a few pieces of value. Pick and shovel crews throw the material out where it can be sorted and only an occasional piece is saved. A visit to such a digging will reveal broken pots and decaying bones scattered over the soil. No record is kept of the sites and thus valuable scientific material is absolutely destroyed. In the valley of the Illinois the looters—for they are such—have destroyed log tombs and most of their contents despite the fact that scientific men are tremendously interested in this extension of an Ohio culture into the Illinois region.

In many instances mounds have been destroyed by the owners of the land for what have seemed to them sound practical reasons. A farmer finds the mounds on his property obstacles to cultivation, and levels them off. In other cases the growth of cities has led sub-dividers, factory builders, and others to destroy the mounds to make way for buildings.

With all these forces working toward the destruction of the mounds, what hope is there of saving them? Some have urged drastic legislation to prevent the opening or destruction of any historic site except by qualified archæologists. But experience has taught that legislation is of little avail unless those affected by it are convinced of its desirability. If we are to save what remains, we must convince the people of the Mississippi valley that the mounds are valuable and should only be excavated by, or under the direction of, trained field workers.

Are the mounds valuable to the community? Near Lewistown, Illinois, is a large mound situated near a farm house. A son of the family started to excavate it and soon found that it contained many burials. He was determined to leave these in place, but realized that to preserve them he must protect them with a building. Consequently the mound was enclosed by walls and roofed over. As bodies were encountered, the earth was removed but the bones and articles associated with them were left undisturbed. (See Figure 4.) Over two hundred skeletons have been thus exposed and with them are the pots and articles of bone and stone which they used in life. (See Figure 3.) The expense in time and money of opening the mound was considerable. The monetary value of all the articles found would not have repaid the work of excavation, but when people learned of the unique exhibit, they began to come from near and far. The fame of the place spread, and this past summer 20,000 persons visited the site, although it is some distance out in the country. Ask the owner of this property if he is sorry he preserved the record by careful excavation. Ask the merchants of the town if the mounds are an asset.

The State of Ohio has long recognized the value of its Indian monuments. It has carefully excavated those which were doomed to destruction. It has purchased and preserved other sites and today it has a number of state parks which attract thousands of its own people. The fame of Serpent Mound and Fort Ancient has spread far beyond state borders and each year hundreds of visitors tour the state to see these prehistoric sites. Wisconsin has been active in the preservation of its mounds, and Illinois has taken steps to care for the great Cahokia pyramids and the historic site known as Starved Rock.

Near Nashville in Tennessee are some of the most perfect pyramids in North America which need only to be made known to prove a great source of revenue to the state and community.

I have cited only a few instances, but every state in the Mississippi valley might have been included.

Up to this point I have stressed particularly the value of the mounds as an asset—something which will bring visitors and funds to a community; but there is another side which is of even greater importance.

We of America have been very close to our aboriginal people. In Europe the early steps toward civilization are buried beneath centuries of struggle and advance. But on this continent, and especially in the Mississippi valley, we still have with us men who had intimate contact with Indian life. We still have groups of Indians; and we still have some undisturbed mounds and other monuments of Indian culture. But we are fast Americanizing the Indian. For good or ill, we are forcing our civilization on him, until in a short time his life will be preserved only in books and museums.

We owe it to posterity to preserve as fully as possible all information pertaining to this life. We owe it to our children and to future generations to preserve the mounds and other prehistoric sites, or to hand on to them the most nearly complete record possible.

Today every historian, every student of human affairs, deplores the fact that the Spaniards destroyed the records of the Maya and Aztec. Some of the early explorers felt it their duty to destroy the old in order to establish the new; and so documents and records relating to rites and religious ceremonies were burned, and the temples smashed. To them it was a duty, but no one today raises his voice in their defense. So effectually did they erase the ancient records that today we are unable to decipher the vast store of information hidden in the Maya glyphs. Now our scholars like Morley and Spinden are spending years trying to decipher this record; our great institutions are devoting funds to the recovery of this lost history, and only recently General Dawes has provided the Smithsonian Institution with funds to make a search of the monasteries of Europe in the hope that some bit of this material may have escaped destruction.

Today we are destroying the record of Indian life in this region; we are not doing it through a sense of duty, but through carelessness and neglect; and our children will not hold us blameless. The historian of the future will charge us with permitting looting and destruction of valuable historic documents, for every advance made in the archæological record adds just that much to history.

[14]

It is the aim of archæology to make the past live again. Do we wish to know the steps taken on this continent by early man on the long journey toward civilization? Do we wish to know how his culture grew from simple beginnings to the high cultures of Central America? Do we wish to know the effects of the American environment on man? If so, we have the materials before us. We can preserve them, or we can continue the present methods of indiscriminate digging and destruction— for a short time—and then they will be destroyed forever.

In the papers which are to follow we are to learn of the value of these prehistoric sites to science. In these sites we have an unprejudiced source of information concerning the development of man and his culture on American soil. It is evident to all who know the facts that we have here valuable documents which should be preserved for science, and which at the same time are distinct assets to the owners of the land and to the the community in which they lie. Our aim should be to acquaint our people with the value of the mounds and we should urge them to use their utmost efforts to see that they are not destroyed through ignorance or through thoughtless digging.

I have been asked to speak primarily of the Indian mounds, but in our endeavor to preserve them we should not neglect historic sites, places of great natural beauty, unique geological formations, or forest lands. We wish to serve science, but at the same time we can provide recreation places for our people. There are in every state many such sites which, if protected and made accessible, will serve as state parks and wayside museums. They will add to our pleasure, they will enrich the community, and they will add to our knowledge of man and the world in which we live.

I might add that I went to Mr. Dawes some time ago and suggested that there were certain valuable sites which ought to be preserved, and he suggested that I should go to members of our Conservation Board and report; and now our Regional Planning Commission has suggested that we should plot all over the State of Illinois the historical monuments, the archæological sites, sites which should be preserved because of their beauty, because of their value to the geologist and historian, or to the archæologist; sites which we might turn into wayside museums or wayside parks, distributed well over the state. With these plotted, we can make our appeal to the Legislature and to the people of the state at large to preserve these public grounds as the property of the state; to preserve them as playgrounds for the people, and at the same time as monuments to a past civilization.

DISCUSSION OF MR. COLE'S PAPER

Rufus Dawes

I have been wondering what reason there might be, other than my admiration for Mr. Cole, why I should be discussing his paper in this society of Sigma Xis and this group of distinguished and eminent scientists. My only qualification would seem to be that I was born within the enclosure of the mound builders at Marietta, Ohio. It was there, as you all know, that the first settlement under the auspices of the Federal Government was made in this great Northwest territory. And I am sure that the men who made that settlement, knowing that that land had been occupied by some mysterious civilization before them, were not surprised that they found the site which they had selected was covered with a magnificent and significant display of the mound builders' remains; because it was true, as they well knew, that every civilization back of them had been a civilization which had received its form and character from its methods of travel and transportation. The white man, when he went into the country, added nothing to what the mound builders or Indians had of such methods. Both of them, at the outset, depended upon water for transportation and travel. It was then no matter of surprise that these mound builders' relics were found upon sites afterwards selected by the white man for his own cities. St. Louis was the site of a great collection of mounds. In Ohio, at Marietta, Newark, Circleville, Lancaster and the sites of many other cities, there had been formerly the cities of these mound builders. That, I think, accounts for the destruction of a great many of these mounds.

And I want to say this much for the men of Marietta. When they went there to make the settlement, they did not destroy these mounds out of mere curiosity. They destroyed them because the use of the ground was necessary for the advancement of a new civilization. But before they destroyed them they made the most thorough examination and surveys of these mounds, under the direction of Rufus Putman, himself an engineer, and called the "Father of Ohio." They made very accurate maps. From the hand of Mannassah Cutler, a learned man, they left a good description of all of those mounds and elevated squares and parallel walls, which formerly existed on the site of this beautiful city at the confluence of the Ohio and Muskingum rivers, and thus located in such a place as to be most convenient to those dependent upon water to carry the products of their industry.

I have wondered that one circumstance in connection with the old map they left has not attracted more of the attention of archæologists; to wit, their description of a well within the greater enclosure at Marietta, said to have measured sixty feet in diameter and to have been at least twenty-five feet deep, and to have been provided with steps down which those wishing to get the water would go with their receptacles. That well, accurately mapped as it is, it seems to me ought even yet to be explored in the hope that it might produce, perhaps not so much, but something along the same lines as the well at Chitzen Itza. For there, surely, you would find examples of pottery and of all the things commonly used in the lives of those people.

Now it seems to me that when we consider such a situation as that at Marietta, where we find one civilization building its structures upon the remnants of another long since disappeared, and both of them adopting the same place because both depend upon the same means of transportation, we have at least this question before us—whether or not the rapidly changing methods of transportation, altering fundamentally the civilization in which we live, may not present to us the opportunity to accomplish the very things that we all want to accomplish. Not so much is to be gained in making philosophical discussions of the advantages of maintaining traditions before great masses of people. Tradition is more than the reason and intelligence of man provides. It is something more than the written pages of history. Reason is but a part of human nature; and tradition is made up of human nature with all its emotions, and all of its pride; the appeal to tradition must be made in some subtle way rather than by arguments, perfectly convincing to a group of intelligent men like this, but not sufficiently powerful to stir the sentiments of the great mass of people.

Now there is a sentiment which controls the feelings of great masses of the people everywhere, to which we can appeal. We have come into the age of the automobile, and of good roads, where the masses of the people move over distances in hours which a generation ago could not be covered in days. Consequently there has been created a measured demand for recreational spots. In Chicago we have for many years been proud of the fact that far-seeing men there anticipated the need, and before the day of the automobile we had provided for an outer fringe of parks entirely surrounding the city, from Wisconsin clear around to the Indiana line. They felt that there at last was a city which had taken means to provide the recreational spots for a population of indefinite size—some thirty thousand acres they provided, not in one compact tract, but scattered through all that great area upon the basis of preserving the forests for coming generations.

They had hardly begun that acquisition of land when we witnessed the change in the habits of the people brought about by the use of the automobile. And now, at a recent Conference on National Parks, we find that authorities have measured the demand existing at the present moment, and have found that there should be about ten acres of park or recreational grounds for every one thousand people. Lo and behold, we find that in spite of all these efforts by which we thought we had provided for generations in advance, we have today in Chicago only the amount now regarded by those who study the question as indispensably necessary for the accommodation of our present population. We have calculated that to provide for Chicago's rapidly increasing population, there must, within twenty-two years be, immediately adjacent to that city—and by that I mean within a hundred and fifty miles—at least twice as much park land as we have now.

In our parks and in the forest preserves about the city we have about thirty thousand acres, but if we are to provide for the estimated population in 1950, ten acres of ground for each one thousand people, there must be provided at least sixty thousand acres more.

Now, how simple it is to appeal to the interest of the public in providing this indispensable necessity of modern civilization, to call upon them in selecting these lands which, in order to achieve the purpose for which they will be acquired, must be scattered and not concentrated, to select them upon the basis of preserving tradition, of maintaining interest in those things which we all know to be worth while. So far from thinking that there will be difficulty about it, I believe that the men who are active in achieving this particular object of public convenience will welcome as an added force to the influences which they exert, your efforts to demand that these Indians mounds should be included in the reservations which are made for the comfort and convenience of modern civilization.

Dr. Cole was speaking to me about this very matter not so long ago, and I suggested to him that he should speak to Mr. Kingery, who is the secretary of the Regional Planning Commission for the district about Chicago, to see if there would be any response whatever to the suggestion that some thought ought to be given to the preservation of archæological as well as historical points of interest, in the efforts of these men. And I received, just before I came down here, a letter from Mr. Kingery referring to Dr. Cole's visit. He said:

> Recently Dr. Fay-Cooper Cole called on us and explained this situation in the vicinity of Powell Mound near East St. Louis, which was about to be razed by a syndicate of individuals; and we took up with S. D. Thomas, County Superintendent of Highways of St. Claire County, the matter of requiring the owner to dedicate the mound for park purposes and surrounding it with subdivisions, and in addition the bringing of a branch of a State Highway into it.

[18]

Through Mr. Thomas' efforts it now appears the plan will be followed, the owner having seen it will be to his advantage in connection with his new subdivision.

My own suggestion that even about Chicago there might, in these forest preserves, be some particular attention paid to Indian trails and things of that kind, was welcomed by Mr. Kingery and his board as being an opportunity for them. I quote again from his letter:

> In the forest preserve plan of Cook County we have taken advantage of certain historical sites, such as the old Indian cemetery near Irving Park Boulevard, the Portage near Lyons, and certain other sites with history attached to them. And in the final publication of the plan we believe these should be marked. Mr. Moreland, the landscape man, with Mr. Endicott, is locating these and is sketching the definite routes of Indian trails which were known, so that there may be an added attractiveness and interest in the use of forest preserves.

I think that suggestion is very practical. I believe that at last we can see a prospect of preserving all of these mounds which you have lately discovered in Stephenson County, and throughout the state, by making use of this demand for recreational spots for picnics, and for motorist recreation, at some or all of these sites. . . . .

THE IMPORTANCE OF SYSTEMATIC AND ACCURATE METHODS IN ARCHÆOLOGICAL INVESTIGATION

F. W. Hodge

(Read by Mr. R. B. Dixon)

A discussion of the subject indicated by this caption would not be necessary if it could only be borne in mind that, in all archæological investigation, specimens are of prime importance only when they illustrate something besides mere handiwork. Because of the lack of appreciation of the part that specimens themselves really play in archæology, they have long been held to be of such paramount importance that most digging has been done for the sole purpose of collecting them.

Specimens of whatsoever character, whether artifacts or otherwise, are of importance and interest chiefly for two reasons: (1) as an index to the culture of the people or peoples they represent, and especially when they reveal varying stages or periods of occupancy of a site, and (2) as illustrations of the product of man's handiwork or of other uses by man.

When an aboriginal site is carefully excavated and artifacts as well as other materials are gathered in such manner that their relations one to another and to the site itself are revealed, then they become highly important evidence as to the character of the culture or cultures of the occupants. Probably more village-sites, mounds, cemeteries, and other aboriginal remains in the United States have been ruthlessly dug for the purpose of looting the specimens hidden therein, or of leveling for utilitarian purposes, than now remain to be excavated by scientific methods. In many cases such digging has been done at the instigation of organizations that should have known better, but in many more instances the exploitation has been the work either of seekers of supposed " hidden treasure" or of amateurs who often with an eye to the main chance, have gathered the better objects with a view to making what they perhaps believe to be an honest dollar. It is thus that untold thousands of specimens of pottery, the most important means of culture determination, have been discarded because broken, while entire pieces have ultimately become scattered to the four winds, with no information respecting the condition or circumstance of their finding. In consequence, private collections and indeed our museums, great and small, are filled with ancient Indian objects with no more information than is recorded on a lonely label giving the name of the state from which derived, and sometimes not even that. Archæological specimens innumerable in public and private collections, thus derived, are often not worth the valuable space they occupy.

Until within a comparatively few years scientific methods of research had not been developed. It was only through long and patient plodding by serious students, eager to make the most of the story which archæology had to reveal, that systematic methods were gradually devised and ways found to wring all available information from every object and from every circumstance associated with its finding. By pursuing such methods, contributions to the knowledge of American archæology in the last two decades have been greater than during all previous time.

By reason of the slow progress in the development of archæological research, and because also of the mistakes made by the pioneers in this field, wherever their work was conducted, we must not hold too severely to account the amateur who has conscientiously endeavored to do his best, handicapped by lack of knowledge of the progress made in the various fields of activity, for the greater part unequipped to meet the problems often presented, yet faithfully recording, and sometimes making accessible by publication, the subjective results obtained. But although innumerable sites have been despoiled, with not even a surviving word of description, the serious student whose ire has been aroused by archæological depredations in this country may be placated perhaps when he recalls that only a few years ago priceless skeletal material found by Egyptian expeditions operating under the guise of archæology, was relegated to the dump-heap. Pot-hunting has not been confined to America by any means, nor to the amateur digger.

We have said that pottery vessels are often the most important of all artifacts recovered, because, as long ago stated by Holmes, "their adventitious records are deciphered with a fulness and clearness second only to that attained in the reading of written records." Pottery, moreover, is the master-key, above everything else made by primitive man, to the determination of multiple occupancy through stratification, and by its usual fragile character it commonly did not find its way very far from the place of manufacture. It stands to reason therefore that it is of the greatest importance that careful note be made of the conditions attending the finding of every example—actual and relative depth, relation to other objects, including those of Caucasian origin if present, and particularly those that seem to be of other types in form or ornamentation, together with many other desiderata apparent to every wide-awake and conscientious observer. Indians selected sites for occupancy with some good reason, such as convenience to potable water or to tillable lands, for purposes of defense, etc. Often these sites were abandoned, to be reoccupied in course of time, sometimes after the lapse of long periods, either by the same people or by others. The importance of distinguishing such periods of occupancy is therefore manifest; and this may be done only by the closest scrutiny of every feature of the site, including the layers or strata of accumulated deposits, and careful observation of every object in its association with every layer or with every burial, as the case may be.

Not all Indians made pottery, to be sure, for some were low indeed in the culture scale, subsisting on the products afforded by a not too prodigal nature and making little in the way of utilitarian, ceremonial, or esthetic objects that have survived to the present time. Other Indians gained their livelihood by hunting, moving hither and yon, never settling long enough in one spot to establish even fairly permanent residence, burying their dead on scaffolds or in trees, of which nothing remained after a few years' exposure. In the areas once occupied by roving tribes archæological remains are necessarily sparse.

Almost everywhere along the streams and sea coast throughout the eastern half of the United States, sites of former settlements are found. Many of these are known to have been occupied in historic time by the same people, who moved from spot to spot, establishing new settlements to which the names of the abandoned ones were often successively applied. In numerous cases these have been dug into by pot-hunters or treasure-seekers and practically destroyed, the salvaged artifacts ultimately finding their way into museums as strays, or into homes where they have been given place with other curiosities on the family whatnot. In the Southern states generally, and especially throughout the Mississippi drainage, aboriginal mounds have been preserved by their owners as places of refuge during times of flood, although even this necessity has not always saved them from the promiscuous digger. Along the Atlantic, from Maine to Florida, shell-heaps, some of them of prodigious size, refuse of the feasts of generations, have been hauled away for the use of the material in road-building or for burning into fertilizer. Aboriginal monuments they, more worthy as memorials of America's earliest history than for the enrichment of a few of us.

So much destruction of noteworthy archæological remains is still in progress that a week scarcely passes without evidence of it coming to our attention. Yesterday we heard of one individual, with purely sordid interests, rifling every Indian grave he can find within motor reach of his home in western Pennsylvania; and today word comes of a farmer in Scott County, Kansas, who has leveled the ruins of the only Pueblo Indian settlement in his state. Does it not behoove our state and local organizations to commence to realize what an asset to education these aboriginal remains really are?

Caves and rock-shelters were nearly always utilized by Indians, for many of them were ready-made abodes in times of stress and in some parts were occupied for long periods, while others were used for burial, for sacrificial deposits, or as shrines. In the Southwest many important cliff-dwellings have been rifled by pot-hunters who did not hesitate to use explosives to make their nefarious work more easy. Fortunately in some of these cases only the exposed artifacts or those lying slightly beneath the surface were gathered, so that it has been and is still possible by thorough investigation to reveal highly interesting and instructive culture stratification. Contents of caves and rock-shelters wherever found have been rifled and their significance lost to science for the sake of a few specimens which proved well-nigh worthless in the hands of the ill-informed. Nowhere has the study of the stratification of remains been conducted with greater acumen than in the ancient Pueblo region of the Southwest, with the result that several periods of culture have been plainly revealed in ruined pueblos and in caves and cliffs, an outstanding achievement in the investigation of ancient American culture history made possible only through the employment of scientific methods. Research of like character may be conducted in every part of America where undisturbed archæological remains still exist.

The untrained digger has usually little interest in anything except the more striking objects. If a skull is in good condition it may be saved as a curiosity; but often skeletal remains found in the eastern half of our country are beyond preservation except by the use of expert methods, and hence are discarded, although to science they are of prime importance, being one of the means of determining relationships and of possible migrations or tribal shiftings, not to speak of the testimony they may offer on social and religious customs. And so with the bones of mammals and birds, and other faunal remains, for their identification, coupled with their interrelation with other objects, may shed important light on the subject of the food quest, and certain peculiarities of their disposal may reveal customs having to do with religious and other beliefs and practices. Kitchen-middens or refuse-heaps, often rich in such animal remains, are generally so poor in the loot sought by the average pot-hunter that he often abandons them in disgust, glory be!

Petroglyphs, if accessible, are often wantonly destroyed by vandal hands, or are so obscured by that public nuisance who carves his own name as of so much greater importance, that few such remains are now found in the East. All petroglyphs that have survived vandalism and the ravages of time should be preserved in the form of paper squeezes from which plaster casts can be made—a very simple and thoroughly effective process. In some localities this should be done at once, for in the progress of our industries vast reservoirs have already caused many such archæological remains to be forever submerged.

A more difficult problem, perhaps, is that of the study and preservation of aboriginal mines and quarries which yielded the materials useful to the early tribes in the manufacture of earthenware, as well as various other useful substances such as copper, mica, salt, hematite, catlinite, turquois, etc.; for such materials are all of present-day value, and when they occur in profitable quantities they have been taken over and operated commercially, so that relatively little of the aboriginal processes of mining and quarrying, simple though they were, can now be learned.

We may speak forever of conducting archæological work only by accepted scientific methods developed after much floundering and waste, and while the amateur may heed appeals to seek the advice of an experienced archæologist before entering on what might prove to be only depredation, there are those who from purely selfish motives will continue their ruinous work. The Federal Government has put a stop to this, so far as possible, by requiring permits for excavation or the gathering of objects on lands under its jurisdiction, and some of the states have enacted laws, not always judicious, perhaps, with the view of preventing similar ravage within their domain. But, after all, the local pot-hunter and collector in the course of time either joins the choir invisible or tires of his hobby—too often, alas, after the field of his devastations has become exhausted—and disposes of his collection to a museum, which most unfortunately is usually a ready market for pot-diggers' loot. When American institutions cease the practice of purchasing collections of which nothing is known save perhaps the general localities whence they come, then will dawn the day when such archæological remains as may then be left, will be subjected only to systematic study and publication.

DISCUSSION OF MR. HODGE'S PAPER
M. W. STIRLING

It is rather difficult to make any comments on this subject that have not been made before. However, repetition perhaps serves a useful purpose in crystallizing our thoughts and enabling us to proceed in a more definite manner in the future.

It has occurred to me that the problem of education as related to systematic and accurate methods of procedure, is of great importance. Before we may have systematic methods applied in the field we must properly educate the men who are doing this work at the present time, as well as those who will carry on in the future. Education in archæology may be acquired in several different ways. Most important for the future of the science is the academic education furnished in our universities, particularly those having separate departments of anthropology.

Education as applied to archæology has evolved hand in hand with the progress and evolution that has taken place in field methods. At the present time the system followed by our leading universities is immeasurably superior to that employed fifteen to twenty years ago when only a very few of our larger institutions were seriously concerned with the subject. Increased popular interest in anthropology, generally, has greatly increased the number of students in this subject. Instruction in our universities is probably the greatest factor in spreading the gospel of correct procedure in archæology.

We have also our great museums, which serve not only to educate the specialist, but the lay public as well. Exhibition methods in our museums have progressed in exactly the same way as have methods of academic instruction. It was formerly the custom in our museums to place on display only the showy specimens from advanced culture areas, whereas regions with a paucity of cultural material were frequently neglected entirely. However, museum men today are beginning to realize that from the educational standpoint it is the typical specimens, rather than those representing the highest artistic development of the region under display, which are of the greatest educational value. It is not necessary of course, to minimize the value and interest to be found in the finer specimens from any region. These are also of great interest, not only because of their intrinsic artistic value, but because they show the capabilities of the people whom we are studying. They do not, however, have the same value to the student as do the common, every-day articles illustrating the culture. A true picture of the life which the student is attempting to reconstruct may only be obtained from an impartial study of this sort.

Mr. Hodge has drawn a rather complete picture of field work in archæology as it is now pursued in the periods with which we are most familiar. The comments I shall make are concerned with the framing of this picture; in other words, a discussion of the two extremes of archæology. On the one hand is the tying up of archæological research with the historical period concerning which we have definite information, and on the other hand the projecting of it backwards to that period of which we may be able definitely to say that there was no human occupancy of this continent.

I cannot stress too strongly the importance of the former branch of research because, as archæological information goes, the data are relatively full. It is extremely valuable when working out the archæological history of a given region that a study of this nature be made a point of departure. It is possible to determine rather definitely the dates of the introduction of certain types of articles of European manufacture which may have been found in an archæological site. We know when and where certain varieties of trade beads were made; we know rather definitely the period during which certain smoking pipes were manufactured and introduced as trade articles among the Indians, and there are innumerable other examples of the same sort which may aid greatly in giving us something definite from which to project backwards a chronological sequence. Specimens of this description have as a rule been given too little significance by archæologists. Many a field investigator has suffered a real sense of disappointment upon finding himself dealing with a post Columbian site. After excavating a number of articles of native origin, and feeling rather triumphant about it, upon encountering a string of glass beads or a silver ornament he is very likely to experience a feeling of depression. There is no justification for such a reaction, and in most instances the archæologist should feel rather a sense of elation. Where an association of this sort is discovered it becomes possible by a process of overlapping to carry a native culture throughout its successive stages of development well back into the prehistoric period. In this manner we may link one site with another and eventually by this method of correlation we may hope to work out a complete and rather definite picture of the culture of any given region.

There is one more point which should be mentioned in this connection. Frequently seemingly unimportant specimens will turn out to be of the greatest significance. There was a time when it was customary for archæologists to collect only the more conspicuous artifacts which were very definitely and obviously related to the material culture under investigation. In this way a great deal of invaluable material has been discarded by field workers. As an example I might mention the necessity for collecting wood which has been preserved in archæological sites, even though it consists of unworked timbers such as are occasionally found in the lower levels of the mounds, and which are found in considerable quantity in the arid regions of the Southwest.

Most of you are no doubt familiar with the work of Dr. Douglass in correlating the growth of tree rings as determined from beams found throughout the pueblo region. By this method he has worked out an accurate chronology of Southwestern cultural development which it is hoped will be completed after a few unimportant gaps have been filled in by field workers this summer.

This tree ring growth correlation has been tied up with the Spanish mission or historic period and carried back to the earliest pueblo sites we know. There is no good reason why work of the same nature may not eventually be carried on for the mound area. What may appear to be simply an uninteresting unworked piece of rotten log which had been used in the construction of the sub-structure of the mound, may in the long run turn out to furnish information of far more value to the archæologist than does the beautiful pottery vessel or carved stone pipe recovered from the same excavation.

I will now discuss the other extreme of our picture which is the ultimate beginning of human culture on this continent. A systematic attempt should be made to establish the latest period during which there were no people in the Americas. In this study we should call for the assistance of the trained paleontologist and the geologist. There should be a system of training whereby we might tie together the specialized work of the geologist and the archæologist, for it is only by bringing these two together that we will be able to determine these early beginnings.

It is perhaps too great a risk to mention any personal opinion as to the probable time when man first entered the Americas, and I shall carefully avoid making any comment as to my opinion on that particular subject. I feel sure, however, that no item connected with archæology attracts more popular interest at the present time, and no point produces more controversy. It is obvious that the reason for this difference of opinion lies in our lack of concrete information concerning these early periods. Recent finds in Florida and New Mexico have produced interesting new discoveries which may shed light on this subject. It would seem that in instances where human cultural remains are found in association with the bones of extinct pleistocene fauna, the problem to be solved is that which will explain to us the length of time that this fauna persisted in these localities. Here the problem seems to be for the paleontologist. In spite of our present lack of knowledge it is not too much to hope that in the future we or our successors may definitely connect the geologic past with the first coming of man into America.

Any other remarks on Mr. Hodge's paper must be random in nature.

I think that the method pursued by Dr. Cole in his systematic survey of the State of Illinois is a model which might well be followed by all state organizations as a method of procedure in this important work. In addition to the actual survey, he has been locating through the aid of his students all of the private collections that have been made in the state, collections which otherwise might never have come to light but which assist greatly in increasing our information upon the archæological history of the state.

Only recently I was in Macon, Georgia, at the site of Old Ocmulgee town which was the traditional founding place of the Creek Confederacy. I went there at the request of a citizen of Macon who reported that one of the great mounds upon that site, which many years ago had been bisected by a railroad cut, was being looted. He, as a man of influence in the community, had temporarily stopped this pot hunting, hoping that the mound could be scientifically and systematically excavated. Previous to my arrival, two skeletons with numerous accompanying artifacts had been unearthed at the base of the mound. The specimens had been collected by the workmen and numerous citizens of the town whose curiosity had been aroused by press notices, so that the material fell into the possession of about two dozen individuals.

A week before my arrival, an Indian medicine man answering to the name of Chief Deer Foot had arrived in town. He had documents in his possession demonstrating that he was chief of all the American Indians, and he made a public statement, published in the press, that at the time of the treaty between the whites and the Indians, wherein the Indians deeded their lands to the whites, it had been specifically stated that all Indian burial places were to remain permanently in the possession of the redmen, and that all articles in these burial places would continue to be the property of the Indians. Hence, as chief of the Indians and therefore their legal spokesman, he demanded that all of the specimens found in the mound be turned over to him. After locating the several individuals who had obtained these specimens, he approached each in turn, demanding that the specimens be turned over to him, threatening legal action in the event of refusal. Curiously enough, his efforts were uniformly successful so that virtually the entire collection obtained from the lowest level of this great mound came into his possession. A few days prior to my arrival he had disappeared to parts unknown. It strikes me that this might be suggestive of a workable method of procedure for archæologists who have had difficulty in obtaining specimens from private collections.

I should like to mention one more point. I think it rather unfortunate that the terms archæology and ethnology have become separated, because after all there can be no division of the two studies. One cannot be a competent archæologist without ethnological training. Archæology is not merely a matter of digging and careful observation, but it requires an ability to interpret these observations accurately. Without knowing the customs of the people who formerly occupied the site on which he may be working, or without knowing the early descriptions left by early travelers of this same region, where such knowledge is available, the value of the archæologist's work will be considerably lessened. The ethnologist for his part cannot competently pursue his studies unless he is familiar with the prehistoric periods of his subject.

I think that is about all that I have to add to the more complete summary of Mr. Hodge. There can be no question that we are arriving at a definite method of procedure in attacking our archæological work. An increased popular interest in the subject has made it easier for institutions conducting field work to obtain funds for research, and our universities are turning out men fully competent to conduct this work in accordance with the most advanced methods. The future of archæology has never looked more bright.

GENERAL DISCUSSION

MR. BARRETT: Both of the subjects treated here this morning are extremely live. There is hardly a state in the Union to which these considerations do not apply intimately. Every one of our states has within its borders archæological sites of one kind or another. They may be mounds, they may be cliff dwellings, they may be old village sites or trails. Whatever they may be, they should be preserved, and may in some cases yield, as has been pointed out, a real monetary return to the general public. That, however, is the least important phase of the subject.

Any one of our cities, with its teeming population, requires and will require as time goes on a greater and greater amount of breathing space in the form of public parks. As we are turning our attention to the general public health, we are going to give greater and greater attention to that particular phase. We are entitled to the interest and the support of our conservation commissions and other bodies of that sort in making these parks serve the triple purpose of beauty spots, health resorts, and the means of preservation of these important archæological remains.

Systematic and accurate methods of investigation are of the utmost importance. As has been pointed out in these papers and in the discussions, we have innumerable cases all over the country of vandalism and destruction. Mr. Freeland mentioned to me just now the instance in which certain individuals in the Southwest used dynamite in the cliff dwellings, wrecking and ruining these great monuments, which should be preserved, and destroying all records.

I have in mind an excellent example of conservation work in the site of Aztalan, of which a large part had been plowed over for seventy-five years before we began to work there. Ten mounds were practically all that were left of that site. We have induced the state to take over a small part in preserving these ten mounds, and we hope to be able to extend that and perhaps reconstruct some of the old wall of Aztalan, so as to give a little idea of the conditions that formerly obtained there. From the archæological knowledge we are able to accumulate and preserve, I think it entirely possible to take some of these old sites and actually reconstruct some of the former conditions. I believe that even the man who has no intimate knowledge of archæology would be much interested, and even more so if he has that knowledge.

MR. MOOREHEAD: I have two comments to make: First, we do not want to be criticized ourselves, and it seems to me the field men should be very careful to restore the monuments. Both the University of Illinois and Harvard University and Phillips Academy have spent a third of their appropriations in rebuilding mounds upon which they have worked. It is not necessary to destroy mounds while studying them. I make a respectful plea that we put all these mounds back regardless of what the owners say.

Second, for three years we have worked on a classification of cutting tools. We have reached 30,000 in our tables, and hope to bring out more in the future. The assignments of dates will be very difficult, but it would be a step in the right direction.

MR. BLOM: I am going to bring you into a field that has nothing to do with North America, but lies in the field of the Maya Indians. Down in our department of Middle American Research we have a card index system of Maya ruins. It consists of a map constantly kept up to date. It includes an alphabetical index of sites, with references to the state and country in which the ruins are found. Then there is a large index giving the names of the groups of ruins, the translation into English of the name, the nearest route of access to the site, and a condensed description of what has been found in that group of ruins. This is followed by a bibliography, giving the names of authors who have written about the place, with complete references. There is also a list of photographs, and a complete set of maps.

A large corps of persons from various institutions is working in this field, and in many cases these institutions have given us their unpublished material. We wish to be of service to anybody going to explore in that country. If an expedition is going out, we hope they will write to us first and indicate their route. Then we can furnish them with information on whatever has been done already. In case we have material that has not been published before, we do not release the photographs we have received in trust, but tell the directors of expeditions that they can get their original material from such and such institutions or individuals.

There is in this index a complete bibliography of all the sites mentioned on the main cards; and there is also a list of all inscribed monuments in each group of ruins. As far as possible the lists contain all dates that have been read. With this goes another index, in which the dates are arranged by their position in the Maya calendar. If one finds a new monument, he can go to the date index, see what has been done already on that particular date, and on which sites he will find monuments having the same date. Then he can go back to the main index, and from the card of such and such a monument can obtain a complete description.

This system might be of help in classifying and sorting the monuments of the area in which we are interested on this continent. If any of you should have prospects of expeditions to Latin America, I want to assure you that we will be very glad to furnish all the material we have; and we hope that when you come back you will turn in your new discoveries so that our index will be an increasingly better clearing-house of information for all expeditions going to that particular area.

ADJOURNMENT FOR LUNCH.

THE VALUE TO THE STATE OF ARCHÆOLOGICAL SURVEYS

ARTHUR C. PARKER

The subject of American antiquities had attracted the attention of numerous students and writers as early as the first quarter of the nineteenth century, but not until the work of Squier and Davis [1] appeared in 1848 did the importance of American archæology receive proper emphasis. "Ancient Monuments of the Mississippi Valley" by Squier and Davis was regarded as important enough to become the first contribution of the Smithsonian Institution; and many of the facts presented by its authors hold good today; in particular, their warning that the sources of information, the monuments themselves, were fast being destroyed.

There have been those who have bemoaned the destruction of the ancient libraries of Egypt and of Mexico, and who have dwelt upon the infamy of the men who burned those records, but little popular interest has been aroused at the thoughtless destruction of the archæological remains of North America. That earthen monuments, ancient village sites, fortifications and burial places might constitute libraries of priceless records never seems to have become a part of popular consciousness. The personal right to dig up relics and traffic in them transcended all feeling of moral obligation and regard for scientific investigation. That each archæological site constitutes something unique and irreplaceable has never been a part of public knowledge. Entire burial places have been destroyed for the relics they contained and not a single observation or record made. It is as if the uninstructed had robbed a priceless library and torn its books to shreds for the illuminated initial letters or for the decorative tail pieces. It is as if these same unknowing persons had failed to conceive that the relation of those things to the text is of vast importance,[2] and that the text itself means infinitely more than the incidental things that come from its signatures. It is so with American archæology. The relation of the specimens to the strata, to each other, to the skeleton with which they were buried has meant nothing to the relic hunter. The result has been a vast and distressing destruction of sources of knowledge.

American archæology presents a most inviting field for scientific investigation. The problems which it adduces have an important bearing upon the history of mankind. Because the American aborigines are not a dominant race at the present time, and because the current civil history

[1] Ancient Monuments of the Mississippi Valley, Vol. 1. Smithsonian Contributions to Knowledge.

[2] A. C. Parker. Methods in Archæology. Ontario Provincial Museum Report, 1923.

of the United States is that of a people of European descent, is no argument that the study of America's prehistory is not essential. There is much that is colorful and inspiring in the study of ancient America. The archæology of the old world owes much to the interpretations that the new world has afforded. *The whole story of mankind is bound up in what ancient America has to tell.*

Increasing interest in the study of the human race, particularly in its origin, migrations, specialization and reactions to environment, makes every fact of value. The enlightened world of the future will not excuse our present-day ignorance and carelessness. So far as America is concerned these basic facts may only be secured during a short period; soon many sources will be utterly destroyed, and this without record.

The facts and principles already stated are well recognized in most scientific circles; but even within the vast body of our intelligent population, to say nothing of the uneducated, there is little realization that the archæology of our own continent is more than a collector's hobby or a museum venture.

American archæologists have a heavy task before them. It is complex because of its very nature. It is concerned with the responsibility of discovering, recording, preserving, and interpreting the material evidences of aboriginal culture in America. It must determine the differences in cultures, the origin of specific cultures, the rise and decline of groups and cultures; it must discover and tabulate all the types of utensils employed by each group and determine the use of each, it must compare similar artifacts and establish the range of each form.

Those concerned with these objects, it is true, have done considerable work to achieve the end sought, but there has been little systematic or concerted effort to explore exhaustively and analyze definite geographical areas, especially by the people of those specific areas. Museums as a rule have sought for sites that promised a striking yield of objects, and, for the most part, the methods employed by trained museum men have been satisfactory. As a result, a number of our great institutions have creditable exhibits and have issued illuminating publications.

Because our great museums have secured so much from regions remote from their doors, local communities have viewed this extraction of local prehistory with feelings mixed with regret and resentment or even helplessness. Some have been openly hostile while others have afforded hearty cooperation. The feeling that some of the recoveries should remain in the localities where found, therefore, has been growing. It would appear that there is some justification in this desire, but up to the present only a few states have prosecuted vigorous archæological surveys primarily for the benefit of the people of the state.

The Commonwealth of Pennsylvania is now undertaking a state-wide survey of its archæological sites. Its experiment will be viewed with interest by other states and its success will considerably influence the future. The states of Ohio and New York have gone to considerable expense to locate and study the aboriginal sites within their borders. Ohio has its mounds and its "mound builders" to attract public interest and support; New York has its striking earth-works and its Iroquois to stimulate and sustain effort. These attractions, however, are not exceptions. Almost every state may find specific or broad themes that will appeal to the imagination of supporting sources.

From what has been intimated, it will be observed that some states have made, or are making, an effort to know and understand their aboriginal archæology. Localization of effort by the localities involved is becoming a practice deserving of further study. It brings up the whole problem of organized effort by the various states, and the duty of these political units to examine their own prehistory.

Archæological surveys, of course, must be justified and must have the support of the cultural agencies within the group; and they must be made to appeal to the intelligence of the people whose support is sought. To promote them requires systematic public education. Key men or organizations must be reached and an endorsement that can scarcely be overlooked, secured and used as argument. The preliminary work of a survey must, therefore, be largely educational.

If all the states have failed to make the archæology of their respective regions a matter of public concern, it is because the public mind has not been organized to perceive the value of archæology. Those who have been led to know and appreciate the meaning of archæology, especially its importance to the localities involved, must, therefore, assume the burden of convincing the several states of the necessity of making an effort to institute surveys. Can archæology be made to seem of value to the state? We believe that it can; and for the following reasons:

1. *Archæology explains the prehistory of the state.*—The recoveries from ancient sites constitute visual exhibits of the people who occupied the state before the coming of a population of European origin. Aboriginal remains are the least permanent of all records when the hand of civilized man interposes. Indian sites are places once occupied by human beings; these sites in most instances will prove of utility to modern man and be sought as areas of occupation. The result is that the Indian remains become destroyed and obliterated. Without these ancient monuments the state cannot explain or illustrate its prehistory.

2. *Archæological remains constitute a vast reservoir of valuable knowledge.*—Judged by every moral standard the state is bound to conserve and protect its resources. The aboriginal sites within each state constitute unique and fundamental sources of archæological facts, highly valued by the scientific world. It would seem that each state, through an organized survey or through the instrumentality of its museums, should authorize and support the systematic attempt to secure archæological information, at the same time protecting the sources from vandalism.

3. *Archæological remains are monumental exhibits.*—The marking of prehistoric Indian sites and their protection from promiscuous digging would not only attract the attention of the sight-seeing public, but would stimulate the investigation by scientists. Archæological monuments should belong to the people, or be protected for them. They are things of public concern, of value to history, education, and art. States that have protected, marked, explored and featured their archæological monuments have found them valuable assets that in many cases have attracted world-wide attention.

4. *Archæological collections are exhibits of lasting worth.*—Wherever archæological collections have been made by trained students of prehistory the resulting exhibits and publications describing them have constituted genuine contributions to knowledge. Their value to science and art is recognized even by those who are neither scientists nor artists. Aside from this value, these recoveries have a market value that frequently overtops anything else within a museum. The state making such a collection is also making a sound financial investment. This sordid fact, however, should not blind one to the more significant values of archæological collections.

AGENCIES OPERATING THE SURVEY

Among the possible agencies for operating a state-wide archæological survey are the following: (1) the state itself through some constituted department, such as a state museum, archæological commission or geological survey; (2) a quasi-official organization, such as a state-supported historical society, a state university, or a state-supported museum operated by a society; (3) an organized group of interested persons, such as a properly recognized archæological association having sufficient funds; (4) a public-minded private individual of means.

It matters little what institution or agency promotes the survey so long as its operating force is composed of trained archæologists familiar with the problems to be met or capable of meeting these problems when they occur. The ideal institution, other things being equal, is a state museum, for then there will be a centralized repository for the specimens, and at least a certain amount of clerical and professional help. A specially constituted commission cooperating with local groups may have difficulty in meeting the problem of distributing the recoveries, especially when it has invited the aid of numerous local historical and scientific societies.

Where there is any reason to believe that the several localities within the state may argue over whether the "relics" shall stay in the town or county where found, great precaution must be taken to explain the significance of the specimens and the real object of the survey. It should be understood that the repository or repositories of specimens must be under the surveillance of those especially trained for such duties. If these precautions are not taken, most of the material will disappear within a short time. It will either be misplaced or stolen. Numerous examples prove this statement. Safety and availability are far more important than the local pride of any community desiring to receive and exhibit relics in an unguarded place where the hazards of theft, fire, and careless removal or displacement may render worthless the effort made to obtain the articles in the first instance. For these and other reasons that might be adduced, a state museum, state historical or scientific society having a safe and permanent building, or a state university museum with proper equipment and a permanent staff of guards, are obviously more efficient custodians than numerous small societies with temporary or ill-equipped rooms and cases, and only occasional supervision.

Under certain conditions, however, duplicate objects, casts, and even unique pieces may be arranged as exhibits and placed by the survey in the keeping of local societies. Loan or permanent collections, selected and distributed by a central office, have their place and value in the scheme of dispensing knowledge. These exhibits may be fastened within standardized cases and thus be made fairly safe. It should be understood that the scientific and educational value of such exhibits does not directly depend upon the presumed monetary value of the specimens. It will take considerable education in some instances to bring this fact home; but if the highest scientific results are to be obtained, the value of the facts and the security of the material evidence must come before any quarrel over the custody of the evidence. When a centrally located state institution inaugurates and supports the survey, there is apt to be little question of the custody of the recovered material. When, however, a large number of local societies unite to prosecute a survey under the leadership of some recognized institution, this serious question may arise. It is best at the beginning to avoid its entanglements. At the outset the agreement should be reached.

PLAN OF THE ORGANIZED SURVEY

Once the survey body is constituted, it must conduct a search for facts. There are three general methods of securing these.

First, citations and records found in county histories or other works descriptive of the state may be consulted. Frequently county histories contain many valuable references to Indian monuments and ancient sites. The accounts of early travelers and missionaries may also be valuable, as in the instance of the *Jesuit Relations.* The journals of military officers may also be found productive, as in the instances of *DeNonville's Journal* and the diaries of General Sullivan's officers, both describing the invasions of central and western New York.

Secondly, actual field agents may travel throughout the territory making examinations and inquiries, and listing their information.

Thirdly, circular forms requesting information may be mailed to representative citizens. From the replies secured, a second series of questionnaires may be prepared and the way paved for a publicity campaign. The New York State Museum in inaugurating its survey in about 1910 found that informants took their circular letters to the local newspapers in numerous instances and secured a wider distribution of the request for information. To the circular letter, therefore, we may also add the preparation of well-written articles for the use of local papers. In the instance of New York, it was found wise to head each press article with local references and to give recognition to local historians and amateur archæologists. (See Exhibit A of this paper for suggested form.)

SUPPORT FOR STATE SURVEYS

The problem of securing adequate support for state archæological surveys will deserve serious consideration. In some instances it may prove a heavy task requiring the help of able financial engineering. To secure funds, therefore, the body attempting to organize the survey must have definite plans and a definite budget of estimated expense. Three general sources of financial support may be considered; first, that of the state legislature; second, that of organizations; and, third, the support of wealthy individuals. Each expected source must be convinced of some valuable return. The state must be convinced that its citizens generally wish the survey; institutions must see that to support it will bring prestige to them; individuals must be convinced that the project will yield them that form of personal satisfaction which their nature requires.

States with organized education departments or well budgeted science or history divisions may ignore the necessity of popular support and press the work as an educational or scientific project. As few, however, are thus equipped, special appropriations must be sought. It takes considerable work to convince legislators that digging for facts about aborigines is a justifiable burden upon taxpayers. The state archæological survey is thus safer when sponsored by another established department of government, or by an institution with considerable state support. Respective examples are New York and Ohio.

Archæological and historical societies, and societies of natural science with considerable funds at their disposal, may sometimes be in position to inaugurate and support a state survey. Such societies, by demonstrating the scientific and popular value of their work in this direction, may be able to attract additional legislative support and perhaps also secure the financial backing of wealthy patrons.

Archæologists, once having outlined the plan for a state-wide survey, may occasionally prevail upon public-spirited men and women to endow the project. This method is perhaps more certain if some personal compensation can be assured, such as attaching the endower's name to the resulting publication series, or to the exhibit halls of the museum acquiring the specimens.

THE STATE SURVEY VERSUS THE OUTSIDE MUSEUM

Up to the present, the greater amount of archæological work in the United States has been conducted by institutions located outside the particular states where the work has been done. There are some exceptions to this statement, as in Ohio, New York, Wisconsin, and possibly Michigan. The observation remains true, however, for such important archæological areas as the Gulf States, the desert region, and the mound area of the Mississippi Valley.

Great institutions, such as the National Museum, the American Museum of Natural History, the Peabody Museum of American Archæology and Ethnology, the Field Museum, and perhaps certain others, seem to have ample justification for a sphere of influence that is more than merely local. If these institutions had not extended their efforts, archæological knowledge would be meager indeed. They have amply justified their work, wherever it has been done, but they have been unable to do all that can and should be done. Much still remains to be done. The question then arises as to who shall do it; the uninstructed relic-hunter or an organized state agency determined to rescue and preserve its prehistory. The relic-hunter digs only to destroy and his recoveries are often abortive things with undetermined parentage. Who can say what they are? The systematic work of an organized survey is quite the reverse in its results; it seeks to present ascertained facts and to correlate them.

Whether the relic-hunter will continue to ruin the field, or whether state-supported agencies shall preserve the field and draw from it the information that an enlightened age demands, depends very largely upon the citizens of each state; but it depends most of all upon how thoroughly archæologists who understand the importance of their quest are able to present it to the public. Archæology must advertise and it must seek thereby to stimulate such a desire to know more of prehistory that support will follow.

METHOD OF CONDUCTING FIELD EXAMINATIONS

Field examinations may be merely preliminary and made for the purpose of determining the character of the site, its specific culture and its excavation possibilities. Information thus secured is made a matter of record.

Preliminary examination may be made either by an amateur or by a trained archæologist. If the work is entrusted to the amateur he should have previously been supplied with a manual outlining the field, its problems and possibilities. This manual should give sufficient information to afford identification of the various cultures and other important facts required by the survey organization. The recognized archæologist, of course, will have the training, experience and knowledge that will enable him to attribute most of the sites which he examines. If the unusual occurs he will readily understand what the unusual features are.

The field survey should be supplied with record books or cards for transcribing the data. Needless to say, the forms used by the various members of the survey should be uniform.

Wherever possible, the field man should take photographs of all prominent sites. Good views should be made of earth circles, walls, mounds, depressions, village sites, burial places, and all other Indian localities or remains worthy of investigation. There should be note-book records of each photograph, and if possible a pencil sketch of the scene or object photographed. Sketch maps are also of importance, but they should be fairly accurate and the orientation should be correct. The preliminary field description of all localities examined should not only be informing as to position and character but should give some inkling as to the probable culture and length of occupation. Frequently, few artifacts are to be found about fortifications and mounds that otherwise look promising. This should be noted.

When preliminary examinations have indicated the desirability of a site for intensive examination, preparations should be made for active operations. The number of men needed should be known, the excavation equipment should be available and all arrangements completed for the housing of men and material. If it is not feasible to board the men with neighboring farmers, ranchers or townsmen, camp equipment should be provided.

The expedition should be in charge of a single head who ought to be an experienced archæologist. His workers and subordinates should know something about the purposes of the work and what it hopes to achieve. However, in numerous instances, a competent leader has been able to instruct and train intelligent students, farm hands and even ordinary laborers so that the manual part of the work was done with excellence. Many expedition laborers develop considerable skill and most archæologists will testify that outside of the constitutionally lazy, there are few failures among expedition hands. It is well, however, to choose honest men who have little desire to collect things for themselves.

Once the site is reached and its limits ascertained, camp may be pitched and the ground staked out. Work maps on grilled paper should be ready in an office tent, where also the data sheets and records should be kept.

If the precise nature of the site is as yet unknown, the site may be post-holed to discover the area of disturbed soil. Once this is known the richest part of the site may be dug. Post-holing gives considerable knowledge about the task at hand. Trench lines or squares may now be marked in readiness for productive excavation.

The present author has found it convenient to run parallel trenches from 10 to 16½ feet wide and to disregard squares except for plotting purposes on the map. The depth of the trench, of course, depends upon the depth of the disturbed earth. His general plan of work may be found in his "Excavations in an Erie Indian Village," Bulletin 117 of the N. Y. State Museum, 1907.

For the purposes of an archæological survey, special field cards should be provided in addition to the field record, the latter frequently being called the "trench book" because all the work done in the trench and all the discoveries made are reported in it. As each object is found a record is made on a card indicating the depth, soil, particular surroundings (as fire pit, refuse pit, grave or surface soil). It is frequently found advisable to place a pencil number on the specimen to identify it with the data card. The specimen, if not too friable, may now be wrapped and placed with the card in the specimen tray.

If interesting pits are discovered they should be described in the notes, measured, and photographed in cross-section. If stratification occurs, the trench cross-section should be diagramed at frequent intervals, say every two and a half feet, or as the strata and pits occur.

Graves should be carefully opened and the skeletons revealed with the skill of an artist carving out a cameo. No specimens should be disturbed or bones moved until good drawings and photographs are made. All skeletons should be saved if the bones can be removed. Fragile bone may be treated with a solution of hot white glue, or possibly ambroid. It is not wise to be in a hurry to remove the contents of graves. Many valuable burial objects have been broken by the carelessness of haste. Wooden articles should be allowed partially to dry and should then be treated in a solution of hot gum acacia; antler objects, such as combs, frequent in New York and Pennsylvania, may be treated with glycerine and then glue, or with glue alone.

Every pit or grave worth opening is worth doing thoroughly. Records should be meticulous but not confusing. The excavator should keep in mind that archæology is hungry for facts, and that there are some which the specialist would give much to discover.

Mounds should receive the same careful examination, and particular care should be taken to find out what is under the mound as well as within it. When the examination is completed, the mound should be restored and covered with sod or seed; or at least left so that the elements will not destroy its form. The same is true of any earthwork. When any archæological work is finished, the conscientious worker, of course, will restore the land to its original contour. Farmers and landowners have frequently been greatly incensed at the vandalism of relic-hunters who left the ground full of holes; and legitimate work has often suffered the handicap of prejudice thus aroused. It is also recommended that copper or leaden plates be buried in all sites that have been excavated, or that some marker be left to indicate that work has been done upon it.

EXAMINING THE SPECIMENS

Expedition records and recoveries are packed and shipped to the headquarters of the survey. Where several agencies have been working, there should be no division of the specimens until all have been studied and their significance ascertained. All should be kept together until it is wisely ascertained which may be considered duplicates.

Specimens are finally cleaned in the survey laboratory, numbered and catalogued, the field records being the source of the principal facts. Recoveries from graves may be kept together, regardless of character. This may also hold true for pit material, but the objects from general trenching may be segregated and classified. In the Rochester Municipal

Museum the "trench run" is tabulated not only by kinds, but by depths, this being useful where there are two or more superimposed cultures. A specimen found ten inches below the surface may not appear anywhere at a five foot depth. This museum is thus able to construct curve-charts showing from what levels the various implements are found and to judge their relative frequency at any and all points. This has been helpful in sifting differences in culture.

When specimens have been studied and the record of the site reviewed some conclusions may be reached. These should be embodied in a monograph. Every site worth excavating is worth describing in a well-written report.

The publication of reports depend largely upon their nature. Certainly all should be available in manuscript form in some central institution. An editorial board may select such as should form the publication series.

This suggests the need of a publisher. Usually this is the survey itself, or the institution with which it is affiliated. Examples of state agencies are the New York State Museum, the New York State Archæological Association, the Ohio State Historical and Archæological Society and the Wisconsin Archæological Society.

EXHIBIT A

DATA ON INDIAN SITES AND MONUMENTS

Please return to the Standard Survey, of the State of Standard, Standardville, Sd.

Information supplied by ....
Address ....
1. Where have Indian relics been found ....
....
Give definite locations of the following classes of Indian remains and indicate position on the topographic map:
MOUNDS ....
FORTIFICATIONS ....
ENCLOSURES ....
VILLAGE SITES ....
BURIAL PLACES ....
Name and locate other evidence of Indian occupation known to you............
....
Where has pottery been found?..
Please list collectors of Indian relics known to you...........................
Please list those interested in Indian archæology and history other than collectors
....
Where are the relics from your region exhibited?.............................

DISCUSSION OF MR. PARKER'S PAPER

GEORGE R. THROOP

It is perhaps rather futile for one who is not primarily an American archæologist to discuss the practicality of matters presented in a paper of this particular kind. But I do think there are certain matters which appeal to the ordinary laymen as well as to those of us who have had some experience in another field. I have had the advantage of a small amount of first-hand observation in the field of American Archæology, and yet I am more impressed by the similarity of method and procedure between classical archæology and American archæology than by any other individual factors that have come to my attention during this particular meeting.

This meeting is primarily held to consider what can be done to educate the public. It seems to me it should be clearly understood what is the procedure in American archæology, and by what particular methods our problems here can be best furthered and advanced. The matter, as it seems to me, divides itself rather clearly into two factors. These are factors which appear in classical archæology, but they also ordinarily come up here. I think they have, to a certain degree, been brought out, but they need considerable emphasis.

The procedure of excavation has to do with two things. One is the actual excavation; the other is the interpreting of material which may be gained in the excavation. That is primarily true in classical archæology as well, and to such a degree in the countries with which I am most familiar—in Italy and in Greece in particular—that there the two divisions can only be successfully handled through the medium of men who are particularly trained in the profession.

In Italy, as you may know, permits for excavation of any kind whatsoever, are not issued except to recognized officials of the Italian government. Despite the fact that America maintains a school of classical studies and that Germany does also, these schools in Rome are never permitted to direct excavations; they can only participate as helpers under the direction of some Italian government official. That means that no mistakes are made. It means that all the materials of value discovered are kept in Italy, and that all of these things are managed in such a way as to further to the best possible advantage the value of the work in hand.

In Greece the strictness is not quite so great, but yet it is rather well watched. There, excavation can be carried on by the different countries that maintain their schools in Athens, by the Americans, the Germans, French, Austrians, Swedes and so forth; but, at the same time, the Greek government exercises strict supervision, and must issue a permit for any kind of excavation which is made anywhere in the country. And, of course, nothing can be exported from the country without the express permission of the Greek government.

It is perfectly natural to say that they can in Europe place restrictions upon private excavation which would be entirely unpractical and impossible to attempt in this country. For instance, if we were to pass laws in the State of Missouri which would forbid a farmer to excavate Indian mounds on his farm in whatever way he would choose to excavate them, I am afraid there would be a revolution, because everything of that kind seems to be the common property of anyone who can get to it first.

It will take a long time to get away from that idea. But it is one of the handicaps to be overcome. The important thing is to start with number one of these factors first; that is, that we should train, so far as we can, the public, the people, farmers or anyone else on whose territory these mounds may be, to understand that if they do excavate, they should excavate in such a way that the materials are not disturbed and are preserved for future use and for identification by experts.

It is, of course, as you know very well, much easier to train a man to be a helper than to train one who can satisfactorily interpret the data which are to be got from the excavation.

We may say that the second procedure is by far the more important, because the material gotten is not of consequence unless it is satisfactorily brought together and interpreted. But I do not feel that that is, perhaps, at the present time, the important thing. What we need is a campaign of education, so that those who open these mounds will not disrupt the data and confuse them in such a way that it is impossible for them to be used later.

It is only a rather chimerical idea, but I wonder if it is not possible to arouse interest in matters of this kind in young high school boys of sixteen or seventeen years of age, by inviting them to go on archæological expeditions, to go into the field, and thus train them up from the very beginning. That is being done now with the Reserve Officers Training Corps. Of course, that work is carried out by the Government. But if a half dozen boys from high school could be added in summer to each expedition that goes out to work, it would possibly be a way of breaking them in. We do not have enough American archæologists in this country. That is perhaps so apparent as not to require mention. But I believe that if more of the young men of the country, perhaps in college as well as in high school, could be interested in the practical aspects of a profession of this kind, the results would be rather far-reaching.

It is not only by the bringing together of a few of the distinguished men in American archæology in the United States that this particular kind of subject can be followed. There must be a kind of spreading of the gospel, as I would call it, which is perhaps, after all, the most important work you can do, because the field is so large that a few men—forty, fifty, a hundred or two hundred—can make almost no impression on the country at large. For instance, it is said, whether truly or not, that there are worthwhile Indian remains in every county of the State of Missouri. If some kind of organization could be formed by means of an archæological survey in each particular county through the medium of citizens who are especially interested, and if we could train others from the high school in an interest in this thing, we could undoubtedly accomplish a great deal.

I feel that it is very pertinent to refer to the State of Missouri because this meeting is held here, and because I feel that it offers perhaps the very best opportunity of any of the Mississippi Valley States to American Archæology at the present time, despite the fact that a certain amount of early work has been done here. I really believe that American Archæology has been exploited less in the State of Missouri than in almost any other state. There is a distinct advantage in this, despite the fact, we might say, that no great amount of scientific work has been done, and that consequently the gross amount of opportunity left is very great.

I wish that through any one of the numerous methods suggested by Dr. Parker and by others, we could achieve the result of getting started, because a matter of this kind needs a starting point from which it may pull, and I hope that this particular conference may have a very considerable amount of influence in bringing that about.

ARCHÆOLOGY AS A HUMAN INTEREST
Clark Wissler

The preceding speakers have shown the importance of the problem to the science of archæology, and have indicated that when the facts are in hand a synthetic treatment of the information from the various states of the Mississippi Valley should give a clear picture of what went on in this area before Columbus brought Europeans upon the scene. With the scientific importance of this I am in full sympathy.

At this time, however, I propose to speak from the human point of view. I spent the first twenty years of my life in one of your states, in touch with the life of your people, and I still return every year to spend a month or two in this same setting. I know the archæological problem as many of you know it—that is, as a layman.

I still remember the thrill of finding the first stone relic. There was in the neighborhood of my boyhood home a farmer, long since passed on, who had a fair-sized collection. From him I learned the names and supposed uses of the common run of stone implements. I am now amazed at the fullness of his knowledge, for during the years that have followed, I have found it necessary to unlearn little of what he told me.

I have dwelt upon this bit of autobiography not because it is unique, but because in the main it can be duplicated over and over. The interest in mounds, stone implements, etc., is universal and spontaneous. Every man is interested in the past of the human race. The farmer or boy scout picking up an arrow-head is spontaneously carried back in imagination to a life different from that of the present. "At this spot," he says, " a man once hunted the deer, or perhaps launched an arrow at his enemy." In brief, the old, whether it be historic or prehistoric, makes a spontaneous appeal. So by the nature of the subject, State Archæology touches one of the basic interests in human life.

The truth of all this comes home to us when we take note of the lure of the past in its concrete manifestations. For example, everyone understands the passion for antiques. It is not merely the idea of monetary value that motivates the buyer of antiques; that is his excuse for what he secretly admits may be a weakness. When Carter discovered that tomb of a Pharaoh in Egypt, the whole world was thrilled: Why? Because it represented the past, it revealed the life of that time in a new light. Ask your newspaper men; they will tell you that every little archæological discovery anywhere will make copy. Many a time I have been rung out of bed by a newspaper office to listen to a telegram from somebody in Ohio, Illinois, or perchance Missouri, telling of a few bones, a pot, or a stone ax, uncovered by accident, or by the spade of a local archæologist.

There is another angle from which the human interest in antiquities may be viewed. That is the collecting interest. Almost everyone either collects something, or hopes to do so. That this is not an abnormality is indicated by the fact that many of our outstanding men were collectors from boyhood. Such collecting is indicative of a tendency to learn by dealing first hand with things. This is the most fruitful type of learning. Everyone knows how the modern museum, supported by governments, states and cities, grew out of this collecting urge; for the museum is but the pooling of collections so that they may be readily accessible to all. Education was once available only to the few, the well-to-do; but the situation in this case was met by public education. In the same way, the public museum has met the situation respecting the collection of specimens. It is now common-place to say that the museum is an educational institution; a place where one learns from things rather than from books. It is well to remember, however, that first there were but individual private collections of antiquities and scientific materials, and that these gradually evolved into the modern museum. The collecting interest is basic in this evolution. What one sees in respect to state archæology is a widely spread collecting interest, with all collectors, young and old, interested in learning more and more about the objects they prize. Here is where the state can function by providing museums and conserving materials, and through proper personnel collecting and making available adequate information. This conserving and spreading of information is to my mind the primary necessity, and it is in this direction that there lies the justification for a State Survey.

The functioning of the state in this way is neither new nor unusual. It has long been the custom to deal in this manner with the assets of the state; for example, it is the rule for a state capitol to house departments for mining, geology, entomology, forestry, etc. Most of these touch important economic factors in life, and for that reason have received support; in other words, they are closely associated with money-making activities, such as the production of iron, lead, coal, oil, etc. History, on the other hand, is supported for educational and patriotic reasons, in some cases perhaps as a concession to the universal interest in the past. But archæology is also a part of your state's history, as has been said over and over; and it should be a part of the background in whose light we see the present. The fact that interest in history and the past is so spontaneous, is evidence enough of its necessity in the scheme of life. You doubtless recall that when Lincoln made his famous address, he said something to the effect that mankind would never forget what was done at that place. I thought of these lines the first time I stood on the ramparts of Fort Ancient in Ohio, and again when standing on the great Cahokia mound just across the river from here: we shall never forget what these ancients did here. These monuments stand as the great achievements of their time; the ancient people must have put forth the best that was in them. Today we also are inspired by great undertakings; we achieve our greatest being in these efforts. So why should we not treasure the past of our adopted country; these earthworks are hallowed spots where men have sweated and perchance died in the line of duty.

I am fully aware that this is sentiment, but what is human life? What is national pride? What is all that makes the world a good place to live in? Man always seeks knowledge, but expects it to increase the joy of life, not to destroy it.

[46]

In their public education programs, the states have spent freely to enlighten the young as to the phenomena of nature. I am a firm believer in nature study, in the necessity for keeping the individual alive to the out-of-door, the natural environment in which he should live, but I see no justification for dwarfing the historical and the human relations aspect of life. It seems to me that the great emphasis put upon nature study was due to resistance upon the part of our children. The history of the nature study movement seems to reveal a certain stubbornness in human nature. Our schools have made a hard fight to interest our children in nature and fundamental science. They have succeeded very well as it is, but teachers seem to find it much easier to lead their charges in the pursuit of historical subjects. Indians, Eskimo, Cave-men, Moundbuilders, thrill them at once. So, it seems to me that knowledge of the past about what men, women and children have done, has a fixed place in the scheme of life, and that society will make no mistake in cultivating its historical and archæological resources. States and governments do spend money to enclose in parks and reserves the work of nature, and there are good reasons why they should. But on the other hand, they neglect the human element; what men have done; an element which makes a more intimate appeal to all of us than even the works of nature. Mounds, earthworks, village sites, etc., are suitable materials for state parks; they add the human touch. The automobile is the most striking feature of contemporary culture. In fact, the whole story of civilization is a progressive triumph over distance. It is because the plain citizen can on a week-end, ride from one end of the state to the other, that the history and archæology of his state as a whole appeals to him. Also the whole Mississippi Valley is occasionally accessible to him, and so he has an inter-state interest. When each state has in its own way made its archæology and history known and accessible, it will have contributed to national solidarity and broadened the outlook of its citizens.

Before closing, I return again to the collecting interest. Some have raised objections to supporting archæological surveys on the ground that whatever the state did would play into the hands of collectors. The mercenary commercial collector is an evil; he may derive some advantage from a survey in that it would increase his knowledge; but on the other hand it should be borne in mind that everyone is a collector in tendency. The educational value of the Scout Camp organizations was placed on a secure footing, when a man in New York had vision and sense enough to build up a course of nature study based upon the collecting interests and instincts of boys and girls. The idea was first objected to on the ground that these young people would be unduly stimulated to destroy wild life and scenic spots. But this was a short-sighted view; the result has been just the opposite. These young naturalists and archæologists soon become real conservationists. I merely make the suggestion that through competent state leadership the scout organizations of these states might become the most effective conservation and survey agency. The right person in the office of State Historian and Archæologist could accomplish a great deal in this way. My point, however, is that collecting is but a manifestation of a deep, spontaneous human interest; it is this interest that seeks an outlet and points the way to state service.

Finally, to summarize what seems to me the point of view of the layman. We see, first of all, that archæology deals with what comes before written history; history and archæology taken together present the facts explaining how human life came to be what it is. The archæology of a state is just as great an asset as its history; in fact, one is not complete without the other. Pride in citizenship, stable social life and fulness of life depend largely upon an understanding of the past. Our civilization is not wholly European. The Indian has contributed his share to it, as no doubt the mound builder contributed to the culture of the Indians who lived here in pioneer days. It seems obvious, therefore, that some regard should be paid to the cultural past of each state. Further, the development of your archæological resources will contribute to adult education, one of the recognized needs of the time. The universal use of the automobile presents a condition calling for parks and exhibits, and these can be made effective because they can be central, and still readily accessible to the whole public.

Society in our day is struggling to become conscious of itself. America is becoming aware of itself as a culture unit in the world at large, and is seeking to guide itself into a more highly rationalized behavior. To some professional critics of our time this seems like trying to lift one's self by pulling up on one's own boot straps. Nevertheless the process seems to work fairly well. However this may be, it seems clear that the probability of success in improving the life we live, depends upon the clearness of insight into the nature of what we call national life or culture. Such insight comes from a knowledge of our own past as well as of that of other cultures. Here is where the archæologist and the historian come to our aid, holding up lenses for our short-sighted social eyes, that we may see ourselves in our true relation to mankind as a whole.

DISCUSSION OF MR. WISSLER'S PAPER

WM. JOHN COOPER

One of our enterprising Washington newspaper men talks most interestingly on the topic, "It is never too late to discover America." Anyone who has participated in this program, who has visited the mounds in this great valley, or who has listened to Director Shetrone's lecture can hardly fail to agree with my news-gathering friend.

Meetings of this character are of great value to our schools and colleges. From such discussions as these will come an interest in a civilization that has been superseded and largely supplanted by our own. To the reasons advanced by Dr. Wissler for interesting ourselves in the culture of the natives, namely, that our own economic greatness has foundations in that aboriginal culture, and that from that culture we take much of value in character training for our Boy Scouts and Camp Fire Girls and similar organizations, I am glad to add further suggestions: First, it seems to me that here is a rich cultural field of study for many adult groups. There has been much discussion within the past year or two about the use of the leisure time of our people. Of what use is this leisure if people know not how to utilize it? Some, of course, do utilize it for study which increases their earning power. A few are studying the problems of citizenship. But what are those graduates of our liberal arts colleges whom Dr. George Herbert Palmer delights to call our "amateur scholars" doing? If there really are some of these amateur scholars abroad in the land, and if any of them have some leisure time and I doubt not that many of them have time to waste, would it not be a splendid thing if we could get them to use some of that spare time studying these primitive civilizations? What a field for amateur as well as for professional scholars! What an opportunity for professional scholars to render service to their fellow men! And finally, what a field for study that would enrich one's life, broaden his horizon, and really liberalize his education! I hope that some of our more progressive women's study clubs will accept the dictum that it is never too late to discover America and launch many campaigns of conservation and scientific expeditions of exploration.

In this connection I think it entirely proper to commend to any such prospective groups of citizen-students Dr. Wissler's monograph entitled "State and Local Archæological Surveys," published in 1923 by the State Historical Society of Iowa. It furnishes a splendid introductory textbook for a kind of laboratory work which I am sure will result in strong movements to preserve these mounds undisturbed until such time as they can be studied scientifically. I should like to offer another suggestion at this time. The United States Department of the Interior is attempting through its National Park Service to offer help to those who would understand our national parks and national monuments.

In some of these parks and monuments are to be found remains of the prehistoric civilization we are discussing here. Why not interest people in spending part of the summer vacation in such a place as Mesa Verde National Park in Colorado? Regarding this we read in the bulletin issued by the U. S. National Park Service the following statement:

The Mesa Verde National Park is one of the few large tracts of land in the United States which have been taken from the public domain to preserve the antiquities it contains. It is the most extensive reservation for this special purpose. Its purpose is educational, and its ruins are object lessons for the student of the pre-history of our country.

We are accustomed to regard the Indians of the United States as a race of wanderers, living in temporary habitations made of skin or bark. The Indians are supposed to live by hunting or fishing and to eke out their food by the cultivation of maize or Indian corn, beans and a few vegetables. While this is true of some Indians it does not hold for all, for there were many different kinds of Indians inhabiting what is now the United States when Columbus landed at San Salvador.

In addition there is much to be learned in those areas known as the National Monuments. Among those of especial interest in this connection are Montezuma Castle, Waputki and Navaho in Arizona, Gila Cliff Dwellings, El Morro, Bandelier, Chaco Canyon and Gran Quivira in New Mexico, Hovenweep in Utah and Colorado.

For preliminary studies in these fields the handbooks published by the Government Printing Office furnish satisfactory introductory textbooks for the laymen.

From such conventions as this, then, should result:

(1) A popular interest in the peoples who lived here before our ancestors came.

(2) Organized classes or study groups of men and women who have collegiate training and are ready to give some of their leisure time to consideration of this primitive culture and to enlisting interest in it.

(3) A popular demand that the remains of the prehistoric culture remain intact and protected from the ignorant and the vandal until they may be studied scientifically.

(4) A renewed interest on the part of the governments of the various states in promoting this scientific study.

(5) A realization on the part of chambers of commerce that these remains are community assets worthy of their careful consideration.

GENERAL DISCUSSION

MR. LEIGHTON: One good result that the archæologists of Illinois have achieved has been that of getting the geologists of Illinois interested in archæological work. I mean to say that the geologist is beginning to see a gap in what we may call the history of the past, in the present human history of Illinois. There has been a lack of investigation into the part played by those people who preceded us here on this continent. The geologist is interested in bringing the history of the past clear down to the present time, and he is therefore interested in archæological work.

The Geological Survey of Illinois became interested in what was being done in archæology in the state when the Cahokia mounds began to be studied by Dr. Moorehead for the University of Illinois, because at that time the opinion of geologists in general had been that the Cahokia mounds were natural features; that they were produced by those natural agencies other than man. So I shall never forget the impression I received when my predecessor (who was then director of the survey) and I, in company with Mr. Moorehead, visited the Cahokia mounds. I could not see how it possibly could be that these mounds, especially Monk's Mound, with its quadrangular form, should be referred to as a product of erosive forces in nature, or of depositional forces in nature, other than human. I am thoroughly convinced that they are man made, in spite of all that has been said regarding them as perhaps a mere terrestrial remnant or erosional remnant within the valley. As soon as I looked at the other mounds and considered the possibilities of their being great kames or sand dunes or erosive remnants from former filling of the valley I became skeptical, and so it was with great interest that I followed the excavations of Dr. Moorehead and studied the materials that were found to be present.

One of the contenders that these mounds we saw yesterday were simply natural features came down to visit the mounds in the course of the excavations, and he and I became much interested in examining the materials together and discussing whether or not this could possibly have been deposited by water carrying those materials floating down the valley, later to become erosional remnants. Finally this man said, "Well, I must confess that I have been converted again. I do not think these mounds can possibly be natural features."

Now, in the course of that work I came to see how it would perhaps be a small contribution to archæology to look into the question of the sources of materials used in the construction of those mounds and other mounds, not only to find what kind of materials are present but to make a precise identification of those materials. It is possible to discover whether or not they were all local materials, or whether for some reason or other materials were brought from distant sources. I do not have reference to the gulf shells or the copper or things of that sort that were used in trade, but to the clays or sands that were used in the building of the mounds themselves.

I also became impressed with the possibility that it might be a small contribution if attention was given to the character of the soil profile that may be found on the mounds; as to how deep a soil there is, as to whether or not there had been formed those natural divisions of the soil profile which in geology we speak of as "A," "B," "C" and "D," or "I" zones, or if they are so young that a soil profile has not developed. Then we could learn whether or not the soil profile of different groups in the state differ in the depth to which they have been developed, so that by this criterion there might be some conclusion drawn as to whether or not they are of different ages.

Furthermore, we could undertake the examination of so-called soils that may be found through the mounds, to find whether those are soils or simply dirty streaks; and the examination of the old soil profiles that may pass under the mound.

And so when we found in the case of the mounds that soil profiles of the valley floor, developed before the mounds were built, passed under those mounds, I think the last skeptic of the origin of those mounds had been converted to the human hypothesis.

Then we might study the actual structure of the mounds; the layers and their relationships, with the possibility in mind that it might throw some light on just how the mounds had been built. You can see that we geologists in Illinois have become interested in lending what little help we may to this study of the Indian mounds of the state, and in tying up our work with that of the archæologist in bringing the geological history of Illinois down to the present day.

MR. KEYES: During the present meeting we have had before us to consider some things just a little bit depressing: the destruction of many of the ancient sites, the commercialization of large sections of the field, and topics of that kind. This was inevitable of course, since there is no profit in not looking at facts. This afternoon, however, the discussion,—and it seems to me to be a very happy thing,—has taken somewhat the constructive side, and I hope I may say a word that will add to that side.

We all know something about the acquisition by the United States government (the slow acquisition, still in progress) of the bottom lands of the Mississippi River, from Rock Island and Davenport to Wakasha, a distance by river of about 300 miles. A large part of those bottom lands and islands have already been secured, as we know, by the Government, but I wonder if you all have noticed an interesting development that has just been consummated in the direction of adding some bluff land to those bottom lands of the Mississippi. The original legislation,

[52]

and the original Congressional appropriation, had to do with the islands and the river and the lowlands only. However, a year ago Dr. James B. Munn, of New York, a well-to-do physician there, who formerly lived in the little town of MacGregor, Iowa, in the ravine just opposite Prairie du Chien, expressed his wish to turn over nearly 500 acres of bluff land to the United States Government, either as the nucleus of a national park, or as an addition to the upper Mississippi wild life and game refuge.

This, of course, was something not contemplated by the original movement, and had to be sanctioned by act of Congress. That action was secured about a year ago, and the addition of these hill lands has recently been made to this upper Mississippi refuge.

Now that is a precedent of value. Some gentlemen living in West Union, Iowa, saw possibilities opened, and they went to Mr. Cox, the Superintendent of this Upper Mississippi Wild Life Refuge, and asked him whether there was anything in the way of legislation to prevent the acquisition of other lands in the bluffs. They were told there was not, and moreover he personally favored such acquisitions. He realized that the bluff lands along the Mississippi had many kinds of wild life, and so far as he was concerned, Mr. Cox favored the acquisition of bluff lands if Congress could be persuaded to make the necessary acquisitions. An organization was effected then in northeastern Iowa, and to make a long story short, it includes ten different counties, and they have secured not only the cooperation of the superintendent of the Refuge, but they have secured that of Iowa Congressmen from northeastern Iowa, and from other parts of Iowa. They have secured the endorsement of the Iowa Board of Conservation, and they are trying to so perfect their organization as to make it likely that legislation in the coming Congress will make possible the realization of their aims. This would not require a large appropriation because the Mississippi bluff lands are not extremely valuable.

What does this mean for archæology? It means a great deal. The whole area in the four states of Wisconsin, Illinois, Minnesota and Iowa is rich in mounds, as we have heard during this meeting. All of the three groups of mounds from Iowa of which Mr. Shetrone showed you pictures last night are in that area. The Indians lived in the high bluffs or terraces below the bluffs, close to the river for the most part. The groups that Mr. Shetrone first showed containing respectively five, three and nine mounds, are all preserved. And the two other groups, perfectly preserved to this day, are on the bluffs next to the river. These are only two or three of the groups of mounds in Iowa alone.

The possibilities of this movement archæologically are immense. The acquisition of the bluff lands would not need to be continuous along the river, but everything on the bluffs would be continuous with the refuge below. Now it struck me that here is where we could all have a part. This matter is going to come up, I hope, in the next Congress, and so far as I know the project has nothing in its way legally; nothing is needed except a very moderate appropriation from Congress to acquire these lands as part of the upper Mississippi Valley refuge. It might not be a national park in name, but it would be in fact. That is, after all, what we want. I hope that our committee, or the National Research Council, will perhaps take it upon itself to inform us as to what the status of the project is, later on, so that we can get concerted action.

MR. PEARCE: I have been interested in all the accounts of and references to preserving the relics of the old Indian life we have had in this meeting. I will tell you one little incident that occurred in Texas which will give you some clue as to our difficulties. One of our citizens offered the Legislature of Texas 21,000 acres in Davis Mountain as a gift to the state for a state park, and the Legislature refused to accept it or do anything with it. The explanation lies in the fact that the Legislature was afraid some money would have to be expended on that park, and it lies in a section of the state upon which they are not interested in spending money for such purposes.

I will say a word about the archæology of Texas. I am working in the central portion of the state where we have some unique remnants of the old Indian life. Not far from Austin I have discovered three distinct cultural levels. Those of you who are particularly interested in mound culture will be interested to know that the upper levels contain many artifacts obviously of the mound builders' culture of the Mississippi Valley. We are in doubt about the invasion from the East over central Texas—whether it was simply a cultural invasion or a popular one. Of course, Texas has many geographical environments within it, but when the archæology of the state has been more adequately worked out I think we will get some interesting lessons in the effect of environment on the different cultures found there.

In historical times the eastern section of the state was occupied by tribes that were sedentary, and had agriculture in a higher state in general than the tribes on the plains. When the forest tribes came out on the prairies and plains they were unable to maintain their forest culture. I judge the explanation of this lies in the fact that those wide prairies made it easy for villages to be seen long distances. The whole situation was radically different, so that from that time on nomadic culture only spread over the plains to the west.

MR. FOX: Speaking for Michigan (having been concerned with the Archæological Society there for some time), when we began studying we went first to our neighbors, Minnesota and Wisconsin and to others who had made great progress, and adopted as nearly as possible their methods. We work, in our archæological society, hand in hand with my friend, Dr. Hinsdale, who is in the anthropological museum of the University of Michigan. In fact, he was one of the founders of our society, and backed it from its inception. The archæological work being done in Michigan is more or less in the hands of the University, and our society through its secretary has been collecting data. We realized as soon as work began that little had been done in our state since about the year 1870.

Dr. Hinsdale did a great deal of work on the mounds of southern Michigan and the peninsula. Dr. Henry Gillman also did a great deal. Just what the future holds for Michigan, I cannot say. My impression, from traveling over the state, is that the culture in Michigan is distinctive, and not as advanced as that of Ohio. Nor do we have any such great number of antiquities. Furthermore, the few we do have, apparently, are practically all in the southern end of the state—many of them located in my own county. We have located some thirty sites, practically every one of which was unknown to the collectors.

The feature of the meeting here this afternoon that has most surprised me is the matter of educational work that can be done. I think if our society is of any value at all to the state in its present form it is in the amount of educational material we are spreading, by means of our meetings throughout the state.

MR. CHARLES E. BROWN: I will speak briefly, and not of Wisconsin, if you will forgive me, for the reason that Dr. McKern and Dr. Barrett have given you an idea of what is going on in our state. But I would like to mention the body of men I met in the year 1904 when I came to St. Louis in connection with the Philippine exhibits of the great Louisiana Purchase Exposition. I was just a young museum man at that time, with much to learn. I came to a strange city and it was to my great pleasure and profit that I fell into the hands, almost immediately, of a great group of men here in St. Louis, one of whom was the late Dr. Henry M. Whelpley, whom many of you knew so well. There were also Dr. Paschal, and D. I. Bushnell, whom many of you knew; there was P. D. Stetson, another remarkable man, old Mr. Chouteau, Judge Dudley, Dr. Wiley and Mr. Jarred Cole. Working here at the Exposition, and meeting these men nearly every evening, I got quite an insight into Missouri archæology. I not only became acquainted with local collections, for these men knew where all of these collections were, and they took me everywhere, but I became acquainted with the very wonderful collection, as I thought even at that time, in the Missouri Historical Society. Likewise, I made my first trip to Cahokia under their direction, and also to the great flint quarries at Crescent, Missouri.

Mr. Fowke was also an inspiration to me. I think he has been that to almost every man who has come in contact with him. He was then engaged in work at the Exposition, but he also was doing a great deal of exploration work in this state along the Missouri bluffs, over in the Ozarks, and in many other places. These men at that time had already accumulated a most valuable amount of both data and specimens in regard to Missouri archæology. They, I think, were the ones who laid the foundations of that work, or at least built on earlier foundations in this state; and I mention this at this time in order that you may also understand that, despite the fact that one or two of the speakers might give the impression that Missouri has been little surveyed and little explored, a great deal of work done has been done in Missouri by a very admirable body of men, and the results, I think, are here to demonstrate the statement.

MR. CALVIN BROWN: Our problems in Mississippi are just about the same as those that I hear from other persons all around about us in the Valley. We have a large number of mounds, and we are only beginning the work of exploration. Mr. Clarence B. Moore and Dr. Moorehead have done some work in our state. The State Geological Survey is doing a little work from time to time, but much still remains to be done; and I hope the interest, both locally and throughout the country, will be increased.

We have certain groups which should be, by all means, preserved as national or as state parks. For instance, the rather famous Nata Highway mound, which is the sacred mound from which the Choctaws sprang from the earth, according to their own story, is still in an excellent state of preservation. How long it will remain that way, unless some official action is brought to bear, I do not know.

There are some excellent opportunities, as I believe Dr. Moorehead will bear me out in saying, for local parks and preserves of that kind. There is much of interest that should be protected.

ADJOURNMENT, AT 4:15 P. M.

PART II
PAPERS AND DISCUSSIONS PRESENTED AT THE OPEN MEETING
OF THE
COMMITTEE ON STATE ARCHÆOLOGICAL SURVEYS

The meeting was called to order at 10.00 A. M., Friday, May 17, with Mr. Carl E. Guthe, Chairman of the Committee, presiding.

MR. GUTHE: We hope that the program of these two days will satisfy all of you, and that you will feel that the trip down here has been worth while. The purpose of this open meeting of the Committee on State Archæological Surveys this morning is to present statements concerning the most recent work that has been done in the archæology of the Middle West, and then to discuss some examples of methods which are being used and studied.

There are four papers on field work, and two on methods. The first paper is by Mr. W. C. McKern, of the Milwaukee Public Museums, who will describe his work last summer in mounds in western Wisconsin.

EXCAVATION OF THE NICHOLLS MOUND OF WISCONSIN

W. C. McKERN

A definite object in view is apt to increase interest in the work of any scientific field expedition. In the summer of 1928, after examining a few potsherds from western Wisconsin, the author led a Milwaukee Public Museum party into the local archæological field with the well-defined idea in his mind to be on the alert lookout for evidence of the occurrence in Wisconsin of the Hopewell Culture of Ohio. The results of that successful search comprise the subject-matter for this paper.

The several sites selected for examination, situated in Trempealeau County, on the shores of the Mississippi, included four groups of mounds and one camp site. The four of these five sites first examined produced only evidence of fairly well-known cultures in Wisconsin, namely, the Effigy Mound and Grand River cultures. However, towards the latter half of the season, work was begun on the remaining site, which produced materials and information of an entirely different character.

The Nicholls Mound, the largest in a group of relatively large conical mounds, is ninety feet in diameter and twelve feet in height. For the Wisconsin field, this is a very large mound. The method of excavation to be employed, as dictated by the owner of the property upon which the mound is situated, was that of trenching, rather than the more desirable method of removal. In order to make the most of an inferior method, a trench ninety feet in length and twenty-three feet in width was cut through the center of the mound.

The entire body of the mound was made up of pockets of soil, clearly apparent in vertical cross-section as lens-shaped bodies. These represent the individual loads of earth as dumped by the builders. The base of the mound was marked throughout by a thin black line of humus, the remains of the original surface turf upon which the tumulus had been erected.

Some rather extraordinary specimens were encountered soon after excavation was started. Scattered centrally throughout the mound, from the top to just above the mound floor, were a series of chipped-stone artifacts of a type and size foreign to any of the previously classified cultures of Wisconsin. These included objects classified as lance points, from eleven to thirteen inches in length, a chert knife fourteen and one-half inches in length, and other knives and projectile points fashioned from quartzite, chalcedony, jasper and obsidian. In similar placement were found a disc-shaped fragment of thin sheet copper, a fire-clay pipe bowl of concave-based platform type, and about one hundred tubular beads rolled from thin copper sheeting.

The beads were placed a few with each of the larger objects, and so disposed as to suggest that they had served as pendants attached to wrappers with which each of the fine large artifacts had been covered. In any case, the regular association of beads with objects indicates intentional placement of these features in the mound material as it accumulated under the hands of the builders.

As the central floor of the mound was approached, the highly decayed remains of strips of bark were encountered. These were disposed parallel and without interval, extending north and south, to cover an area twenty-five feet in diameter and marginally raised two feet above the mound floor. A structure of light poles had supported this bark shed, which rested marginally on banks of earth taken from the sub-floor pit which the shelter was designed to cover, but the entire center of the structure had caved in due to the accumulating weight of earth upon it in course of mound erection.

[58]

The materials of this shelter were identified as the barks of a number of locally indigenous trees.

Beneath the bark shelter was an angular, flat-bottomed pit, two feet in depth, entirely lined with bark, in which had been placed the remains of six adult individuals and one small child. Four of the adults and the child had been interred in the flesh, extended and prone on the back; the remaining two were represented by reburials of bundled bones, probably the remains of primary scaffold burials. All skeletal materials were in the last stages of decay, to such an extent that they crumbled under the delicate touch of a paint brush, and had acquired a rather bright orange-brown color from contact with the bark which lined the pit.

Associated with the bones were: six copper celts, of which the largest weighed two pounds eleven ounces; a large chalcedony artifact of the scraper type; two copper plates, one of which was definitely of the Hopewell breast plate type; about forty pearl beads; four ear ornaments, fashioned of balls of wood, covered with sheet silver and perforated for purposes of attachment, found two on either side of one of the skulls. Two of the copper celts and both copper plates had served to preserve associated pieces of cloth, made entirely of fine nettle-fiber cord. Two techniques of weave were represented; a coarsely woven open twined technique, and a finely woven twined technique, with what appears to be crossed warp.

In a smaller tumulus adjacent to the Nicholls Mound, associated with burials, a large chert lance point, a broken stone pipe bowl of concave-based platform shape, and a pottery vessel of pronounced Hopewell type were found (Figure 6). Amateur excavations in a third mound had previously produced, among other typical objects, copper ear spools of the Hopewell pattern.

The materials and data obtained from these mounds (Figure 7) define a new archaic culture for Wisconsin, new in the sense that it has never before received recognition in our classifications. No single trait of previously defined mound builders' cultures is evidenced in these finds, with the doubtful exception of pit burials, doubtful since this particular type of pit burial differs in all respects from the widespread Wisconsin type.

Of the culture traits illustrated in the finds in these mounds, all are compatible with the Hopewell culture of Ohio, and at least half of them are recognized Hopewell markers; such as the concave-based platform pipes, pearl beads, copper breast plates with cloth adhering, wooden beads covered with sheet silver, type of copper celts, type of pottery and copper ear spools.

The only immediate conclusion to be advanced is that a group or groups of mound-building Indians in locally prehistoric Wisconsin possessed a Hopewell-like complex of culture traits which can hardly be explained from any other consideration than that of dominant influence of Ohio Hopewell culture.

DISCUSSION OF MR. McKERN'S PAPER

MR. SHETRONE: I think Mr. McKern's findings are those of a rather typical Hopewell development in Wisconsin. Some of you may recall that in my paper at Evanston I outlined the broad extent of this culture, finding evidence of it in eight states, prominently in six of them. In so far as Wisconsin is concerned, aside from two or three local trait deviations, it appears to me to be quite typical. In looking over the development in this area to the west, I found the most interesting thing, perhaps, of all, in the fact that we have a trait deviation, particularly as regards pottery. The vessel which Mr. McKern has shown is a typical Hopewell vessel, but that does not hold true for this entire area adjacent to the Mississippi. In Davenport, where I spent two days this week in looking over pottery from the Iowa mounds, I found a very definite and decided influence on the ware, presumably as a result of contact with some of the western culture. Moreover, it is of particular interest that certain characteristics which we regard as basic and ever present in the Ohio area are rare or entirely absent in this area to the west.

The spool-shaped copper ear ornament which is ever present in Ohio is not entirely lacking out here, but has almost disappeared. The same is true of the copper gorget or breast plate, which with us is ever present but is quite rare farther west.

I made a plea in the paper I read at Evanston for concerted effort on this Hopewell culture in the several states in which it occurs. My reason was that with the interest which is so much in evidence just now in the Middle West, it might be well to devote considerable study to particular phases of the subject. The fact that the Hopewell culture is so clean cut in its characteristics and of such significant interest, led me to feel that it might serve as a pivotal consideration for those of us who are fortunate in having Hopewell remains in our areas. The variation in Hopewell traits as they occur in the several areas or sub-areas might very well give us ultimately information bearing on the very interesting questions of the migrations, chronology and interrelations of these various Hopewell developments in the entire area.

THE WILLIAMS AND GLOVER SITES IN CHRISTIAN COUNTY, KENTUCKY

William S. Webb

(Abstract of a paper by Wm. S. Webb and W. D. Funkhouser, read at the St. Louis Conference on Midwestern Archæology by Wm. S. Webb, and to appear July 15, 1929, as one of a series of publications by the Department of Anthropology and Archæology, University of Kentucky.)

Each of these sites in Christian County, Kentucky, some three and five miles respectively southeast of Pembroke, Kentucky, consists of a mound in the midst of a prehistoric village site, and is surrounded by a cemetery of stone grave burials. The Williams site mound and cemetery was excavated in the summer of 1928. The Glover site had been partially explored previously.

The Williams mound was excavated by "slicing," each cut being ten feet wide and extending east and west across the mound. Earth was shoveled out and hauled away by a scraper, thus allowing a very thorough examination.

The mound was shown to have been built on the site of a wooden structure, formed by driving posts into the earth and weaving between them coarse grass and stems to form a wattle-work wall. The building was destroyed by fire, leaving the stumps of posts in the earth. The level of the mound was then raised by bringing up on it more earth and another structure, similar to the first was erected. When investigated, the mound, in the center, showed at least three distinct occupation levels as revealed by post molds made by the decay of the stumps of posts in walls. (See Figure 5.)

It is suggested that these two sites may indicate an extension into Kentucky of the prehistoric people denominated by Meyer (41st Ann. Report, Bureau of Am. Ethnology) as the Gordon culture, because of certain very definite similarities, which are listed below.

(a) The Williams mound certainly, and it is believed the Glover Mound also, was erected on the site of buildings which had been destroyed by fire.

(b) These buildings had rectangular walls of posts and wattle-work.

(c) Pottery sherds in the Williams mound showed textile impressions and were characteristic "salt pan" sherds, such as are always associated with the stone grave people.

(d) Stone graves in the Glover cemetery contained salt pan sherds.

(e) The double lug rim sherds, described by Meyer, were common in the Williams mound.

(f) Corn was used in the building, as shown by charred corn cobs.

(g) Periwinkle shells were found in abundance.

(h) Although the amount of animal bone found was not great, the remains of the buffalo were conspicuous by their complete absence.

(i) In the Glover mound, a water bottle of painted ware was found decorated with the so-called "four world-quarters cross and encircling sun symbols."

(j) Both the Williams and the Glover mounds were adjacent to large cemeteries typical of the stone grave people of the Cumberland River Valley.

DISCUSSION OF MR. WEBB'S PAPER

MR. NELSON: I would like to ask whether the pottery found was near the bottom?

MR. WEBB: The pottery seems not to be stratified—simply sherds throughout the whole mound. There seemed to be no broken vessels or pieces which might have been related. There was nothing at the bottom of the mound except ashes four to six inches thick, with flag stones.

MR. NELSON: I asked because some years ago I had an opportunity to work in the entrance to Mammoth Cave where there had been found human deposits. The bulk of them had been removed in making way for tourist traffic, but at least the basal portions of two fairly extensive deposits were left. I was unable to find any pottery in that material, although outside, up and down the Green River, were stone box graves, and elsewhere there was plenty of pottery. In other words, my conclusion was that here in Kentucky, as in regions in New York, in New Jersey, and even up as far as the St. Lawrence River, there is evidence of a culture level antedating the appearance of pottery.

MR. BLACKMAN: I would like to ask what was the method of tempering.

MR. WEBB: It seemed to be shell tempering, and some of it gravel and even sand. It was almost impossible to identify it certainly. I might add that we have found this pottery used as flooring in stone box graves in nearby cemeteries.

MR. MCKERN: Was any interpretation made of the walls or post molds?

MR. WEBB: It suggests a wattle-work wall inasmuch as the matting was lying both inside and outside of the line of posts, and therefore, presumptively, was representative of the wall which was burned. It fell and was, apparently, covered over even during the time of burning, because its remains were charred. That may suggest intentional burning. There are evidently three levels of occupation, upon each of which post molds were found, showing impressions of the bark in the wall of the mold.

MR. MCKERN: Was there any indication that these represented the walls of houses?

MR. WEBB: Yes. This wattle-work material does not seem to be plastered at all—no evidence of it. But the pattern made by the posts was rectangular, and the lines run east and west, and north and south, seeming to follow the cardinal directions.

MR. BARRETT: Was there any possibility of determining the original form of these mounds? Were they simply round mounds, or was there anything in the way of a truncated pyramidal form? My point in asking this question is that some years ago, in our work in Wisconsin, where we had to deal with the truncated forms of mounds to a certain extent, we found the same type of post mold structure.

In the southwestern corner of a pyramid, we found a complete square formed by post molds, apparently representing a structure of the same type that Professor Webb has found. But we also found surrounding this whole site a complete stockade wall, with watch towers at regular intervals.

I am wondering if the work has been carried far enough in Kentucky to determine the existence of such a stockade, and also the original form of the mounds.

MR. WEBB: The form of the mound was not ascertainable, because it had been under cultivation for many years. It is located in a very fertile valley. I talked to the old negro man who had been on the property for many years, and he told me it was now about half as high as he could remember it. Evidently it would have been easy for it to have been plowed down a good number of feet. In fact, there may have been higher levels of occupancy, entirely destroyed. It was very difficult to tell where the mound began because of this cultivation. We could find no evidence that there had been a stockade wall. The work is not yet finished, and will be continued for a couple of months this summer.

SOME RECENT NOTABLE FINDS OF URN BURIALS IN ALABAMA

PETER A. BRANNON

Recent discoveries made along the Tallapoosa and Alabama Rivers in central Alabama have greatly increased our interest in the customs of the aboriginal people who inhabited that section of the Gulf States. Members of the Alabama Anthropological Society, within the past six months, have investigated several localities formerly occupied by these natives, and an outstanding contribution to our knowledge of the early history of America has been made. This section is particularly rich in evidences of a burial custom which has interested historians as well as scientific investigators, and burials in urns have been recently noted in large numbers.

The floods of the past March made possible many rich finds, but even before that time, sites on the headwaters of the Alabama River and on tributary streams in Lowndes County had yielded fine results to the archæologist. By carefully studying these remains in their original position, an excellent opportunity is given to note economic and cultural conditions at a date certainly long before the coming of DeSoto, the first white man in this part of the South.

Social customs are indicated in the manner of wearing ornaments and in the placing of possessions with the loved ones who had gone on to that Great Beyond. Two shell hairpins found over the left ear at one place, and four over the right side of the head at another, suggest different modes of dressing the hair. Many of the burials are accompanied by ear ornaments, but none by nose-pins. Nearly all of them wore breast gorgets, the larger number of which show incised designs in conventional forms, but many clearly illustrate the ivory-billed woodpecker, the rattlesnake, the hand, the eye and the sun.

The custom of placing the dead in pots at interment is said to have been a Choctaw culture indication; if so, these people extended their influence as far east as the source of the Alabama River. The traditions of these people say that they put the bodies out on pole racks or brush arbors when death occurred, and then when the flesh had sufficiently decayed, they gathered up the bones and buried them. The finding of a group of vessels suggesting that they were all placed in the grave at the same time clearly corroborates these traditions.

Recent finds of pottery washed by the rains of early spring (1929) from their original deposit place, at a site known in later years as Autosse in Macon County, indicate these people as having been far above their later descendants, so far as their cultural status went. The vessels are of a heavy earthenware, shell tempered, glazed with charred grease, and some of them of a capacity of eight gallons. One recent day's work by five members of our Society resulted in the taking out of eleven of these fine pots, all in perfect condition. A number in fragments, beyond recovery, were also found. These had no skeletal remains in them and do not indicate a use other than economic. I believe they were used to store walnut oil, a commodity much prized in this section.

[64]

Less than thirty days ago, Edgar M. Graves, Dr. R. P. Burke and Howard H. Paulin of Montgomery located in a cache-like arrangement twelve urns, every one covered with a bowl, and all containing skeletal remains. The largest is twenty-six inches in diameter, about two feet deep, and had in it eight skulls and the larger number of the bones of these skeletons. Several were adults but there were also children and babies. Mr. Graves considers them as all of one family or clan. Several of the other pots or urns had more than one skeleton in them. The smallest of them all is just eight inches in diameter, but in it was the complete skeleton of a baby. The arrangement of this group may have been intended to represent a constellation. The vessels were very close to the surface, in fact recent plowing had carried off the cover of one of them.

The first indication of this kind of an arrangement of vessels noted in this state was on this same stream, Pintlala Creek, but nearer the mouth than those found in April of this year. Several years ago we found nine urns grouped around a central one. In this case a vault-like placing had been attempted. A hole about twenty-five feet in diameter was apparently first cut into the solid red clay. Into this was poured quartz gravel, then periwinkle and river mussel shells from the kitchen middens or refuse piles, and into this ashes.

The vessels after arrangement were surrounded with layers of gravel, shells and ashes, and then covered with clay. This had been hardened by burning, indications of fires on the pile being very evident. You will readily see that a very good attempt at permanent preservation of the remains had been made.

Frequently interments in the earth alone accompany those within the pots and are apparently contemporaneous. In most cases these are flexed; that is, bent up with the knees under the chin and sometimes with the elbow over the head. Occasionally, bark or wood slabs were used in covering vessels, and in casing the loose burials, though usually an attractive bowl was used to cover the vessels. Burial-urns are nearly always of thin, poor quality of earthenware, and suggest that they were made altogether for this purpose and had rarely served any previous economic need. The bowls which we find serving as covers are nearly always works of art, many having the ornamentation on the inside of the lip. No bowls and few pots have handles. Whenever a vessel does have handles, it is more apt to have six than four. In no case have we ever found a burial urn with legs.

The conventional roll-forward and loop-back serpent scroll design, and the design in some manner suggesting the rising sun, are the most common from central Alabama, while the woodpecker and the hand and eye are found most often in our Moundville culture. An attractive as well as economic design of ornamentation consists of upright parallel embossings of clay attached immediately under the lip and on the neck of the vessel, serving to reinforce the edge as well as to beautify it.

Mr. Clarence B. Moore, of Philadelphia, first noted our urn burials. Those most prominently figured were located at Durant's Bend on the Alabama River. In recent years, most of our finds have been in Lowndes County at a site passed by DeSoto in September, 1540, and noted by one of his chroniclers as "an old abandoned town." No European contacts have ever been suggested in connection with urn burials, indicating that the custom was obsolete here before the explorers passed through.

Though not extensive, the shell objects, bead ornaments and pendants are indicative of a high civilization. The ability to express ideas on shell in the form of pictures is prominently demonstrated. The squatting figure inserting a sword through his tongue, which protrudes to an exaggerated length, is frequently seen on the gorgets or cameos. This suggests to students of Mexican cultures a contact with that region, and perhaps a verification of the legend of the people found here in historic times, that they came from Mexico. All our Alabama natives were Muskhogean-speaking peoples.

Strange though it may seem, we do not find any suggestion of the influence of the flora of this section in their art, though the beauty of the primeval forest was doubtless evident. The fauna, however, did impress these prehistoric people. Several of our southern animals are pictured. The gopher, the mink, the turtle, the fish, deer, raccoon, and the ever-present serpent are shown. Their pipes often represent animals and birds. The stars, the sun with its rays, and a design that may be intended to represent rain are prominently in evidence.

While the historic period of the Gulf country overlaps the pre-Columbian, our archæological investigations clearly indicate a demarcation of cultural influence. Of course, it is obvious that trade objects are found in our burial places, but there are few instances where the post-Columbian and the prehistoric overlap. The civilization of the older people was advanced far beyond that of the natives found here after DeSoto. However, early man in the South was not nomadic. He was raising corn several hundred years before Columbus touched our shores, and his mounds are the material evidences of that local condition which forced him to build earth lodges where they erected pueblos in the Southwest. Poles in the thickly wooded country of the South, along the swamps and streams which provided an additional food supply, made house-building easier than in localities where stones had to be found.

[66]

DISCUSSION OF MR. BRANNON'S PAPER

MR. SHETRONE: Mr. Brannon, I would like to ask whether the material culture that goes with these urn burials is similar to that of Mr. Moore's Moundville Site?

MR. BRANNON: Exactly the same thing. I think the Moundville culture extended as far east as seventeen miles east of Montgomery.

MR. COX: Mr. Brannon, may I inquire if these urns containing the skeletal material had the bottoms broken out—did they appear to be "killed"?

MR. BRANNON: No, sir; we have never seen a killed vessel in that part of the state.

MR. COX: Do you find the bones of animals with these burials?

MR. BRANNON: Many of them have bear's jaw-bones, and sometimes they have a deer's bone made into an awl, or they may have a bird's leg made into some sort of perforating tool.

MR. FOX: Were these pots built up by coiling?

MR. BRANNON: I do not think they had any other than the coiling process. Frequently we find the bottoms of broken vessels showing the small coil.

MR. BROWN: Are the covers always inverted?

MR. BRANNON: Yes, sir, they are always inverted, and the covers are invariably of a much better material than the urns. I think they must have originally been domestic utensils. Nevertheless they always have the sun design, together with the similar rain design.

TRAILING DeSOTO

JOHN R. FORDYCE

(Abstracts from paper as presented, full publication to be made elsewhere)

I have been interested in the expedition of DeSoto primarily because his was the first expedition of Europeans which penetrated into the interior of the southeastern United States. The members of his party were the first white men we know of who saw the great Mississippi River far inland. The first accounts concerning the Indians who lived in the regions back from the coast come down to us from the historians of DeSoto's explorations.

On different trips connected with my engineering duties I have seen mounds and village sites, strewn with broken pieces of pottery, chips of flint, and fragments of human bones which the plow had turned up; and I have traced old moats and embankments through plowed fields.

Old maps show trails marked out that were long ago great trade routes crossing the continent. Some of the early maps which our land surveyors made show faint, dotted lines described as "Indian Trading Paths," or "War Paths," or "Traces"; and some of our old histories tell how the road which runs from one town to another was at one time an Indian Path which the early white settlers cleared and made broader. Now our great Government highways and many of our railways are located almost exactly along these old Indian trails.

DeSoto and his men found the Indians living as they had been living for centuries, gathered together in fortified villages, the Chief living on a mound, the temple located on another, and on still another the great council and guest house. The mounds of today have a great interest for us, but what a still more intense interest they would have if we could see them occupied by the Indians as DeSoto found them, or see the various savage dances and ceremonies which he and his men witnessed, as they wandered over the states of Florida, Georgia, Alabama, Mississippi, Arkansas, Louisiana, Oklahoma and Texas.

I wanted to visualize these old scenes and the Indians who had once lived in these villages and traveled these old War and Trade Trails, and so I gathered together all the books and maps that I could find which told of the travels of the early explorers in this southland of ours. From these sources I began gradually to piece together the route of DeSoto. Afterwards I got in my car and actually went to various places to see if the topography filled the description given by the old historians.

In order better to understand the old accounts of DeSoto's Expedition, I made a study of the Indians who lived along his route; their friends and their enemies; their customs and their religion. I have studied the changes which civilization must have brought about in the land—the drainage of swamps, the changes made by the construction of levees and the cutting away of the timber and the changes in river beds by cut-offs.

History shows that on May 30, 1539, DeSoto with 570 men and 224 horses had landed on the west coast of Florida, worked his way through swamps, rambled across the rolling country of Georgia as far east as the Savannah River below Augusta; thence up into the hills of northern Georgia where he came in contact with the Cherokees; over into the country of the Creeks, down the Coosa River, down the Alabama River to a point near the present city of Selma, where he crossed and marched into the country of Tascalusa. In a fight with this Black Warrior of the Choctaws the Spaniards lost most of their baggage and all of their powder, but killed over 2000 of the Indians and burned the village of Mauvila which has been located as just opposite the present city of

[68]

Demopolis in western Alabama. DeSoto's route from this point turned northward up the eastern side of the Tombigbee River to about Aberdeen, Mississippi.

On the 16th of December, the exploring party crossed the Tombigbee River, and after a victorious encounter with the Indians settled in a town that appears to have been just west of the present town of Egypt in Chickasaw County, Mississippi. Here they spent the winter, and from here started westward again on the 26th of April, 1541. During the winter and during the first few days of the spring march they were frequently attacked by the Indians, losing most of their clothes and their weapons. Near the present town of Houston, DeSoto attacked the Indians, hoping to find provisions, but none were found. From here, on the last day of April, they set out and marched for seven days through a deserted country, hilly and swampy, and still deserted at the present day.

The Old Trading Trail from the east comes out of the hill country of central Mississippi near the present city of Grenada in Grenada County. It winds over to the crossing of the Yalobusha River just below the mouth of the Tippo Bayou and crosses the deep Tallahatchie River near Minter City. This town is located on a ridge between the Cassidy Bayou and the Sunflower River, which is thickly dotted with Indian mounds. Here DeSoto found villages where corn was plentiful, and easily captured them. Then he and his followers went on up the Sunflower River, through the present towns of Webb and Tutwiler, and on Saturday, May 21, 1541, came to an opening in the trees and saw the great Mississippi, which at that time flowed more than twelve miles to the east of its present bed. So it was at Clarksdale, Mississippi, now far inland, that DeSoto first saw the river.

Timber was built into barges under the direction of the engineer Francisco. The Indians hovered about in their war canoes but did not attack. These Indians were of Siouan stock—" fine looking men, large and well-formed, and what with their awnings, the plumes and the shields and pennants and the number of people in the fleet it appeared like a famous armada of galleys."

Early one morning, before daybreak, the first boats built by white men to navigate the Mississippi River began to cross over, and two hours before sundown the whole party was on the other side. It is evident that they crossed to a point or sandbar on the Arkansas side, because it is related that the cavalry jumped out and waded ashore. The first town they came to was called "Aquixo," which may have been the original name of Arkansas. After wandering around and bridging a small river or bayou

[69]

they waded through a swamp and came to Casqui. This, they said, was a high, dry country in which pine trees grew. This description alone would fix the point of crossing the Mississippi River; for Crowley's Ridge, near Helena, Arkansas, is the only place in Arkansas on the western shore of the Mississippi River where such a description fits.

The Chief of Casqui was most friendly, for he needed these Spaniards to help him punish his ancient enemy, the Chief of the Pacahas. DeSoto, in order to demonstrate his religion and convert the Indians, joined his men in a religious ceremony. This being concluded, DeSoto and his men marched northward, up the river toward Pacaha. The Indians of Casqui preceded him to build bridges and clear the trail, and their main army followed him.

In two days they reached a plain, and came in sight of the town of Pacaha, surrounded by a man-made moat. The army of Spaniards and Indians did not attack at first, and most of the people of Pacaha escaped by boats. Later Pacaha was induced to return to a grand " get-together " dinner to which DeSoto invited both chiefs.

There is a picture in the Capitol at Washington which shows DeSoto discovering the Mississippi River. Evidently the artist did not read his history very carefully, for he has shown the Spaniards in all their glory with banners flying and armour brilliant in the sun. Contrast this picture with the actual facts—" numbers of soldiers who had been a long time badly covered, clothed themselves then. Of the shawls they made mantles and cassocks; some made gowns and lined them with cat skins as they also did the cassocks. Of the deer skins were made jerkins, shoes, stockings and shirts, and from the bear skins very good cloaks such as no water could get through. They found shields of raw cowhides out of which armour was made for horses."

DeSoto stayed here at Pacaha for over a month, then marched back down the river, passing the village of Casqui, whose chief went with him to the end of the trail which led to lower White River, helped him to cross over through the connecting cut-off to the Arkansas River, where DeSoto and his men landed on the south bank. Marching still downstream, on the western bank of the Mississippi, they came to a large town, which must have been, by the description of its site, located on Lake Chicot in the southeastern part of Arkansas. From here they marched northwest through country whose description fits the low, swampy region in Chicot and Drew Counties (soon to become the spillway for overflow waters of the river), and arrived at a town on a river that is evidently the Saline, near the place where now the Cotton Belt Railway crosses, in Cleveland County. Marching on they crossed some high ground, found a salt spring

[70]

(on Bayou de Sal in Clark County), and came to the Ouachita River. Staying there several months, they discovered the present site of Hot Springs. During this stay, the Spaniards heard that back in the hills upstream there lived a fierce and strange tribe of Indians called the "Tulla." DeSoto attacked and conquered these Indians, who lived along the Ouachita River near what is now the town of Cedar Glades in Garland County.

Looking now for a place to spend the winter, the party turned southeast to the Caddo River and down to a site near the present city of Arkadelphia where they crossed the Ouachita and going on, stopped for the winter near the present city of Camden. During the winter DeSoto's interpreter died, and on March 6 he decided to build a ship and send to Cuba for reinforcements. Wandering as he thought toward the Rio Grande, he came to "Ayays" a town on the bank of the Ouachita just above Monroe, Louisiana. (Mr. Moore, of Philadelphia, has investigated this village site and found many beautiful specimens of pottery from its mounds.)

The Spaniards built a pirogue and crossed the Ouachita, then traveled for three days through a low, swampy country, undoubtedly the Bœuf River bottoms below Monroe. Coming to the town of Tutelpinco, they found it deserted and without corn, and near it the waters of a large lake flowed into the river with a swift current—Turkey Lake in lower Franklin Parish. Finally two friendly Indians showed the Spaniards the way to cross on rafts of timber and reeds, and they arrived on March 29 at Nilco, undoubtedly either at the present town of Jonesville or just below it at Serena, on Black River. There are large Indian mounds now in the town of Jonesville. The cemetery is on one, the hotel on another, and another was an island of refuge for stock in the 1927 overflow which covered this country. The town of Deer Park on the west side of the Mississippi River may have been the site of the town of Guachoya, but I believe that it was the present town of Clayton. DeSoto determined to go to Guachoya himself, for he wished to find out if the sea was near and if it was a good place for building ships.

He arrived there on Sunday, the 15th of April, 1541, and had his second glimpse of the Mississippi. He marched into the town, took possession and inquired vainly for information about the sea. From here he sent one of his own men, Juan de Anasco, to see if he could reach it. At the end of eight days de Anasco returned reporting that he had only traveled fourteen or fifteen leagues on account of the great bogs and cane brakes. This report fixes the location of Guachoya in the Red River delta, where DeSoto died on the 21st of May, 1542.

After his death a vote was taken and it was decided to abandon the ship idea and to march westward in hopes of reaching Mexico by land. Luys Moscoso, whom DeSoto himself had chosen, was made leader of the expedition, and on Monday, the 15th of June, the whole party started west.

The French maps of this region show that in the early days there was a well-defined trail which ran from Vidalia opposite Natchez to the westward. This trail crossed the Black River at Jonesville and reached the high ground at Rhinehart; it then went west to Packton, where it forked. The southwestern branch went to Natchitoches and the northwestern branch went north of Saline Lake, a point well known to the Indians, for there they made salt. The Spaniards arrived on the 20th of June at the province called "Chaguate." On this route they had crossed "a desert country."

From the fact that salt was made from a lake, and that the Indians had customs similar to those of the Tulla, it is highly probable that the Spaniards were marching westward along a trail which led up Red River into southeastern Oklahoma. Coming at last to a river (which must have been Red River) they were unable to cross until eight days later, when the water had gone down. This point of crossing I believe to be just north of Paris, Texas.

After wandering in Texas, seeking in vain the route to Mexico, starved and almost exhausted, they returned to Nilco, only to find that there were not sufficient provisions. Through many difficulties, they arrived at last at a well provisioned town near the present town of St. Joseph in Tensas Parish, where they set immediately to work to build ships. This village was still an Indian village when La Salle visited it 150 years later, and his men found some Spanish guns in the Temple.

In June the brigantines were ready and as the river began to rise they were floated off. On July 2, 1543, the Spaniards sailed down the river to the Gulf.

I have checked these places carefully by the stories of the old writers and by the actual conditions existing and described. In conclusion, I have proved to my satisfaction that wherever the old historians said there were mountains, the route goes by or over one. Wherever swamps were mentioned I have found one today or showed that it has been there. The rivers are still at or near the places where they should be. The salt springs are still running today, and though the Indians and buffaloes are gone forever, their old village sites and mounds still stand as reminders of the vanished people. We should preserve these monuments and each state or neighborhood of our country ought to help to do it.

[72]

DISCUSSION OF MR. FORDYCE'S PAPER

MR. BLOM: I think your calculations of time and miles are correct. It is very rare in open country that one can cover twenty-five miles; and in many cases it is impossible to travel even a half mile in a day. You mentioned a desert country. In Spanish " deserto " does not mean desert as we understand it, but an abandoned country, a region deserted by the population.

MR. MOOREHEAD: I want to say a word on this interesting paper. In exploring in the South in the past few years, we have all searched for physical evidence of the presence of DeSoto. Having read the journals in English and noticed how cruel he was to the Indians, I was in hopes we would never find any physical trace of DeSoto's trip. We did not at the Etowah site where we worked for four seasons. Thirty-seven miles north of Etowah we find the same culture, and in the top of a mound we found a skeleton not as decayed as the others, and surrounded by a crude crib. With it were three objects that had the appearance of sword blades four to five inches long, highly oxidized. Experts in the American Museum thought they probably were Spanish, but this is not certain. However, they were not early French.

MR. FORDYCE: The La Salle narrative, I think, reports that at this point Aminoya where the ships were built, there were some Spanish guns found in the temple. I think Mr. Brannon has a fragment of one of the DeSoto cannon. Mr. Brannon, could you tell us where that came from?

MR. BRANNON: There was an old man connected with our institution from Mississippi, who identified a breech block in the possession of the State of Alabama as being positively similar to those used by the Spanish Armada at the time of the attack of some forty years before DeSoto got to the Southern Gulf country. So far as our records indicate, and so far as history records, that was the only expedition that brought any cannon of the Armada type to this country. We do have one of the breech blocks that were found twelve or fourteen miles from Montgomery.

We have another specimen which may be a DeSoto relic, however. We find indications in the city limits of Montgomery that here was one of his stops. This town is spoken of in the DeSoto narratives, again some sixty years later by the Charleston travelers, in 1714 by the French, and in 1804 by Andrew Jackson. It was abandoned in 1836, and must have existed practically 400 years. The narrative states that DeSoto stopped there from the 6th to the 13th of September, 1540, and that he swapped thirty pocket knives for thirty women. The Anthropological Society has thirteen of those pocket knives, we think. We found a large old knife blade, which was sent to the leading cutlery establishment in America, where it was identified for our chemist. This company commented that the knife contains the highest purity of iron which had ever reached their laboratories; that it was not latter-day iron, and that it indicated a very fine grade of a type much older, and of European origin.

Now, the fact that we found European iron of such a fine grade would indicate that it was a pocket knife that DeSoto traded off at that time.

MR. KEYES: May I ask Mr. Fordyce whether his results have been published?

MR. FORDYCE: No, they have not. I have been studying this subject for about twenty-five years, but I am an engineer and have only worked at it as I went into one section or the other, and as I found time.

MOUND AREAS IN THE MISSISSIPPI VALLEY AND THE SOUTH

WARREN KING MOOREHEAD

During the past forty years various institutions and individuals have carried on extensive explorations in the mounds, village sites and cemeteries of central and southern United States. These operations have been extensive, have involved expenditure of a large sum of money, and the public may properly demand an account of our stewardship. In my informal remarks I shall attempt to deal with essentials. Beyond question we have several distinct mound builder cultures in the twenty states comprising the area mentioned. That these overlap, or in some instances present certain features in common, no one who really understands this subject will deny. My audience will realize that my remarks are not critical of any individual or institution, when I state that it appears to me, if to no one else, that we have gone entirely too far in extending the boundaries of certain of these cultures.

There are available a large number of reports on field operations and very considerable collections in some forty or fifty museums. Obviously I refer to collections from the mound area and not elsewhere. These should be considered in a broad rather than local aspect. Therefore our first proposition is to the effect that there was a general and prevailing custom of mound building throughout the entire region between the Great Plains and the Hudson valley. Within this enormous tract lies a territory roughly estimated at 600 by 800 miles in extent in which mound art—if one may use the term—is rather highly developed. Surrounding it in the greater area, mounds and their contents indicate less complex cultures. This geographical area of 600 by 800 miles is characterized by high local development in certain places. We have not time to enter into detail, but we might remark that chief of these is the famous Hopewell culture of the lower Scioto valley, to which could be assigned thirteen of the nineteen type units which for my own convenience I have assumed in measuring the cultural status of mound-building people. The high Etowah culture of north Georgia and of the Tennessee-Cumberland valleys of Tennessee is assigned eleven points in this scale. When Professor Mills coined the term "Hopewell" for the lower Scioto, he also employed the term "Fort Ancient" for a more widely spread culture. As I have remarked elsewhere, this term "Fort Ancient" has been accepted although it is rather unfortunate. It means neither high mound builder art nor yet an exceeding low status but might be roughly compared with the term middle class, commonly employed to differentiate the bulk of individuals from those who are extremely well to do or very poor. Fort Ancient, then, applies to the average village population of the Kanawha, Illinois, Arkansas, Wabash, Savannah and other valleys. Beyond question that body of our Indians of the middle class (Fort Ancient) occupied most of the mound area under discussion. Within that body, as I have stated, are these highly developed local centers. In addition to the two named is the site (probably Creek) at Moundville, Alabama, which has been explored by Mr. Moore and Mr. Brannon, and which although less than one hundred and seventy-five miles from Etowah to the east, is quite different.

In the central Illinois valley there is also a high development somewhat comparable to Hopewell but not entirely so, since it presents several other characteristics. To that Illinois group, surrounded as it is by general Illinois Indian culture, one would assign eight of our cultural points or type units.

In both Florida and south Georgia—particularly in the former—there are an enormous number of shell mounds, platforms for houses or temples, and indications of a very heavy and industrious population. Notwithstanding extensive labors for many winters on the part of Mr. Moore or others, the tribes of Florida and south Georgia do not present a development equal to the centers mentioned. In short, I would assign them but four or five points in our scale.

It seems to me that we have minimized the extent of aboriginal trade and commerce. These great centers, such as the famous Duck river site, certain large sites in Wisconsin, and Hopewell and Etowah themselves were in being for generations if not for several hundred years. The most skilled New World artisans were attracted to communities more or less sedentary. Art developed—as did our own great art centers—in the cities. A study of the collections verifies the statement that many objects from these centers were carried to distant points. It seems to me that the preponderance of evidence is to the effect that although in most of our highly developed local centers we will find objects suggesting a knowledge of, or trade with, some distant dominant culture, yet the great majority of artifacts, and so forth, are of local manufacture and form or concept.

I have often thought of that little colony, speaking Sioux, located at Biloxi, Mississippi. They have been described by ethnologists. There was found a small body, far removed from the great Siouan stock to the northwest, and surrounded by Muskhogean tribes, yet no one would claim general Siouan influence in that part of the South.

As to the origin of the Hopewell culture, I might offer a theory. Years from now, when explorations throughout the Mississippi Valley shall have been completed, more competent observers will probably solve the question of origins. My hypothesis may not be correct, although I desire to have it recorded. It cannot be set forth very briefly.

I have never believed that the Hopewell people originated in the Lower Scioto valley. There is no evidence that they dominated Kentucky to the South, which is a buffer state between the Tennessee-Cumberland and the Ohio. The Kanawha valley has not been explored, but such specimens as are available indicate a considerable divergence from pure Hopewell. The Muskingum in eastern Ohio is probably Hopewell, or closely allied to it. No Hopewell objects were carried down into the South so far as we can ascertain. There may be some in Kentucky, but I am speaking generally, keeping in mind preponderance of evidence. Trade objects at Hopewell indicate a knowledge of the South, and that it is more recent than the Southern works.

Far up in the Northwest have been found a few monitor or platform pipes, log burials occur in the Liverpool district (Illinois), human maxillaries worked into ornaments, and grizzly bear tusks—favorite Hopewell trophies—and some other objects. It may be, as claimed by some, that this indicates an offshot of Hopewell in southern Illinois, eastern Iowa, or central Wisconsin—as in the case of the Sioux at Biloxi. With due respect to my distinguished co-workers who differ with me in this matter, permit me to state that while such objects may have been introduced through barter, or small colonies sent out by the home village, I do not believe that is the correct solution.

My theory is to the effect that a certain band or tribe of Indians—probably very early Algonkin—reached or originated in eastern Iowa. One branch may have worked up into Wisconsin. The other proceeded east-

[76]

ward through Illinois and Indiana to central Ohio. The objection to the Southern theory of origin lies in the fact that the ceramic art so prominent in the South is not in evidence to any extent in the Hopewell tumuli; that is, they have found a few pots, but in the scores of mounds explored from whence they (Putnam, Mills, Shetrone and I) took hundreds of burials, it may be said that pottery is practically absent. On the Nettler farm in 1927 in one tumulus we found considerable pottery, six or seven typical Hopewell axes of copper, cut human jaws, etc. This is in the region where it is now claimed there was distinct Hopewell development.

Reverting to my theory, as these people proceeded eastward they built mounds, but constructed very few characteristic Hopewell earthworks— the squares, octagons, etc. Now and then one finds a crescent or circle. Not until we enter Indiana do we observe true geometric works so typical of these people. On reaching the Scioto Valley, where conditions were extremely favorable for their development, they remained, became sedentary, and attained the culmination of their wonderful development.

Mr. Charles C. Willoughby, who has given some attention to the subject, is of the opinion that the solution to this mound problem lies in a complete study of symbolism, and that there were very highly developed mound cults regarding which, at present, we know little or nothing. He has not perfected his study of the earthwork and cosmic symbols as evinced in copper, on bones, or presented by the earthworks themselves. All of us join in the hope that at some future time he will undertake this important investigation.

I have purposely omitted the great Cahokia group from my remarks. It is in a class by itself. It is distinctly Southern. Five seasons spent at that place in extensive work have not yet produced the mortuary edifice of these people. It is the largest known village north of Mexico, being, by actual tests, about six miles in extent. That so large a population made use of one or more structures for the interment of their distinguished dead no one doubts. Until this discovery is made, it is impossible for us to present conclusions worthy of the name concerning the Cahokians, for, obviously, we cannot study art unless we possess art objects.

I have said nothing as to the origin of mound building in general in our country. That, as the writers say, is another story and too lengthy to be inserted here. One might remark, however, that Mrs. Nuttall has found seven distinct comparisons between early Toltec art and our Etowah finds. Whether this is a mere coincidence, or whether it indicates that the Etowahans worked their way gradually from central Mexico to Georgia, is problematical.

A chief objection to this theory lies in the fact that it is some 1500 miles from the last tumuli of central, northern Mexico to the first mounds of size in eastern Texas. Indians, familiar with mound building, would scarcely traverse 1500 miles and leave no remains. Yet how are we to explain the monolithic axe, idol heads, plumed serpent, seated figures, and other similarities?

These impressive monuments, so numerous in twenty states, are fast disappearing. It is fitting, therefore, that we urge their preservation in state, city, and town parks. That is of primary importance. Once preserved, their thorough exploration can be safely deferred.

DISCUSSION OF MR. MOOREHEAD'S PAPER

MR. GUTHE: We are all grateful to Dr. Moorehead for presenting the general problems and theories which we must always keep in mind while we are working on the specific things we find. One is sometimes apt to overlook the greater problems while studying the little technical details of a given region.

I think Dr. Moorehead's conclusions will certainly be amplified and upheld on certain points, and perhaps disproved on others, as we become better acquainted with the small problems that we have before us, such as the distribution of pottery types, the distribution of copper, types of stone implements, and so on.

MR. COLE: I agree very heartily with Dr. Moorehead on the desirability of the study of skeletal material. However, we must not depend too much upon such studies for this reason: that if we go to any ethnological situation—in California, for instance—we find a similar culture spread over a large number of tribes and groups. If we consider our ethnological field in general we find a similar culture will spread over diverse physical groups.

It is quite evident from the little work we have done in Illinois that there are several physical types in this culture area. While it is important to study skeletal material, the results obtained do not necessarily affect cultural history.

MR. GUTHE: Mr. Moorehead has very admirably taken the step from the reports of field work to the study of methods. This meeting is an open meeting of the Committee on State Archæological Surveys, an advisory body which is trying to bring to archæologists an interchange of methods and technique as well as a knowledge of the results of others. For that reason there have been included in this morning's program two papers on methods. I know that you will all be interested to learn that the Milwaukee Public Museum has developed a form for recording the data obtained in a field survey.

[78]

A FORM FOR RECORDING DATA OF FIELD SURVEYS

S. A. BARRETT

(Abstract from stenographic notes)

Those of us working in the Milwaukee Public Museum have felt for a long time that in our archæological work in Wisconsin we needed some definite form for recording archæological sites and features as uniformly as possible.

We have had working in Wisconsin for many years a large number of members of the Wisconsin Archæological Society, and they have accumulated a vast amount of data on the archæological remains of the state. The work that has been done by that society and its members is most admirable in every respect. However, these data have been accumulated in accordance with the ideas of each individual who made the notes, and nothing in the way of a standard form has been adopted.

With this in mind, Mr. Townsend Miller, one of the members of the Wisconsin Archæological Society who has been much interested in the archæological work and excavations of the Museum, set about to develop a form of several sheets, so that when the data obtained from the various workers is compared, one will know precisely where to look for a definite item of information.

The form that Mr. Miller finally worked out was brought to us. Mr. McKern and I went over it and made a few minor suggestions, but the sheets, four in number, are essentially as Mr. Miller originally worked them out. To these is added one of the ordinary maps of the state, on which the site is located. There should also be an index of the various townships, so that the data sheets can be filed in a notebook in accordance with the key system which Mr. Miller has also worked out.

The sectional map, which is the first unit of the four sheets, presents a section divided into sixteenths. Before the field work starts it is necessary to transfer to this map from one of the county atlases such features as rivers, lakes, railroads, roads, and also the various farms and farm limits; in short, all of the features which will serve as guides to the worker in the field. On this sheet, also, the records already in existence are brought together under each one of the sections. This information is checked in the field and any new features found are added.

The second sheet, which is made to face the first, is a form entitled: " Form for Data on Sites." On this sheet blanks are indicated for numbered sites—1, 2, 3, 4, etc.—corresponding numbers being placed on the map. Additional data concerning each site are entered upon the third sheet, such as features of sites, including mounds, village sites, quarries, etc.

The fourth sheet is of cross-section paper, adapted for making careful drawings of the features of the sites.

I should like to emphasize the point that this form has been in actual use. The question has been voiced from time to time as to the workability of such a system as this, and as to whether it was too complicated for use by anyone who was not an engineer, for instance. The simplest equipment necessary is a camera tripod and a drawing board, a ruler and a simple sighting device, such as a combination of a ruler with a couple of pins. The use of these forms by Boy Scouts is obtaining entirely satisfactory results, and has shown that this method is a perfectly easy and workable one for use by almost anyone. Of course, the better the equipment used, the more quickly the work can be done, and to the better advantage.

Formerly, the results of county surveys were frequently jotted down on miscellaneous scraps of paper and much of the necessary data left to memory only. Unless some fairly accurate and careful recording methods are adopted mistakes are almost sure to occur, no matter how carefully and conscientiously the work is done.

These forms are not, of course, entirely perfect in all respects; but we believe they are a step forward in the solving of problems of recording of archæological field work, and are a positive and definite record, which does not lose its value with the passage of time. The recorded data should be duplicated and two copies placed in different repositories, in order to eliminate the risk of loss. If anyone cares for copies of these forms, we shall be glad to give you samples, or to have copies printed for you in quantity. I present this to you that you may consider it for the future, and possibly work out some set of forms which can be universally adopted, which would greatly facilitate the exchange of notes and information. The systematization of archæological information is one of the things that, in my estimation, we need most at the present time.

DISCUSSION OF MR. BARRETT'S PAPER

MR. GUTHE: I know you are all interested in this matter. As far as I know, it is the first attempt in the Mississippi Valley to systematize field survey records on printed forms. In the Pueblo area such forms have been developed, but as they stand they are not applicable to this region.

MR. SHETRONE: I would like to ask Dr. Barrett, if in selecting the survey station some permanent feature is chosen so that in the future the record may be definitely correlated with the actual site?

MR. BARRETT: Yes. The same principle should be followed in a survey of this sort as is followed by a civil engineer in locating a station for any survey. The survey station must be reckoned from some definite datum point, such as the corner post of a section or quarter section.

MR. BLOM: It is customary to have north at the top of the map. Once the magnetic or true north is determined, it is quite unessential to relocate the orginal point of observation. The magnetic declination changes constantly, so it is much better to orient maps to true north than magnetic north.

MR. BARRETT: That is quite true. The only reason that the true north is not always parallel to the lines on the paper is that in some cases the sketch of mounds of certain forms would run off the sheet if so oriented.

MR. PEARCE: Down in Texas, unfortunately, we do not have this system of land survey. Our state was laid out under the old Spanish method, by walking ponies across the prairies in any direction, or following the streams. I have been much puzzled to know how we can get anything as accurate as you evidently can get up here. The best we can do is to get the landowners' maps, and the contour maps, in the few places where they have been finished. We are up against a serious problem.

MR. BARRETT: I would say the only thing to do under those circumstances is to follow the topographical work done by civil engineers. They have to depend as I understand it, upon natural features. A point so many feet to the east or north of a natural feature is chosen as a starting point for the survey.

MR. HINSDALE: Over in Michigan we are mapping the state in some detail, and we have two systems of maps which we use. One is the Hutchins system, which has been adopted by the Government, laying out the country in six mile townships, and the other is the system of the U. S. Geological Survey topographical maps. I would suggest reducing this form to the basis of the Geological Survey topographical maps so far as obtainable in the different states.

MR. BARRETT: There is no question but that where a Geological Survey map can be had, you are in luck, but in a great many parts of the country the quadrangles have not been worked out. And there is another thing to be said against the topographical maps. They are generally on such a small scale that while they are excellent for trail work or features extending over a large area, we should not be able to indicate mounds or small features. I think that this system for detail work would prove very satisfactory, if used in connection with the topographical sheets.

MR. COLE: I want to say a word of appreciation of the suggestion Dr. Barrett has given. We have been doing something of the same kind in Illinois but not as elaborately as they have in Wisconsin. We found the township map very desirable indeed. We use the geological maps where available, and transfer the necessary data to township maps, which are nearly always available, and which anyone would be able to use. The Geological Survey map is much more difficult to read, and does not permit the accurate placing of small sites.

MR. FORDYCE: One of my most interesting sources of information about the old trails has been the group of old maps on file in nearly all of our state capitals.

MR. BLACKMAN: I get about half my archæological information in Nebraska from the old deeds.

MR. HINSDALE: There are about twenty-four or twenty-five hundred township maps filed in Lansing, the capital of Michigan. Every one of these has gone through the hands of my secretary or through my own, and we have found a great deal of valuable information.

MR. BUTLER: Mr. Chairman, I think we have been favored with a very valuable suggestion in Dr. Barrett's paper. The Eastern states, the old original states, are all in the same condition as Texas. The Hutchins system was not applied there. It came too late. But in some of the states they are going over the old surveys. Those who are interested in Florida will be glad to know that superimposed upon the map of that state which shows the old Spanish grants, there is now available the Federal system of land surveys which makes a very fine check for the work of the student of geology or of archæology.

MR. GUTHE: Last year at the annual meeting of the Committee it was recommended that an attempt be made to work out a blank for the inventorying of archæological collections both in museums and in private hands. That work was assigned to Dr. Greenman of the Ohio State Museum, and during the winter he has been getting acquainted with the problem.

A FORM FOR COLLECTION INVENTORIES

E. F. GREENMAN

In view of the increasing activity in state archæological survey work, some attempt should be made to bring about uniformity in the use of terms, and in the methods of describing archæological objects, in order that the work done in one state may be compared with that in adjoining states. At the present time a number of state institutions are using different schemes in classifying artifacts and in plotting their distributions, and when the time comes to compare results the matter can only be worked out by going back to the original objects themselves, which in many cases will be impossible. Distributions common to more than one state can only be worked out by the use of a uniform terminology.

A method is here presented for classifying the so-called arrow and spearheads and analogous forms. The plan is based upon four interrelated columns, each of which deals with certain definite attributes of these specimens. These are suggested by descriptions of typical outlines.

The caption "Hafted points" is used for the purpose of avoiding the uncertainty and confusion resulting from the use of terms with a functional significance, such as "arrowhead," "spearhead," "knife," "drill," etc., a great many of which grade into one another almost imperceptibly. By "hafted point" is meant a flint or stone point which by its form lends itself to attachment to a handle or shaft, whether notched or unnotched. In the first column hafted points are divided into four main classes on the basis of outline—angular, convex, concave and indeterminate. This classification was suggested by Mr. George Langford, of Joliet, Illinois. The unnotched leaf-shaped form has been omitted for the reason that it cannot readily be fastened to a shaft. It belongs with the "knives" or cutting edges.

The second column describes types. By "type" is meant the frequent linking together of a number of features on the same specimen. The forms in this column are true types, and when the name given to each type is set down, no more description is necessary, except when it is desirable to subdivide the types by the presence or absence of minor features such as serration.

The third column is devoted to "single details, intentional, definite and well made." Any one or more of these single features may occur on any point. In this column an attempt is made to enumerate all the important features occurring on hafted points in the eastern part of the United States which are of value in plotting distributions, with the view to describing, by the use of the numbers preceding the descriptions, any point that might be found.

The fourth column records the conditions under which the specimen was found. It is understood that unless indicated by one of the Roman numerals of this column a given specimen or group of specimens was found on the surface of the ground. The fourth column is of first importance, as it will permit the correlation of certain types or single features on hafted points with the different kinds of archæological sites, and together with the notation of the particular locality, *i. e.,* county, township, section, etc., in which the specimens are found, complete distribution data will be recorded.

In describing a given collection, it would be necessary to group the specimens on the basis of this classification, and then to use an arrangement of names, numbers, words and letters, set down for record on a piece of paper separate from the pamphlet containing the classification—a code, if you wish to call it that, made by reference to the pamphlet and deciphered in the same manner. The first numbers would indicate the quantity, the other numbers and symbols would refer to the various columns.

This scheme of classification will not describe every point that will be found, nor will it describe each point completely, but it will accommodate the larger number of forms. It would be possible to so classify hafted points that each specimen would fall in a class by itself. It is the intentional forms whose distributions are significant, and for that reason stress is laid upon the *types* and upon the single features such as those in the third column. Those forms which this classificatory scheme would not " catch " can be described separately with a degree of completeness optional to the one doing the work. The amount of time involved in analyzing and recording the contents of a collection by this apparently complicated system is not as great as would at first appear, for the greater number of specimens in a given collection will fall in the indeterminate class of Column 1, and in Column 2, being thereby summarily disposed of. As use of the scheme proceeds, Column 2, containing the *types,* will receive occasional additions from Column 3, wherein the frequent linking of single features on the same specimen will discover new types.

The principal aims of this method are: first, to expedite comparison of work done in various states, by various individuals; and second, to make it possible to set down on paper with a small expenditure of time and in a comparatively precise manner the facts regarding a collection of specimens, for the purpose of study and record. Such a record, made by one who fully understands his work, will be of almost as much value to the student as the original objects.

A pamphlet of six or seven pages with a descriptive classification of archæological specimens, in plain English, with a short text explaining its use, and outline drawings for illustration, would not cost a great amount of money, and, circulated among workers in the eastern part of the country, both trained and amateur, would stimulate the preservation of data that will otherwise be lost. It will also serve to increase intelligent interest in the subject on the part of amateur collectors. But most important of all, it would obliterate state boundaries by standardizing the method of procedure.

[84]

HAFTED POINTS

Column 1.

Series A, Angular
A. Unnotched
B. Side notch
C. Shouldered
D. Base notch
Series E, Convex
E. Unnotched
F. Side notch
G. Shouldered
H. Base notch
Series J, Concave
J. Unnotched
K. Side notch
Series X, Intermediate forms tending to any one of the classes in the A, E, and J series.

Column 2, Types of frequent occurrence

Bevel
Small notch
Sloping shoulder
Fluted
Wide-stemmed
Round base
Fishtail
Notch stem
Bell-shaped
Tapered sides
Thick blade

Column 3, Single, intentional details, well-made and definite

1. Indented base
2. Pointed base
3. Round base
4. Expanded base
5. Long base
6. Expanded stem
7. Long stem
8. Half stem
9. Diagonal notch
10. Restricted notch
11. "Lock" notch
12. Curved notch
13. Wide, shallow notch
14. Width over half of length
15. Cross-section approaching the circular
16. Flake on one side
17. Serration
18. Ceremonial serration
19. Three notches
20. Four notches
21. Corner notch

Column 4, Where found

I. Mound
II. Enclosure
III. Village-site
IV. Grave
V. Cache

Other artifacts, including points not intended for hafting, celts, grooved axes, pestles, ceremonial slates, pipes, pottery, implements and ornaments of copper, etc., can all be classified and recorded in a similar way.

Constructive criticisms of this scheme will be gratefully received.

DISCUSSION OF MR. GREENMAN'S PAPER

MR. BROWN: I think the paper presents a very nice piece of work. I should like to ask Dr. Greenman why he departs from all previous classifications? Most of the others are based largely on the origins of implements from simple forms and their development to more elaborate forms.

MR. GREENMAN: My only excuse for neglecting these other classifications is that they are too thorough and too good. This classification is admittedly incomplete, but must necessarily be so to be practical. The other classifications are in no way general, and would not describe specifically any type or single feature for which the student might be looking, at least so far as my experience with these classifications goes. I do not believe in mentioning such features as the restricted notch, the diagonal notch, or anything not basic.

MR. NELSON: The very fact that we have so many different attempts to classify projectile points shows of itself that it is a very difficult problem. There are two practical reasons why we need systematic classification. We need it in the museum in connection with the cataloguing; because we do not group into types, we have to show each one separately. Then there is the other reason, which has already been mentioned. For anything like strict, rigid comparative studies, not only within our own country but throughout the world at large, we have to have some such system so that we will know what we are talking about.

For a good many years I have had to struggle with the matter from the museum point of view, in order to shorten the work as far as possible. I finally devised a scheme of my own which, while also tentative, is, I think, a little less conflicting than the one just presented, and a little less complicated than those which have gone before. I have no criticism to offer on any of these previous schemes except this, that they break down of their own weight.

I began this way: I concluded that the point end of the spear and arrow points and knives have no classification value because they are all pointed to a greater or less degree. And while the side may bulge in or out, may be plain or serrated or what not, the classification characters develop in the base or butt end. We have not time to go into

[86]

great detail, and I have not quite finished with it myself. But this is the beginning of my scheme. There are only four possible outlines that a base can take; it is either straight, concave, convex, or pointed. It seems to me that practically all blades were derived from a blank, a rough outline form; and this outline form had one of the four base shapes that I have mentioned. And so I prepare for cataloguing by separating the material into the four different groups and subdivide each group separately.

A specimen may have notches—one notch in the base, or a number of notches, *i. e.*, two which make a stemmed point, or there may be small or big notches running in from the side. These notches occur in all the four different groups, and one can erect a family tree, holding true for each of the four groups, whether with concave or straight base.

Finally, it is necessary to separate the stemless points from the stemmed forms. The latter group has stems which either contract forward, have parallel sides, or expand forward toward the point end. I separate each main group into these sub-groups. Then I divide each sub-group by the character of the barb, which either projects downward, horizontally, or recedes.

That is the basis of my scheme. I believe it is simpler than the one the speaker has presented, but I should like to talk it over with him and see if there could not be an advantageous combination of the two.

MR. KEYES: Anyone who has done what Dr. Greenman has done in attempting to classify the types of implements is deserving of praise, I am sure. I find myself without a judgment at present as to the best system, but the work of Mr. Nelson and Dr. Greenman ought to be published as tentative schemes, so that we could study them in detail. It is pretty evident that any system will have to stay in quarantine until we can criticize it and try to apply it in our actual problems. Ultimately, I believe the best will survive.

MR. SHETRONE: In preparing his scheme, Dr. Greenman was kind enough to bring it to my attention, but I did not have time to give it detailed consideration. We feel that the older classifications are not sufficiently inclusive. In preparing his scheme Dr. Greenman had in mind, as I understood it, something that would be more elastic, which might be abridged or modified in any way. I believe it can be simplified. A certain number of forms that he has listed seem to me to be of such relative scarcity that we might consider them merely accidental. With the help of the experience of each of these gentlemen I believe that something could be evolved which would serve us much better than anything we have as yet.

MR. COLE: May I suggest, Dr. Greenman, that a blueprint or drawing be sent to the various workers in the field, perhaps as an addition to the scheme you are now proposing. It might be well to have the blueprint or mimeographed form quite simple in its preliminary stages, perhaps giving two or three of the schemes most widely used, from which you have drawn suggestions. Then we could each make our comments and suggestions from our own experience. In going over a large number of local collections in our survey work, we have felt that it is necessary to have as simple a scheme as possible. Then as new forms are encountered, drawings in outline of the new type may be made, with record of its percentage and peculiar features. The problem is really very important.

PART III
ADDRESSES AT THE DINNER FOLLOWING THE CONFERENCE SESSIONS, TOGETHER WITH ADDITIONAL RADIO ADDRESSES

MR. BARRETT: In the rush and bustle of our modern American life, as we hurry about over the country, or in our cities, by fast train, automobile and other rapid means of conveyance, bent on those all-engrossing tasks imposed upon us by our business, our profession, our social engagements, even by our pleasures and relaxations, few of us ever stop to consider the foundations upon which our modern life and culture are really based, and when we do we most frequently think of these, either as originating within ourselves or within the modern culture which dominates us. If we ever do look for an historical background we refer it to our forefathers of the Old World. Almost never do we give credit for anything in our modern life to the American Indian, that most interesting individual who inhabited the Western hemisphere from time immemorial, before the coming of our forefathers as the conquering race from the East.

Yet, if we would give pause for a moment we would find that we owe much to this indigenous American culture, for it was a real culture and one of a high order which existed here in America long before the coming of the white man. Whence came the names of many of our states, such for instance, as Connecticut, Massachusetts, Minnesota, Missouri, Wisconsin and others? Whence came the names of many of our great cities—Chicago, Minneapolis, Milwaukee, and those of cities and towns scattered over the entire country? Whence came the name of our principal river, the great Mississippi with its great tributary, the Missouri? Whence came the names of many of our mountains, lakes and other features with which we have to deal daily, names which we so glibly roll from our tongues and which we rarely recognize as derived directly from the American Indian?

There are other features of our daily life equally important, or perhaps even more important than these. Do we ever stop to reflect that many of the vegetal foods which we now consider absolutely essential, are derived from plants which were domesticated by the American Indian long before Columbus set foot on these shores? Among these may be named the maize or Indian corn, and the lowly potato; now two of our greatest crops and chiefest sources of wealth in the agricultural districts of America—in the whole world in fact. There are also the tomato, the peanut, the bean, the squash, the tobacco and many other plants, all of which are in daily use in most, if not all, American homes, and all of which were derived from the American Indian and introduced from his aboriginal culture into ours.

There are also many medicinal plants which are now of the utmost importance to us. Two of these alone need be mentioned, quinine and cascara; again introductions from the aboriginal American culture into ours.

Much has been given us also in the arts and crafts, in the way of beautiful fabrics, in basketry, blanketry, pottery, stone and metal working. Much also has been added to our art in the matter of design. Only recently have we come to recognize that we owe at least some debt of gratitude to the American Indian for the introduction of these.

We travel great distances to visit the temples, palaces, pyramids and other architectural works of the Old World, losing sight of the fact that some of the most beautiful of these same architectural types are to be found right here in the Western hemisphere. I refer, of course, to the stone temples and palaces of Mexico and Central America and to the pyramidal structures of these same regions and elsewhere on the American continents. In fact, in Mexico, at Teotihuacan, stands the greatest pyramid in the world; one larger even than the great pyramid of Ghizeh in Egypt.

The American Indian is commonly thought of as belonging to the Stone Age, and it is true that he was an exquisite worker in stone; flaking, chipping, grinding, polishing stones of various kinds into objects of both utility and beauty in a manner hardly, if ever, to be surpassed by stone workers anywhere in the past history of the Old World.

He was, however, at the same time a metal worker of no mean ability. Certain tribes had mastered the art of handling gold, silver and even platinum, and we find beautifully fashioned vessels and exquisitely wrought jewels and ornaments in these metals. One of the most interesting instances of metal working is the fashioning of copper into objects of utility and ornament which was carried on in our own Great Lakes region. The source of supply, so far as we now know, for all of this copper was Isle Royale in Lake Superior, and the immediately adjacent shores of Upper Michigan and Wisconsin. Here these primitive miners extracted this, to them, precious metal, with most painstaking care and unstinted labor. Not only are the locations of many of these mines now known, but recent research has brought to light the methods used by these primitive miners in extracting this red metal from the solid rock in which it was, in most instances, deeply imbedded. From this source copper found its way in aboriginal times to most, if not all, of the tribes of the Mississippi Valley and even beyond these limits, just as at the present time our modern trade routes supply this metal from this same locality to many of the markets of the world.

These ancient workers, who formerly inhabited the Great Lakes region, were also builders, not of stone temples and palaces like those south of the Rio Grande River, but of works almost, if not quite, as interesting: the great earthworks dotted throughout a major portion of the Mississippi Valley from the Gulf to the Great Lakes. Such earthworks are among the most durable of monuments, and we have still left to us, despite the vicissitudes of time and the ruthless hand of the white man, such great mounds as those at Cahokia near St. Louis, those of Hopewell and other groups in Ohio and neighboring states, and the effigy earthworks so characteristic of Wisconsin and vicinity. Associated with these are, of course, camp sites, Indian trails, mines, quarries and other features of great interest; all of which are worthy of the most careful consideration and intensive study. The mounds at Cahokia differ materially from those of the Hopewell type, and they, in turn, differ materially from the effigy forms already mentioned. Do these three types of mounds, representing, as associated material shows, probably three distinct types of culture, indicate different periods of occupation in point of time or were they concurrent? If they represent different periods what was the order of their succession? These and myriads of other interesting problems such, for instance, as the problem of the territorial distribution of these types of culture present themselves for solution. Until recently, for instance, it has been presumed that the Cahokia mounds marked the northernmost limits of that southern type of earthwork and culture, but at Aztalan in Wisconsin, unmistakable evidence of the penetration of this southern type at least as far north as this point has been found. Recently, also, it has been found that the Hopewell type of culture extended as far north at least as the region of Trempealeau in Wisconsin. Thus gradually by the careful study of these problems and the systematic accumulation of data, it is possible to arrive at a solution of one after another of these questions and to build up the prehistory of the region, a history which is just as fascinating and just as important as is the modern and current history of the white man which we preserve so carefully.

So important have these problems of the prehistory of the Middlewest become that they have enlisted the active interest of the National Research Council, and there is now in session in the City of St. Louis this conference called by the special Committee on State Archæological Surveys of the National Research Council, which is devoting itself to the consideration of these important questions of prehistory, and out of whose deliberations we may hope for great progress in the solution of some of these highly important problems not alone on this region but upon the prehistory of the whole country, and in fact of the world at large.

From these and similar conferences, there should come a fuller appreciation, not only upon the part of students of American prehistory but upon the part of the general public, of the importance of carefully surveying and recording all data of every kind concerning aboriginal history in America and especially, we hope, the appreciation of the importance of the preservation of the ancient works of our predecessors on the American continent wherever these may be located and whatever may be their nature.

Thus by the careful study of such remains and by the gradual accumulation of data and information concerning our earliest Americans, will we come to a full appreciation of the truly intimate, historical relation existing between us, with our special culture, and the American Indian with that great early culture which flourished in the Western hemisphere so long before our coming as a conquering race. Thus may we come to realize the importance of the American Indian in our every day life.

MR. GUTHE: In the few minutes at my disposal, I would like to talk to the thousands of people in the Mississippi Valley who have taken an interest in the many fascinating Indian relics that are found in mounds and burials grounds throughout the area. These relics are like the words in an historical document. The earth in which they occur constitutes the page upon which the words have been written. In any historical document the information contained depends upon the relationship which one word bears to another. If the words are removed and arranged according to the number of letters they contain, the meaning of the document is entirely lost. Similarly, when the arrowheads, Indian skeletons, and stone axes are removed from the earth in which they are found without any consideration being given to the relationship which one specimen bears to another, the record of the document has been irrevocably destroyed.

To the historian, then, the archæological specimen, be it a beautiful stone axe, or a highly polished ceremonial stone, is of no value without this documentary evidence. It is important, then, for the amateur interested in our ancient Indian inhabitants to remember that a specimen is of no value scientifically without a record. Conversely, a specimen with a record is of great value. It need not be perfect, or even complete.

[92]

Fragments of pottery, fragments of skeletons and broken specimens in general, if accompanied by their records, are of more value than the most beautiful and perfect specimen which has no history.

If you are an amateur, and sincerely interested in tracing the forgotten history of the Indians who once lived in your country, you can be of great assistance to the archæologist by following certain suggestions which I should like to make.

The first point is that no digging be done in mounds or burial grounds until you are quite sure that you understand the kind of information which the historian needs in order to obtain the story which the specimens you find can give. I do not mean by this that you shall not dig. I simply ask that you make a concerted effort to discover the best methods of excavation. Any museum or society in your neighborhood interested in the story of the Indian will be glad to render you all possible assistance in this matter. The only reason the historian is anxious to prevent digging is that through well-intentioned but ignorant enthusiasm valuable records are destroyed.

The second point I desire to make is that concerning the scope of the collections which you might make. There are great museums in this country and abroad which are trying to secure representative collections of the archæological specimens over a large area such as North America or Europe. These great museums, with large staffs and plenty of funds, find it impossible to make such a complete collection over a period of many years. It is logical, then, that a single individual, living in a small town or on a farm, can only duplicate, in a very small degree and very inadequately, these larger collections. A general collection of archæological specimens in private hands usually loses its entire value because of its haphazard and irregular formation. On the other hand, the great museums, and even some of the smaller institutions, are unable to visit every spot of importance in the country, and moreover have no desire to do so. Yet a knowledge of the archæological evidence in very small regions is important in building up the story of the Indian life of our country.

The archæologists can depend, in a large measure, upon the local amateurs. If you will concentrate upon obtaining as complete and perfect a collection as possible of the archæological specimens of your given district, whether it be a township or a county, you can render to science a very distinct service which no other individual can do as well. Instead of feeling at the end of years of labor that you have an imperfect and incomplete collection of specimens from the entire world, you can have the joy of knowing that at the end of half a life time you have an absolutely unique collection from your own county which has not been duplicated by any one else. Obviously if you have such a collection it is not necessary for you to buy or sell specimens. You will find in your county or in your special region certain friends who are also making collections, and with these friends you can exchange material which will be of interest to both of you. Unfortunately a traffic has grown up in archæological specimens so that today an entirely fictitious monetary value has been placed upon stone specimens. This emphasis on the value of the specimens has resulted in floating collections going from the hands of one collector to another and back to dealers which, scientifically, have absolutely no value because there is no record associated with the individual pieces. This has a double effect. In the first place, it puts a premium upon the destruction of historical evidence, and in the second place, it causes the spending of money by the amateur upon material which he cannot sell to any museum or institution.

The fourth point that I desire to mention is the necessity of keeping an accurate record of the specimens you have in your collection. Each specimen should have a number. This number should refer back to a catalogue in which, under the heading of the number, a complete record is given of the specimen. This record should contain a statement as to when the object was found, by whom it was found, where it was found and the conditions under which it was found. Such information raises the specimen from a purely art piece to an historical document.

My last point is one that is rather self-evident, but which is often forgotten by the enthusiastic archæologist. We men who work in museums are frequently confronted with individuals bringing to us material which was collected by their relatives who have recently died. Usually this material is not accompanied by information.

There is also another aspect which we decry, namely, that after a man has spent many years of his life in bringing carefully together a representative collection of his area, upon his death this collection is either destroyed or dispersed because none of his heirs are interested in what he has done. Therefore, my last point is that each of you who have collections should make some provision in your will, or preferably before your death, for the disposition of your collection into hands which are able to care for it in perpetuity and preserve it for the use of students of the future. Any museum or historical organization will be glad to advise you on the best way in which this can be done.

Finally, let me again emphasize the fact that specimens alone are of no value scientifically, and that the commercial value placed upon them is purely artificial. If you are interested in helping the historian to interpret the Indian life of our country, let me urge you to get in touch with your nearest museum or historical society and tell them of your interest. You will find them always willing to cooperate and assist you in any way they can.

MR. DIXON: For countries without written history, an archæological investigation affords the only means to reconstruct the past. The New World has, except in Mexico, Central America and Peru, no history which goes back to the period of Columbus' discovery, and for many reasons our actual history covers little more than one hundred years.

We cannot help feeling a lively curiosity as to the history of our country in the days before the white man first put in his appearance, and we can partially reconstruct this history through the ages by our archæological work, but such work must be intelligently planned and carefully carried out, otherwise the story that the mounds and earthworks, the village sites and burial places, have to tell will be lost.

In this country archæological work has passed through several successive stages. First comes the early period, motivated primarily by curiosity, in which the sites are dug into just to see what is in them. Such digging is generally done by the owners of the property on which the site occurs, or by some local enthusiast who desires to form or add to his collection of Indian curios. Sometime during this early period the professional dealer in curios and antiques makes his appearance and digs with the purely commercial purpose of securing specimens which he can offer for sale. In all such cases the work is done by untrained and uninformed persons whose only appreciation is for the things they can find, and who do not realize that in their haphazard and unscientific methods of excavation they are missing or destroying evidence of great importance; for the detailed observation and recording of data on construction, or on the exact position of the objects found, or on their spatial association, is often of fundamental significance in determining the conclusions to be drawn from the find. The recognition, for example, that a particular object or burial is intrusive and is of later origin than other finds in the mound; or the discovery, as a result of careful sifting of the earth, of a single glass bead indicating that the find must date from after Columbus' time, may be a clue of the greatest value. Archæological work of this primitive character has been carried on almost since the beginning of the settlement of this wide Middle Western area. Later, with the founding of museums and universities, the rise of interest in prehistoric materials and the development of genuinely scientific methods of excavation and record, archæological work was begun by the trained investigators, and the results have thrown a flood of light upon the whole problem of the early history of this region. The time has now come, however, when we should look forward to the greater systematization of such studies. The people of each state should feel that they have a definite responsibility to see that the archæological remains within the borders of the state should be adequately recorded and studied, and that some at least should be preserved as records for all time. Many sites have been destroyed. Many are being destroyed or concealed through the annual plowing and cultivation of fields. Others are being dug away in the course of road building and railway construction or the gradual growth of towns and cities. In many sections, therefore, there is need for prompt action. To that end it is hoped that the Legislature of each of the midwestern states will take appropriate action to organize a state archæological survey. An annual appropriation for this survey should be made sufficient to provide:

1. For the salary of a competent trained field archæologist who should direct the work of the survey.
2. For the making of a detailed archæological map of the state on which would be recorded, by section and township, the location of all known archæological sites.
3. For the carrying out of careful excavations in certain selected sites.
4. For the adequate publication of the results.

The annual appropriation for these purposes would not be large: a few thousands of dollars would, in a few years, produce extremely valuable results.

It is also to be hoped that one or more of the notably fine examples of each type of site may be purchased and preserved by the state as a state monument in accordance with the action of the Federal Government in so preserving as national monuments some of the finest of the ruins in the Southwest. Such a survey might be organized either directly under the auspices of the state or placed in charge of the State University. It is clear that popular interest in archæological work and its results is growing. Certain states, notably Ohio, have for some years recognized their responsibilities and opportunities of this sort and have already done most valuable work—work of which they may well feel proud.

It is to be hoped that as a result of the present Archæological Conference in St. Louis the people of all the midwestern states may realize more fully the value of joining in what should be a great cooperative undertaking to make the best possible use of the remnants still surviving of the works of our aboriginal predecessors, and to preserve some of them at least for our children and our children's children when the last survivors of the Indian (the Original American) shall have been absorbed into our national population.

MR. MOOREHEAD: Your attention was called yesterday afternoon to the Cahokia Mounds. It was emphasized that the rest of that important group should be preserved. I believe that we answered, somewhat superficially, some of the leading questions concerning Cahokia. A report just published as a Bulletin of the University of Illinois will acquaint you with the known facts. I shall deal with the human side.

During four seasons spent at this place, thousands of visitors have come, and there are three questions so frequently asked that we term them "the eternal questions." The first, of course, is: "Where did the Indians originate?" The second: "How old is this mound?" And last and best: "Who is paying for this work?" People seem consumed with curiosity as to who is footing the bills. From the tone in which the last question is asked, I suspect some of the inquisitors doubt the wisdom of spending money in our line of research.

Among these hordes of visitors were many intelligent persons. The majority, as one would expect, were not especially interested. Yet the park feature appealed to them all. Therefore, I quite agree with the remarks of our distinguished friend, Mr. Dawes, that we should emphasize the fact that these parks are a great asset and benefit to the community at large.

You may wish to know a little of the inside history of how Cahokia was saved. This has never been published, but as seven years have elapsed, there is no harm in telling the story. We had spoken to many groups of persons all over southern Illinois, and urged the preservation of the Cahokia mounds, all of which had little effect. A friend of mine mentioned how the late and genial Mr. Barnum saved Stonehenge, and told me to follow that idea, but carry it further and make the people mad. Therefore, I prepared a very severe statement for the press, and it was printed simultaneously in a number of large cities. It cast reflection on the intelligence of southern Illinois folk who preferred filling stations, hot dog stands, dance halls and bungalows to the greatest monument north of Mexico. I stated that in any other state but Illinois the mounds would have been included in state parks long ago. The reaction to this tirade was immediate. I was roundly denounced in the press, a politician hurried from East St. Louis up to Springfield, and a bill to make a state park was passed in forty-eight hours.

Dr. Wissler and two or three other speakers have referred to the achievements of the American Indian. That is quite important, but we must not lose our perspective. We deal too much with the mere materials of archæology and are apt to forget the human interest side. Consider the Indian men and women themselves for a moment. Time forbids that I mention more than a few outstanding characters: King Philip (Metacomet); Samson Occom, in answer to whose plea the Earl of Dartmouth founded Dartmouth College; and Tecumseh, the great character of the Ohio Valley, a wonderful man and a born orator and leader. The last named was one of three distinguished brothers, all of whom opposed encroachment by the whites, and two of whom yielded up their lives in defense of their country. At the Treaty of Greenville, Tecumseh uttered this striking metaphor on being asked to join General Harrison and the other officers on a raised platform, but preferring to sit among his young warriors: "The sun is my father (pointing upward); the earth is my mother (pointing downward); on her bosom will I respose."

On the Great Plains the fighting Sioux, an upstanding people, produced Red Cloud, Sitting Bull, and others. The career of Red Cloud and his young men marks a dramatic episode of American frontier life. Sitting Bull was our Bismarck of the Plains. A man of blood and iron, he feared no one, and despised the trickery of political commissioners. His famous statement will not soon be forgotten: "All white men are liars, and bald-headed ones from Washington are the worst liars of all." In the far West Sa-cah-gah-wea guided Lewis and Clark to the Pacific Ocean, and Chief Joseph directed the longest cavalry raid in the history of our country—eleven hundred miles through the heart of the Rocky Mountains.

In the World War the American Indian with man power of about 37,000 people fit for military duty furnished nearly 12,000 young men. This percentage applied to our whole population would have given us an army of between twelve and fifteen million.

It is fitting that we remember the achievements and the character of this native American race. Through the preservation of these ancient, impressive monuments we do well to memorialize that fine race, the American Indian.

MR. COX: Unwritten history is the kind of history that should be preserved. In my humble opinion, we do not need in the archæological work today quite so much visualizing, quite so much correlating, or quite so much tying up, but we need to apply in archæological work the same kind of plans and ideas that we apply in every other human endeavor and effort.

Therefore, I want to appeal to you, for the benefit of yourselves and the benefit of your people; for the benefit of those who will succeed you;

[98]

to get out in your neighborhood in the Mississippi Valley if you want to find evidence of pre-historic man. Talk to your neighbor about it and try to bring about a public sentiment and a disposition to maintain those particular structures as pieces of unwritten history.

To those of my unseen audience who are looking after the spiritual welfare of mankind, I appeal. We know that a great deal of the actual visible evidence of instruction that we have received from the Holy Book has been proved and established with the spade of the archæologist. What greater duties can you perform than to tell your people of the necessity and the duty and benefit of preserving evidences of prehistoric man in your neighborhood?

To you, Mr. Farmer, undertaking to scratch the earth to make a living—the man who is talking to you has jumped many a clod between the plow handles. I know something about your ordeals and your trials. Do you know that this prehistoric man, about whom I am trying to talk, is the man that produced the corn that makes a living for you and your family? Do you know that these prehistoric men discovered and produced tobacco and made it known to the world? Mr. Farmer, when you go out to dig your Irish potatoes, do you know that the first Irish potato ever heard of in the world was found in America, created or developed by prehistoric man? There are reasons, Mr. Farmer, why you should be interested in preserving for history that structure on your farm, so do not permit its history to be destroyed by some relic hunter who will get into this mound or into this tomb solely for commercial purposes, and destroy every bit of history. If you have such a thing on your place, talk to some of your school teachers, talk to some of your professors, and they will be glad to go there and help you bring out the real history.

To you who are engaged in this mission of educating the nation, may I respectfully submit that you cannot accomplish any better result in your whole experience than by undertaking to educate youth to the benefit of the history that will be furnished by the preservation of the mounds and structures and other evidences of prehistoric man in the Mississippi Valley.

I respectfully make these suggestions for the consideration of the public. When we want to accomplish anything in any state within this Union, whenever we want to accomplish anything with the Federal Government, we know that we must create behind the purpose that we seek to accomplish a public sentiment that will sustain it. Therefore, I sincerely hope that you will take under consideration that plan of development, under the organization which is functioning in your immediate territory, whether it be an academy of science, an historical society, or an archæological society. Develop a committee whose duty it shall be to look after the needs of that particular community, whether it be with the state or with the nation. Let that committee make certain recommendations, and submit its recommendations to the National Research Council, an organization that is doing more to accomplish real good for this nation, in my humble judgment, than any other organization in it.

After you have submitted these plans to that organization, bring them back to your state, submit them to your educational institutions. Then, with their recommendation, submit them to your legislative body, and you have a basis to work from that should be successful.

Finally, my last word to you all is: take this seriously, take it earnestly, and when you go home think about this, talk to your neighbor about it, talk to your friends about it, create an interest in it, and do not forget to tell your member of the Legislature and member of Congress what you want.

MR. KNAPP: I must necessarily feel my inadequacy to this occasion, for I cannot in any way pose as an archæologist, nor even as an historian. But I think that all of us here in the United States, and especially in the Middle West, in the Mississippi Valley—all of us who as boys have roamed the beautiful Ozark hills, or wandered through the equally lovely country of Wisconsin or Illinois or Ohio, have come upon evidences of the early inhabitants of our America. We have been intrigued and fascinated as boys with the flint arrowheads that we found in the corn fields. We have looked with wonder at these mounds that have been referred to by these gentlemen who have done so much to acquaint us with the true, inner meaning of these monuments of early America. We have allowed our imagination to build for us pictures of the charm and beauty of wild America before the coming of the white man; and the thought that through carelessness and neglect, through ignorance and lack of appreciation the remnants of this earlier day still to be found in the Mississippi Valley should be allowed to perish—this thought, I say, is simply intolerable.

These scientists gathered here this evening, these archæologists who really know what they are talking about, realize far more fully than any of us who are mere tyros in the field can realize, how great the loss has been in the past, and how great the loss to science will continue to be unless a real and intelligent interest is aroused in the people of the Middle West, which will guide them to do as the last speaker has indicated; to take means to preserve and to protect these monuments of the historic past of America.

Even those of us who are not archæologists or scientists are horrified at the wanton destruction which has gone on. It was my pleasure to teach for four years at a little college in Wisconsin, just at the junction of the Wisconsin and the Mississippi rivers. The valley there is known as the Prairie du Chien. It was once inhabited by a tribe of Wisconsin Indians, but when the whites first began to settle there, the folk-lore and the traditions of the place tell us, there were some twenty-nine very extraordinary Indian mounds on the prairie, some larger, some smaller, but all of them containing materials of very real archæological interest.

When I went there in 1917, and on hearing of these Indian mounds and on reading of the Wisconsin Historical Society doing something about them, I set out to try to find some of them. Of the twenty-nine or more that once had been there, I was able to locate but a single mound. On the hills back of the college, however, there were two still very fine mounds remaining; one a so-called "totem" mound, in the form of a bear. On the grounds of the University of Wisconsin I had the pleasure of examining another in the form of a tortoise, or turtle.

Why was not this great collection of mounds on Prairie du Chien preserved? Because the people there took no intelligent interest; because the farmer and the railroad and the settler plowed them down, and tore them down, and destroyed and scattered as mere articles of curiosity the remains found in them. Those mounds have been lost to science, and with them we have lost a deal of real evidence on the history of early man here in the United States. If the proper study of mankind is man, then there is just as much real interest and merit and value attaching to a knowledge of primitive man here in the United States as there is to the tomb of Tutankhamen in ancient Egypt.

Therefore, if, as announced by Professor Breasted at the American Historical Society meeting in December, ten million dollars have recently been given to the Oriental Research Foundation at the University of Chicago, it is a great pity that the people of the United States cannot appreciate the value of what they have at their very doorstep, and cannot take necessary and adequate steps to preserve and protect what still remains to them.

In tune with the beautiful remarks of the preceding speaker, I urge most earnestly on all those who are listening to the speakers here this evening that the effect of this conference on Midwestern Archæology be to arouse in the people of the Middle West a real, a determined interest, and a disinterested effort to preserve and to protect and to develop an interest in, the historic monuments of this, our own country.

DR. THROOP: I feel we have lagged perhaps in this particular field. We are dealing very largely with what might be called the practical side. We already know how to excavate in archæology—or you do, and I think we have done very valuable work so far as we have gone, but have neglected the putting of the case of American archæology before the American people.

It seems to me that this could be done without great difficulty. We are classed as archæologists but we need to be propagandists. We should not neglect any opportunity to put ourselves before the public, and before the school system, whether higher or secondary. Everyone knows, from classical history, from ancient history, from Roman and Greek history, and from Oriental history, what Oriental archæology or Greek archæology or Roman archæology means. It is necessary, to my mind, for the public to be educated to what the American Indian means to us. I wonder if someone could not be inspired to write a pamphlet in which he would answer the three great questions which stand out in American archæology.

I think if I were to classify them in my own mind, I would say, first, " Why American archæology? " second, " Where American archæology? " and third, " How American archæology? " I think we have solved the last one, but have not even touched the first one. I think few people in this country realize the " why " of American archæology, and perhaps that explains our being so neglectful.

This question has been touched on in various ways in this particular meeting, as when it was said that we owe the American Indian the corn produced in the United States, or that we owe the American Indian all of the Irish potatoes produced in the entire world, and that we owe other things as well, fruits or vegetables or whatever they may be. It is almost impossible to estimate in dollars the wealth which the American Indian has brought into this country. But we should go farther than that and show that the native Indian actually had an art, and still has an art, which is very important; that in their cliff dwellings, in their arts, and in their manufactures of various kinds, what they made has a significance which is comparable very often to that of the things which we admire in early Oriental or in early Grecian or in early Mycenaean archæology.

I believe these things can be presented, and should be presented; but seldom do I see them presented anywhere. Seldom do I see anyone who really knows about them; but I believe the interest and enthusiasm for American prehistory could be much more generally aroused than interest in civilizations which are across the sea, because of the fact that these things are here at our very doors.

[102]

If some of you men who are writing learned reports and learned surveys of what you have found in this field and what you have discovered in that field would get out popular brochures and pamphlets to interest high schools and college students in the field of American archæology, you would find at the end of a few years that you are not a mere fifty, a hundred, or two hundred, but you would be numbered by the thousands.

I feel the second question is one about which I should not talk, but inasmuch as you have met here in St. Louis, and inasmuch as you call this the conference on Midwestern Archæology, I think I can say again, as I did this afternoon, that the central west is bound to be the home of the next most important investigations in the field of American archæology. This may particularly apply to Missouri and to the adjoining states, because of the fact that they have been less exploited; because there has been less interest in scientific work; and because the field is so rich and virginal.

I believe a pamphlet or brochure which would give an archæological survey of the United States; which would show where work has been done, and where it might be done; which would explain the purposes, methods and manner of work, could be written in ten, fifteen or twenty pages and distributed widely in the schools, high schools and other places, and would be of tremendous value. I do not know whether you think that suggestion is practical or not. Certainly I do, having been in school work all my life and knowing how little the students know of the American Indian.

Despite the fact that there is no subject in the United States about which our people should be so well informed, I believe there is no subject about which the ordinary student at the present day is less informed. That statement is made advisedly, and I believe it is entirely true. I do not believe one student in twenty at the present day knows that the Indians were the real mound builders. Most of them believe the theories of decades ago that the mound builders were antecedents of the Indian, and that they disappeared before the Indians came. I merely cite this as an illustration of the prevalent misinformation and misunderstanding.

I feel, as I said, somewhat of an outsider in making these suggestions, but if we are to have fruit from meetings of this kind, if we are to be able to say that the Mississippi Valley is to become the great center of archæological investigation, we must look not only to the methods by which we work, but also the propagation of interest and to the spreading of news and the arousing of interest in the entire program.

MR. COLE: I think the suggestion of Chancellor Throop is a very good one. We are likely to become so engrossed, so interested in the details of our subject that we find it difficult to let the outside world know what we are doing. I believe there is great need for the popularization of science, for the writing of simple books, which people can understand, getting away from all technical names, and telling the story of the advancement of man.

Perhaps you would be interested in knowing of a development which one of our great institutions here in America is attempting along this line. The American Institute of New York, after a discussion with scientific men in various portions of the country, decided that perhaps their greatest service to science would be to dramatize several of the sciences—to put them on in such a way that they could be shown by film or in a theater, telling their story in such a way that the public would come because they wanted to see the play. It is of interest to us here to know that the subject they have chosen to start that series (which I think will cover six or seven sciences) is the story of the coming of man, in which they are going to portray the story of man's struggles up through the ages—the first appearance of man on this earth, the life of man in the caves, and finally the coming of the modern races into Europe and the building of our own civilization.

It is a wonderful story indeed, and if it can be staged, as some of our greatest dramatists believe it can (and the motion picture men in this country say it is perfectly feasible to film, and they want it), I think we have an opportunity to tell a story that is going to awaken the people as to what the science of anthropology means to a nation. I believe more of our efforts should be directed to letting the people on the outside know what the people on the inside are doing. This will build us good will, which we need and should have.

MR. BARRETT: In our discussions and deliberations we are dealing with a problem which is decidedly live here in the Middle West; a problem which, though it deals with people who have passed, is just about one of the livest problems that the country has to deal with. We have, I think missed one bet in the consideration of the problems of archæology. Archæology is, of course, nothing more nor less than one of the branches of anthropology, which is an all-embracing science with more ramifications than perhaps any other.

Ethnology, the other branch, is analytical; archæology is synthetic anthropology. Your analytical chemist takes a compound and analyzes it, to find out what is in it. We take the living Indians, analyze their cultures and find out the elements and how the cultures are put together. With the archæological evidence we use synthetic processes and make

[104]

deductions as to how those people lived in former times. The two are so closely allied that you cannot separate one from the other. I think we would do well in our archæological work to pay the strictest possible attention to the analytical work of the ethnologist and correlate the two.

We have had here a great deal of discussion of the pros and cons of excavation of village sites and mounds and remains, but one matter I should like to mention a little more in detail. The Chancellor mentioned a moment ago in discussing publicity for our archæological work, the writing of a book or series of booklets which will place before the public the importance of archæology and the interpretation of archæological remains. That is most excellent, but the book needs illustrations and I think that this should be done in the most practical, visual way.

Being a museum man, I am always harping on visual instruction, which is the special province of museums; and if we will take the evidences that we gather from archæological excavations and visualize these by means of artistically arranged groups in our museums where we can bring our classes and teach them to interpret the evidences we have found in these excavations, we will be doing a very great work.

There is one other means by which this interpretation may be done, and done very effectively. Setting aside a park with a lot of mounds in it is an excellent thing to do, and I do not mean to disparage it; but when your average visitor, who is not acquainted with archæology, goes to that park, he sees simply so many little hills of earth. However, if in this park we could establish an outdoor museum, or perhaps a building for the display of exhibits, we should be able to visualize for this visitor the interpretation of finds that have been made in those man-made hills. We could reconstruct actually a portion or all of the village that was at this site, or the temple or whatever these remains show to have been there. We would thus place this mound group or this site, before the visitors in a manner which would enable them to really interpret what they see.

If that can be done in our parks—if we can make these prehistoric people live for the visitors—we will excite the greatest possible interest. And with that general interest we are going to have the support that is necessary to carry on this work and develop similar plans for other parks and localities.

MR. GUTHE: I want to tell you the story of an organization which is very close to my interest, one which I hope will grow and expand through the cooperation and help of all of you.

Some ten years ago various archæologists in this country watched with great interest the growth in the knowledge of the historic life of our own

Southwest—New Mexico, Arizona, and neighboring regions. They saw the work which was being done in Ohio and Wisconsin, and the interest in archæology which was being manifested in Alabama. In looking over the map of the central United States, it became apparent that there were some areas in the Mississippi Valley about which practically nothing was known, speaking archæologically.

These far-sighted men came together and brought into existence the Committee on State Archæological Surveys of the National Research Council. Dr. Dixon was the first chairman, and efforts were started in 1921 to revive local interest in archæology in some of the Middle Western states, particularly in Indiana, Iowa, and Missouri.

Dr. Dixon was followed as Chairman by Dr. Wissler. The work expanded. The Archæological Survey of Iowa was started, the Archæological Survey of Missouri was begun, and in Indiana, through cooperation on the part of several organizations, a beginning was made in gathering together information about the archæology of that state.

The function of this Committee as originally conceived and as directed by the purposes for which the National Research Council was organized, was to act as a clearing-house and advisory board for those interested in archæology in our country. It soon became apparent that the problems which were brought to the Chairman of the Committee were such that they could not be answered off hand, and it was necessary to begin accumulating information and material in the files of the Committee so that these questions of real importance to the local groups could be answered in the best possible way.

Through the years the Committee work expanded. Dr. Wissler was succeeded as Chairman by Dr. Kidder, and after a few years the mantle of chairmanship was thrown upon my shoulders.

The scope of the Committee's interest has increased. It has come in contact with some sixty institutions in this country that are carrying on archæological field work, and that are correlating archæological evidence from books and from old manuscripts, and from things not necessarily found in the ground, so that today the Committee contains in its files information from the institutions working in some twenty-four of our states, as well as the institutions working in Canada.

That is all very well as far as it goes, but obviously it is of no purpose to collect all this information in one set of letter files and keep it there, like archæological specimens. This information is of no value unless used for educational purposes, for bringing together and helping various groups interested in the work.

The result has been that the Committee, which now consists of eleven archæologists, is trying to work out a series of policies, trying to determine some of the ways of solving the larger problems which confront the archæologist.

Obviously, we cannot solve these problems by having one man lay down the law for others to follow. The problems can only be solved by securing the cooperation of all those who work together towards a common goal. For this reason, the Committee has taken upon itself to send out from time to time great numbers of mimeographed bulletins, questionnaires and circulars, in the hope that you will all respond and supply the committee headquarters with the information needed; in order that you, in turn, may profit by the results obtained through our interests and experiences.

We have in mind, for example, in the near future a small pamphlet of some fourteen pages which will contain the very broadest outline of what should be done in studying archæological problems in one's immediate neighborhood. We have in mind bringing out a manual of considerable detail which can be used not only by the enthusiastic amateur, but also by the trained archæologist.

Another scheme is nearly completed; that of bringing together the opinions of various individuals in the field on the best way to work out a blank which will make possible the recording of archæological collections, both in private and public hands.

There is a plan on foot, also, through the committee headquarters, to gain the information that you all have regarding symbols to be used on maps; and again, the best way of recording sites in an archæological survey is being worked on by this Committee, with the cooperation of all of you.

MR. DAWES: I think I shall begin my talk tonight with a text from Deuteronomy: "Thou shalt not remove the landmark which they of old time have set in thine inheritance." I hope that many of you will accept this Biblical injunction literally, and regard it as an inhibition against removing the relics of the mound builders and of the ancient people who once dwelt in the land of our inheritance. And I know that all of you will take it as an injunction to maintain tradition as a guide to the future.

These men who occupied this land before us are a part of our tradition. Their mounds remain. We use the corn which they developed, the Irish potatoes, the peanuts, the tomatoes. And we have their love of nature and the enjoyment of the woods and streams—we profit by their knowledge and we hand it down through our Boy Scouts for the young to enjoy.

And so I think that we may regard it as a part of our duty in preserving tradition to maintain these mounds which adorn these states of the Mississippi Valley.

Tradition is not a thing that may be described in history. It is not a thing which may be defined by reason. It is not a thing of reason, but of human nature, of which reason is but a part and not the greater part. It is made up of our emotions, our feelings, our pride of ancestry, our ambition for the future. It is the instinct of the race and must be preserved. Tradition is the utilization of the best that is in the past to enable us to cope with the present and to shape the future. It is the remembrance of facts and monuments, of signs and symbols, if you please; of all that has gone before us which would enable us to build up for ourselves and to maintain for ourselves those standards of conduct which have been created out of our past, and to keep alive the sense of obligation to public service.

Mr. Cox has just said to me, " In these days we move too fast," and it is true. It is good for us all to stop and to look back once in a while. In the maintenance of traditions there is something in an aristocracy which has a certain advantage over any democratic form of government. The old spirit of *noblesse oblige,* the acknowledged obligation of favors received to perform a public service, has left in itself a fine tradition. In a democracy there is nothing to take the place of *noblesse oblige* except the preservation of ancient " landmarks which have been set in the land of our inheritance " by the men of old; and except in the ritual and the monuments to record the historical events which have marked the upward steps of our progress, and the celebration of historical events or of epoch-making steps in the progress of mankind.

And how fortunate are we who live in this particular era where all that is behind us has been a constant progress towards advancement. Considering the past, whether of the period since the white man has been here or for the ages back of this particular era, we see such a record of continued progress that in looking forward we may safely indulge our hope and our confidence.

Now, as President of the 1933 proposed World's Fair in Chicago, I say that we undertake this task with a full realization of the obligation that is upon living men to recognize the obligation they owe to the men who have created this great era in which we live.

In the midst of changes such as we have seen in the last few years, at the close of a period which has witnessed such an extraordinary alteration in the industrial habits and in the social habits of men, we think that it is time for us to stop and look back and by some sort of an exposition make clear, if we can, what are these conditions that have brought about such changes—what they mean, and in what they may result.

The National Research Council has approved the general outlines of our plan. The National Research Council has appointed committees to carry out this plan; and with the committees appointed by the National Research Council we are confident that we can present something of value and of interest. But I assure you all who are interested in archæological matters that no exposition of this kind could be complete unless it included a well directed exposition of anthropology and archæology. Nor could it be useful unless that exposition was presented in such a manner as to arouse the interest of all the people in this area to the incalculable value of these mounds which now remain as relics of past civilizations. Let them all be preserved, and placed in state parks.

MR. STIRLING: For two days this body has discussed the necessity for cooperation between different organizations working in the archæological field in America, and particularly in the mound area of the Mississippi Valley. We have spoken of technical methods of research and have devised ways and means by which the work in the field will be rendered more efficient and can be done in a more systematic manner. We have spoken of the pot hunter and the archæological vandal; we have heaped anathema upon his shoulders and spoken of ways and means by which we might possibly curtail or stop his nefarious activities. In short, there have been but few subjects dealing with archæology and with the interests of this body which have not been thoroughly discussed.

But it seems to me there is one topic on which I might profitably add a few words, and that is something concerning the history and the nature of the institutions which I represent: The Smithsonian Institution and the Bureau of American Ethnology, which is a part of that great institution.

Almost ninety years ago an English gentleman by the name of Smithson died, leaving behind him what was for that day a very tidy fortune; and to the surprise of everyone he left his fortune not to his normal heirs, but to the then comparatively new United States government for the establishment of an institution for the diffusion of knowledge among men.

It was left to the Congress of the United States to determine what should constitute this diffusion of knowledge; and it was eventually agreed that the establishment of a great government scientific clearing-house would most effectually carry out the wishes of the donor of this fortune. Thus the Smithsonian Institution had its inception—a small beginning, perhaps, but it rapidly expanded. Before long it had outgrown the original foundation, necessitating the annual appropriation of an ever increasing fund by the Congress of the United States in order that it might continue its operations in a manner befitting such an institution, respresenting such a great country as ours.

The study of anthropology and its allied branches was not actively pursued until about the beginning of the seventies. Previous to this time, immediately following the Civil War, Major J. W. Powell, whose name is familiar to all of you, went out into the southwestern region of the United States, a region which then lay beyond the frontier; and there, in the most picturesque section of our entire country, he began upon his own initiative a geographical survey of the region.

Major Powell had with him as assistants a group of men who were enthusiastically interested in this sort of work, and who conducted it for very little recompense except the satisfaction they derived in knowing that they were doing useful scientific work. Although the primary object of this survey was a study of the geology of the region, they found the entire region occupied by tribes of very interesting Indians. In addition, they found the region dotted with ruins, stone and adobe buildings erected in the most picturesque locations on the sides of mighty cliffs or in the valleys beneath them.

It was at this time that the attention of the world first became focused upon the ruins of what is now known as the Pueblo region. Major Powell, returning to Washington, obtained the support of the Government for this work and was made director of the newly formed Geological Survey. Returning to the Southwest, he continued his work with redoubled energy.

In time, however, the interest of Major Powell became diverted more and more from geology to archæology, and to a study of primitive living Indian tribes of the region. Upon his next return to Washington, in 1886, he prevailed upon the Congress to establish a separate division of the Smithsonian, to be known as the Bureau of American Ethnology. The duties of this Bureau were to collect all of the data possible concerning the surviving Indians remaining in various parts of our country, and also to conduct archæological researches in the ruins that had been left behind by other groups preceding them. Major Powell was not only made director of the Geological Survey, but also director of this newly formed Bureau of Ethnology.

At that time there were no departments of anthropology in our great educational institutions. There had been no systematic study whatever of the early remains of the original inhabitants of this land. There had been no systematic study of the inhabitants themselves, so that it was necessary for Major Powell, in building up this Bureau, to collect about him such individuals as had developed, through incidental contact with the Indians, an interest and enthusiasm in this work. Such names as those of Frank Cushing, James Mooney, James Stevenson and Mrs. Stevenson, and many others of that early day, give us a knowledge of those who laid the foundation for the systematic science which anthropology has now become in this country. Their work at first had to be pursued under very considerable difficulties, principally the lack of definite information and of published material upon which to base the field work which they themselves conducted. In fact, their principal responsibility was to lay the foundation for the systematic work that was to be followed in later years.

The Bureau of American Ethnology at the present time has, among its duties, not only the pursuit of field work in various parts of the country, but it has also become, in a way, a court of appeal for the population throughout the country who are interested in matters pertaining to anthropology. We receive in the Bureau every day a very large correspondence consisting principally of inquiries concerning subjects relating to the Indians and Indian remains. Some of these are very naïve, for the simple reason that the great mass of our population has had no opportunity to become adequately educated upon these subjects. I believe, as Dr. Throop has mentioned earlier in the evening, that one of the most important duties we can perform at this time is to put suitable text-books in our grade schools and high schools, giving accurate and up-to-date information upon the native period of American history. When we stop to consider that the few hundred years that have elapsed since the discovery of America by Christopher Columbus constitute but a small fraction of the entire period of the time that human beings have occupied this continent; and that the early periods of its history were far more colorful and eventful than the subsequent events, it seems a pity that the facts we do know of this period cannot be put before our young students. When this has been done, I am confident that the people throughout the country will have the same knowledge of anthropology and of archæology as related to America as they now have of classic archæology and anthropology in the old world, and of our own early history.

I sincerely hope that this, in time, may be done. The Bureau of Ethnology has endeavored throughout the years of its existence to supply, as far as it is able, much of this long felt want. There is probably no organization in the country that has published as many pages or as many volumes dealing with the American Indian and with the subject of Archæology as has our Bureau. Most of these publications are, however, for the advanced student, although it is intended that they be intelligible to any reader. We stand ready to assist at any time, to the best of our ability, any of you who are interested or professionally engaged in the study of archæology. And we invite you all to communicate with us on any problems that may be on your minds concerning what we believe to be the most interesting and live subject of the present day.

MR. COLE: In closing, I want to express a word of thanks to Dr. Terry and Dr. Throop and our other friends in the city of St. Louis and the State of Missouri who have made our visit here so enjoyable; and I want to express to Dr. Dunlap the feeling which we all have of appreciation for the fine work he has done in making this conference possible. Also, I know perfectly that, hard though Dr. Dunlap has worked, he could not have made it such a success had it not been for the equal endeavors of Carl Guthe. And even they could not have been successful without the faithful service of the Secretary of the Division, Mrs. Britten. So, in behalf of the Conference, I want to thank all of those who have taken part, and I hope, indeed I am sure, that we will all leave here, much more assured of the future of archæology than when we came here two days ago.

RESOLUTIONS

On motion of Mr. Cox, the following resolutions were unanimously adopted:

I. *Resolved:* That we are intensely impressed with the purposes of this meeting, and are inspired to do our whole duty.

II. *Resolved:* That we hereby express our desire and anxiety to co-operate with the National Research Council in the accomplishment of its laudable and patriotic purposes.

III. *Resolved:* That we hereby express our gratitude to the Council for having given us the opportunity to attend this Conference, and will hold in kind remembrance the courtesies shown us by the National Research Council.

APPENDIX A
LIST OF MEMBERS OF THE COMMITTEE ON STATE ARCHÆOLOGICAL SURVEYS NATIONAL RESEARCH COUNCIL

Carl E. Guthe, *Chairman,* Director, Museum of Anthropology, University of Michigan, Ann Arbor, Michigan.

Peter A. Brannon, Editor, Alabama Anthropological Society, Montgomery, Alabama.

Charles E. Brown, Director, State Historical Museum, and Secretary, Wisconsin Archæological Society, Madison, Wisconsin.

Amos W. Butler, member of Executive Committee, Indiana Historical Society, 52 Downey Ave., Indianapolis, Indiana.

Roland B. Dixon, Professor of Anthropology, Harvard University, Cambridge, Massachusetts.

Frederick W. Hodge, Curator, Museum of the American Indian, Heye Foundation, Broadway at 155th St., New York City.

Charles R. Keyes, Research Associate, State Historical Society of Iowa; Director, Iowa State Archæological Survey, Mt. Vernon, Iowa.

A. V. Kidder, Research Associate in charge of American Archæology, Carnegie Institution of Washington.

Warren K. Moorehead, Director of Archæological Explorations for the University of Illinois; Director, Department of Archæology, Phillips Academy, Andover, Massachusetts.

H. C. Shetrone, Director, Ohio State Museum, High Street and 15th Ave., Columbus.

M. W. Stirling, Chief, Bureau of American Ethnology, Smithsonian Institution, Washington, D. C.

Knight Dunlap, *ex officio,* Chairman of the Division of Anthropology and Psychology.

Fay-Cooper Cole, Chairman-elect of the Division of Anthropology and Psychology.

APPENDIX B
LIST OF MEMBERS OF THE CONFERENCE ON MIDWESTERN ARCHÆOLOGY

Babcock, Willoughby H., Curator of the Museum, Minnesota Historical Society, St. Paul, Minnesota.

Barrett, S. A., Director, Milwaukee Public Museum, Milwaukee, Wisconsin.

Blackman, E. E., State Archæologist and Curator, State Historical Society Museum, Lincoln, Nebraska.

Blackmar, F. W., Professor of Sociology and Anthropology, University of Kansas, Lawrence, Kansas.

Blom, Frans, Director, Department of Middle American Research, Tulane University, New Orleans, Louisiana.

Brannon, Peter A., Editor, Alabama Anthropological Society, Montgomery, Alabama.

Brown, Calvin S., Archæologist to the Mississippi Geological Survey, University of Mississippi, University, Mississippi.

Brown, Charles E., Director, State Historical Museum and Secretary, Wisconsin Archæological Society, Madison, Wisconsin.

Butler, Amos W., member of Executive Committee, Indiana Historical Society, 52 Downey Ave., Indianapolis, Indiana.

Caulfield, Henry S., Governor of Missouri, Jefferson City, Missouri.

Cole, Fay-Cooper, Chairman (1929-1930), Division of Anthropology and Psychology, National Research Council, Washington, D. C.; Professor of Anthropology, University of Chicago.

Cooper, Wm. John, United States Commissioner of Education, Washington, D. C.

Cox, P. E., President, Tennessee Academy of Science; State Archæologist, Franklin, Tennessee.

Dawes, Rufus, President, Chicago World's Fair Centennial Celebration, Harris Trust Company, Chicago, Illinois.

Dellinger, S. C., Curator, University Museum, Professor of Zoology, University of Arkansas, Fayetteville, Arkansas.

Dickson, Don F., Dickson Mound Exploration, Lewistown, Illinois.

Dixon, Roland B., Professor of Anthropology, Harvard University, Cambridge, Massachusetts.

Dunlap, Knight, Chairman (1927-29), Division of Anthropology and Psychology, National Research Council, Washington, D. C.; Professor of Experimental Psychology, The Johns Hopkins University, Baltimore, Maryland.

Fordyce, John R., Colonel, Engineers Corps, O. R. C., Hot Springs, Arkansas.

Fox, George R., Director, E. K. Warren Foundation; President, Michigan State Archæological Association, Three Oaks, Michigan.

Fox, Lawrence K., State Historian, Pierre, South Dakota.

Freeland, W. E., Majority Floor Leader, House of Representatives of Missouri, Jefferson City, Missouri.

Futrall, John C., President, University of Arkansas, Batesville, Arkansas.

Greenman, E. F., Curator of Archæology, Ohio Archæological and Historical Society, Columbus, Ohio.

Guthe, Carl E., Chairman, Committee on State Archæological Surveys, National Research Council; Director, Museum of Anthropology, University of Michigan Museums, Ann Arbor, Michigan.

Hinsdale, W. B., Custodian of Michigan Archæology, Museum of Anthropology, University of Michigan, Ann Arbor, Michigan.

Keyes, Charles R., Research Associate, State Historical Society of Iowa; Director, Iowa State Archæological Survey, Mt. Vernon, Iowa.

Knapp, Thomas M., Chancellor of the University and Dean of the College of Arts and Sciences, St. Louis University, St. Louis, Missouri.

Leighton, M. M., Chief, State Geological Survey, University of Illinois, Urbana, Illinois.

Lemley, Harry J., Hope, Arkansas.

Logan, W. N., State Geologist, Department of Conservation, Indianapolis, Indiana.

Mason, J. Alden, Member of the Division of Anthropology and Psychology, National Research Council; Curator, University Museums, University of Pennsylvania, Philadelphia.

McKern, W. C., Associate Curator of Anthropology, Milwaukee Public Museum, Milwaukee, Wisconsin.

Moorehead, Warren K., Director of Archæological Explorations for the University of Illinois; Director, Department of Archæology, Phillips Academy, Andover, Massachusetts.

Nelson, Nels C., Curator of Pre-historic Archæology, American Museum of Natural History, New York City.

Over, William H., Curator, University of South Dakota Museum, Vermilion, South Dakota.

Parker, A. C., Director, New York State Archæological Survey; Director, Municipal Museum, Rochester, New York.

Peacock, Charles K., Assistant State Archæologist, 817½ Market Street, Chattanooga, Tennessee.

Pearce, J. E., Professor of Anthropology, University of Texas, Austin, Texas.

Putnam, Edward K., Director, Davenport Public Museum, Davenport, Iowa.

Setzler, F. M., Indiana State Historical Commission, 334 State House, Indianapolis, Indiana.

Shetrone, H. C., Director, Ohio State Museum, High St. and 15th Ave., Columbus, Ohio.

Snodgrass, Richard M., University of Chicago Survey of Illinois, University of Chicago.

Stirling, M. W., Chief, Bureau of American Ethnology, Smithsonian Institution, Washington, D. C.

Taylor, Jay L. B., Engineer, Pineville, Missouri.

Teel, William R., Chairman, Indiana Archæological Society, 420 Guaranty Bldg., Indianapolis, Indiana.

Terry, R. J., Professor of Anatomy, Washington University, St. Louis, Missouri.

Throop, G. R., Chancellor, Washington University, St. Louis.

Webb, William S., Head of the Department of Anthropology and Archæology, University of Kentucky, Lexington, Kentucky.

Wheeler, H. E., Curator in charge of the Museum Department, Birmingham Public Library, Birmingham, Alabama.

Will, George F., North Dakota Historical Society, Bismarck, North Dakota.

Williamson, George, Honorary Curator of Archæology, Louisiana State Museum, State Normal College, Natchitoches, Louisiana.

Wissler, Clark, Curator-in-Chief of Anthropology, American Museum of Natural History, New York City; Professor of Anthropology, Yale University, New Haven, Connecticut.

Appendix C

Figure 1. Monk's Mound, Cahokia Group, in East St. Louis, Illinois; a view of the south end. A state park now includes Monk's Mound and some others of the Cahokia group. (Photo by Gordon S. Severant, 1927)

Figure 2. The "Red" Mound at Cahokia, so called because of the fall color of the sumac which covers it.

Figure 3. Pottery vessels, from the mounds near Lewiston, Illinois. (Photo by Don F. Dickson)

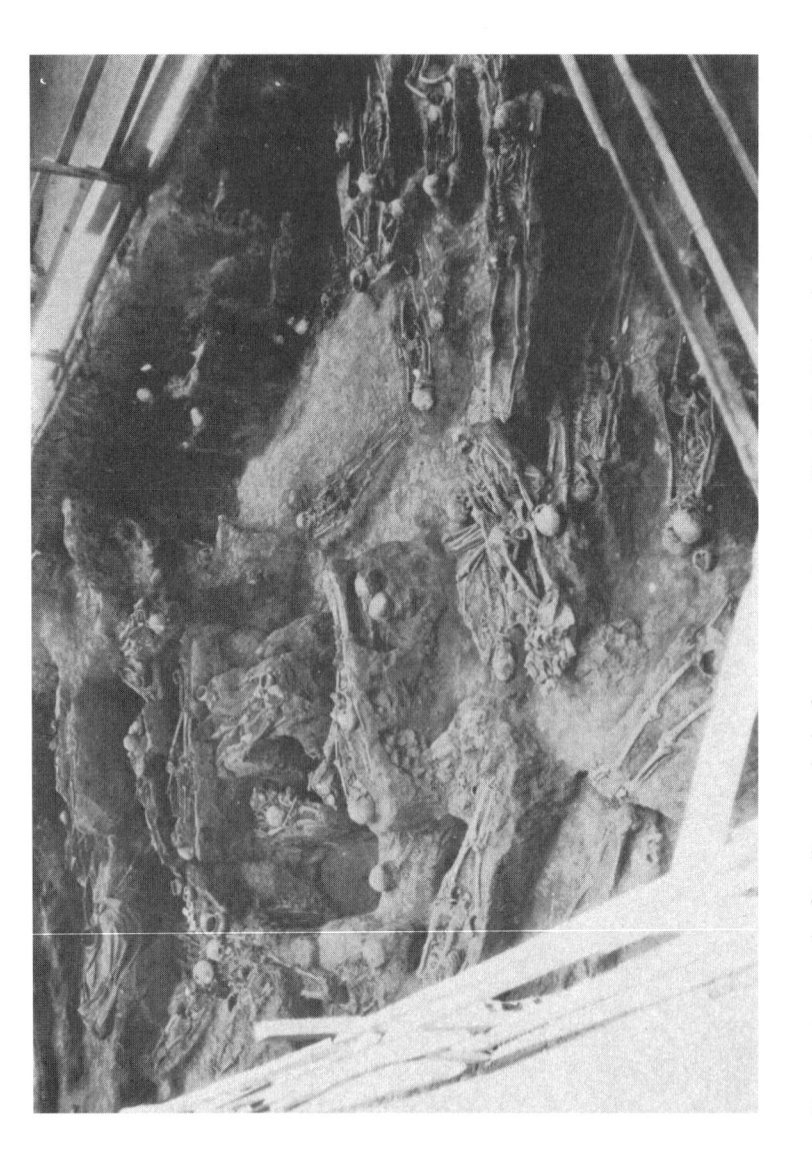

Figure 4. An example of intelligent methods of excavation at Lewiston, Illinois. Dr. Dickson has here excavated two hundred and thirty burials, leaving every bone and implement as it was found. A fire-proof building erected over the excavation preserves the materials for educational purposes.

Figure 5. Post-molds, uncovered by excavation of a mound near Pembroke, Kentucky. A structure erected on this level with posts set in the ground had been burned, and then more dirt piled on the site. The stumps of the posts decayed, but their molds are left. (See abstract of paper by Webb and Funkhouser, page 61.)

Figure 6. A fine specimen of pottery, used by the "Hopewell" people of Ohio long before the advent of white settlers. This pot was found on a mound near the banks of the Mississippi River, north of La Crosse, Wisconsin, indicating the contacts between prehistoric peoples of two regions. (An exhibit of the Milwaukee Public Museum.)

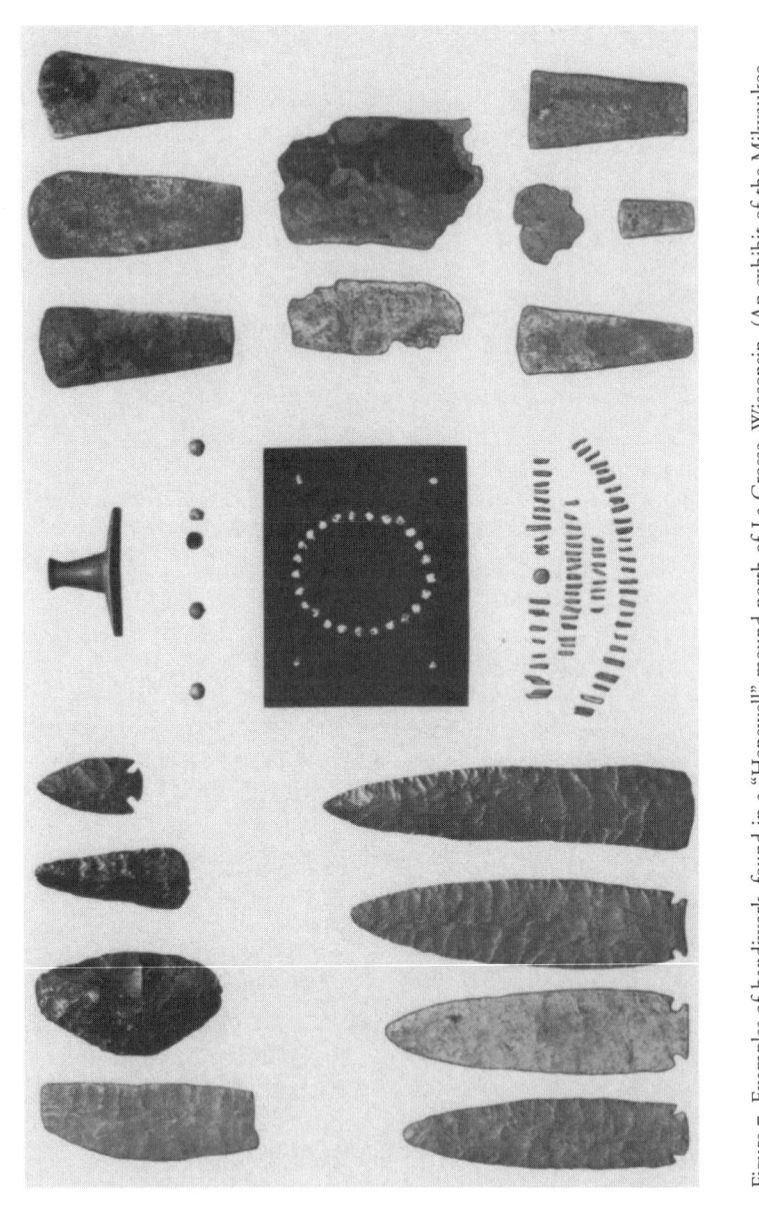

Figure 7. Examples of handiwork, found in a "Hopewell" mound north of La Crosse, Wisconsin. (An exhibit of the Milwaukee Public Museum.)

Conference on Southern Pre-History

Held under the Auspices of the Division of Anthropology and Psychology Committee on State Archaeological Surveys National Research Council

Hotel Tutwiler
Birmingham, Alabama
December 18, 19, and 20, 1932

Issued by the
National Research Council
2101 Constitution Avenue
Washington, D.C.

The Highest and Largest Pre-Historic Mound in Alabama; Mound B at Moundville

Conferees attending the 1932 Conference on Southern Pre-History in Birmingham, Alabama, are: *Standing left to right:* Mr. S. C. Dellinger, Mr. Edward Ginnane, Master John Rothermel, Mr. J. F. Rothermel, Mrs. Paul Pim, Prof. Wm. S. Webb, Dr. Forrest Clements, Mr. James Ford, Prof. Fred B. Kniffen, Miss Caroline Dormon, Mrs. Marion Hale Britten, Dr. Carl E. Guthe, Mr. Moreau B. Chambers, Mr. Gene Stirling, Mr. Henry B. Collins, Jr., Mr. Neil M. Judd, Dr. John R. Swanton, Mr. K. W. Grimley, Dr. A. T. Poffenberger, Mr. E. F. Neild, Miss Clara Berentz, Dr. Knight Dunlap, Dr. Walter B. Jones, Dr. Clark Wissler. *Kneeling left to right:* Mr. Q. B. Schenk, Mr. Paul Pim, Mr. H. E. Wheeler, Dr. W. D. Strong, Dr. Ralph Linton, Mr. E. C. Horton, Dr. J. E. Pearce, Dr. Fay-Cooper Cole, Mr. M. W. Stirling, Mr. H. F. Wenning.

Table of Contents

Program [iii]   213

List of members and guests [iv]   215

First day, December 18

Trip to Alabama Museum of Natural History and
    to Moundville [1]   219

Informal dinner and welcoming speeches [2]   220

Informal discussions of Moundville and Etowah excavations,
    led by Dr. Moorehead and Dr. Jones [3]   221

Second day, December 19

Morning session, Dr. Wissler, Chairman
The Interest of Scientific Men in Pre-History, Dr. Linton [3]   221

Southeastern Indians of History, Dr. Swanton [5]   223

The Pre-Historic Southern Indians, Mr. Stirling [20]   241

Afternoon session, Mr. Brannon, Chairman
The Bluff Shelters of Arkansas, Mr. Dellinger [31]   252

Report on Moundville Culture, Dr. Jones and Mr. DeJarnette [34]   256

Archaeology of Mississippi, Mr. Collins [37]   259

Pre-Historic Cultures of Louisiana, Mr. Walker [42]   265

Archaeological Tasks for Tennessee, Mr. Peacock [48]   271

Significance of the East Texas Archaeological Field, Mr. Pearce [53]   276

Evening session, Dr. Wissler presiding
The Relation of the Southeast to General Culture Problems of American
    Pre-History, Dr. Swanton [60]   283

Third day, December [20]

Morning session, Dr. Guthe, Chairman
Exploration and Excavation, Dr. Cole [74]   299

Laboratory and Museum Work, Mr. Judd [79]   303

Comparative Research and Publication, Dr. Wissler [88]   313

Afternoon session, round-table discussions
Field Methods, Dr. Moorehead, Chairman [93]   319

Laboratory Methods, Dr. Strong, Chairman [94]   319

Research and Publication, Dr. Webb, Chairman [95]   320

Dinner and evening session, Dr. Guthe presiding [96]   321
Appointment of committee to arrange for future conference [96]   322

Adoption of resolutions [97]   322

Appendix

Figure 1. Indian tribes and linguistic stocks in the eastern part of the United States about 1700   324

Figure 2. Relation of Indian tribes in the Southeast to the physical divisions   325

Figure 3. Hypothetical route of De Soto   327

Figure 4. Distribution of certain cultural traits in the Southeast   328

Figure 5. Tribal movements of the eastern Indians   329

Figure 6. Significant tribal locations   330

Figure 7. Archaeological culture areas   331

Program

Hotel Tutwiler, Birmingham, Alabama, December 18, 19, and 20, 1932

Sunday, December 18
9:30 A.M. Trip to University of Alabama Museum and the Moundville excavations

Lunch in Tuscaloosa.

7:00 P.M. Dinner at Hotel Tutwiler, Birmingham

8:30 P.M. Smoker, informal

Welcome to the Conference from the City of Birmingham—Mr. Earle

Response from the National Research Council—Dr. Poffenberger

"The Significance of Moundville"—Dr. Moorehead and Dr. Jones

Monday, December 19
10:00 A.M. "The Interest of Scientific Men in Pre-History"—Dr. Linton

Response by Dr. Wissler, Permanent Chairman of the Conference

"Southeastern Indians of History"—Dr. Swanton

"The Pre-Historic Southern Indians"—Mr. Stirling

12:30 P.M. Luncheon

2:00 P.M. Recent Field Work in Southern Archaeology, Session Chairman, Mr. Brannon

Arkansas—Mr. Dellinger

Alabama—Dr. Jones

Mississippi—Mr. Collins

Louisiana—Mr. Walker

Tennessee—Mr. Peacock

Eastern Texas—Mr. Pearce

(Ten minutes of discussion to follow each paper)

6:30 P.M. Dinner

8:00 P.M. Dr. Wissler presiding

"The Relation of the Southeast to General Culture Problems of American Pre-History"

Dr. Swanton

Tuesday, December 20
10:00 A.M. Session chairman, Dr. Guthe

"Exploration and Excavation"—Dr. Cole

"Laboratory and Museum Work"—Mr. Judd

"Comparative Research and Publication"—Dr. Wissler

(Each paper to be followed by discussion)

12:30 P.M. Luncheon

2:00 P.M. Round-table discussions

A. Field methods, Chairman, Dr. Moorehead

B. Laboratory methods, Chairman, Dr. Strong

C. Research and Publication, Chairman, Dr. Webb

7:00 P.M. Dinner, followed by informal speeches, Dr. Guthe presiding

List of members and guests

Barnes, George D., Dayton, Tennessee

Brannon, P. A., Alabama Anthropological Society, Montgomery, Alabama

Britten, Mrs. M. H., Secretary, Division of Anthropology and Psychology, National Research Council, Washington

Burton, E. M., Director, Charleston Museum, Charleston, South Carolina

Chambers, Moreau B., Department of Archives and History, Clinton, Mississippi

Clements, Forrest, University of Oklahoma, Norman, Oklahoma

Colburn, W. B., 333 Washington Rd., Grosse Pte. Village, Detroit; and Asheville, N. C.

Cole, Fay-Cooper, Professor of Anthropology, University of Chicago

Collins, Henry B., Jr., U.S. National Museum, Washington

Cramton, F. J., President, Alabama Anthropological Society, Montgomery, Alabama

DeJarnette, D. L., Alabama Museum of Natural History, University, Alabama

Dellinger, S. C., University of Arkansas, Fayetteville, Arkansas

Dormon, Miss Caroline, Chestnut, Louisiana

Dunlap, Knight, Division of Anthropology and Psychology, National Research Council

Earle, Samuel L., Birmingham Museum Association, Birmingham, Alabama

Evans, Robert, Shreveport, Louisiana (guest of Mr. Neild)

Ford, J. A., Clinton, Mississippi

Guthe, C. E., Chairman, Committee on State Archaeological Surveys of the National Research Council; Director, Museum of Anthropology, University of Michigan, Ann Arbor, Michigan

Hayes, James, Shreveport, Louisiana (guest of Mr. Neild)

Horton, E. C., Birmingham, Alabama (President, Birmingham Anthropological Society)

Jones, Walter B., Alabama Museum of Natural History, University, Alabama

Judd, Neil M., U.S. National Museum, Washington

Kniffen, Fred B., State University of Louisiana, Baton Rouge, Louisiana

Lindfors, Mrs. D. L., Secretary, Committee on State Archaeological Surveys

Linton, Ralph, Vice chairman, Division of Anthropology and Psychology, National Research Council; Professor of Anthropology, University of Wisconsin, Madison, Wisconsin

Moorehead, W. K., Phillips Academy, Andover, Massachusetts

Neild, Edward F., Shreveport, Louisiana

Peacock, Chas. K., Chattanooga, Tennessee (East Tennessee Arch. Society)

Pearce, James E., University of Texas, Austin, Texas

Poffenberger, A. T., Chairman, Division of Anthropology and Psychology, National Research Council, Washington

Stirling, Gene, Harvard University, Cambridge, Massachusetts

Stirling, M. W., Chief, Bureau of American Ethnology, Smithsonian Institution, Washington

Strong, W. D., Bureau of American Ethnology, Smithsonian Institution, Washington

Swanton, J. R., Bureau of American Ethnology, Smithsonian Institution, Washington

Walker, W. M., Bureau of American Ethnology, Smithsonian Institution, Washington

Webb, William S., University of Kentucky, Lexington, Kentucky

Wenning, H. F., President, East Tennessee Arch. Society, Chattanooga, Tennessee

Wheeler, H. E., Birmingham Public Library, Birmingham, Alabama

White, James, Jr., Montgomery, Alabama (guest of Peter Brannon)

Wissler, Clark, American Museum of Natural History, New York City

[1]

Report of the Conference on Southern Pre-History

Introduction

The Conference on Southern Pre-History held at Birmingham, Alabama, on December 18, 19, and 20, 1932, was called for the purposes of reviewing the available information on the pre-history of the southeastern states, discussing the best methods of approach to archaeology in this region and its general problems, and developing closer cooperation through the personal contacts of the members of the conference. During the past few years, the interest in Indian pre-history of the lower Mississippi Valley and the southern Atlantic states has been increasing steadily, and a number of institutions have undertaken research work in this field. Developments from studies of the same period in the northern part of the Mississippi Valley and from work on certain southwestern problems indicate that as the knowledge of the pre-historic cultures of the Southeast increases, the problems of the neighboring areas will be more clearly understood. It was for the purpose of fostering more rapid increase of this knowledge that this conference of experts in the study of pre-history from all over the United States was called to meet with interested students of the South.

Many of the members of the conference arrived in Birmingham during Saturday afternoon and evening, and in spite of a severe storm through the general area, the majority had arrived by early Sunday morning.

Sunday, December 18

Through the kind offices of Mr. H. E. Wheeler, of the Birmingham Museums Association, a number of Birmingham citizens were at the Hotel Tutwiler Sunday morning with their automobiles, to serve as guides on an excursion to Tuscaloosa and Moundville, which occupied the first day of the conference.

Arrived at the Alabama Museum of Natural History in Tuscaloosa, the members of the conference found that Dr. Walter B. Jones, the director, and his staff had made elaborate preparations for their entertainment. The main exhibit halls had been arranged for a special exhibition of the material culture disclosed through the work of this organization at the nearby famous site of Moundville.

The modern equipment, consisting of both wall and table cases, displayed adequately the artistic and representative materials. The exhibits were arranged by types of material, that is, in groups of pottery, of stone implements and ornaments, and of shell and bone materials. The [2] long series in many cases made possible an analysis of the general characteristics and enabled the visitors to obtain a good perspective of the material culture and the extent of individual variations in the pieces. The artistic arrangement and the lack of crowding in the cases themselves not only emphasized the more noteworthy items, but prevented "museum fatigue." The value of the exhibit was enhanced by hand-printed labels which contained sufficient information to acquaint the average visitor with the relative importance and significance of the several traits illustrated. A model of the site itself gave a bird's-eye view of the locality, which was visited later.

At noon the museum staff served an informal luncheon in the basement workrooms of the museum, demonstrating the traditional hospitality of the South to the grateful members of the conference.

In the afternoon, the majority of the members were taken by automobile to the site of Moundville, a short distance from Tuscaloosa. In spite of the rainy weather of the past week, it was possible to visit several of the mounds and to walk over some of the terraces. A photographic print of a map of the site was given to each person, making it possible to locate the mounds, in spite of the trees which covered some of them. The orientation of the visitors was further facilitated by large signs set up at each point of interest, carrying the same letters as those given on the map. The descriptions and comments by Dr. Jones and his staff, in answer to questions, increased the interest in the site. Late in the afternoon, after another hour or so in the museum at Tuscaloosa, the group returned to Birmingham, arriving about 6:30 P.M.

Dinner session, Sunday, December 20

The first general meeting of the Conference was a group dinner in the Hotel Tutwiler at seven, which was followed by an address of welcome from Mr. H. E. Wheeler, who first mentioned the regret of Mr. S. L. Earle that illness prevented his attendance. Mr. Wheeler pointed out Alabama's importance to southern archaeology in both natural and historical resources and expressed the appreciation of the citizens of Birmingham and the workers throughout the state for the choice of Birmingham as the meeting place.

Dr. Poffenberger responded in the name of the National Research Council and of its Division of Anthropology and Psychology, expressing appreciation for the hospitality of the Birmingham group and of Dr. Jones and his staff, which gave such an auspicious opening for the conference. This had afforded the members an opportunity to become acquainted with one another and concentrated attention on the pre-history of the local region as an introduction to the problems to be discussed during the next two days.

[3] Dr. Warren K. Moorehead and Dr. W. B. Jones were then called on by Dr. Poffenberger for informal comments on Moundville and Etowah and the significance of these two great sites in the culture history of the Indians of the South.

Monday, December 19

Morning session, Dr. Wissler, chairman

The meeting was formally opened by Dr. Wissler, permanent chairman of the conference, who called on Dr. Linton to speak on "The Interest of Scientific Men in Pre-History."

Dr. Linton:
All anthropologists who are working with either the ethnology or archaeology of North America east of the Rocky Mountains have a vital interest in the results of archaeological work in the Southeast. No matter what their individual specialties may be, they find themselves confronted with problems in their own particular fields, which must remain unanswered until more information from the southeastern area is available.

In the territory east of the Rocky Mountains two main types of culture are discernible. In the northern woodlands and northern and central Plains, there are a series of cultures based on hunting. These cultures are characterized by a more or less nomadic life and by relative simplicity of material culture, social organization, and religion. To the south and east of these hunting peoples was a whole series of other cultures, which were mainly dependent on agriculture. These were characterized by a much more settled life and by a richness of cultural content, which elevated them far above the hunting cultures. These sedentary cultures reached their highest development in the Southeast, and everything indicates that it was in this region that the new economic basis of life first became established and from this region that it spread northward and westward until it had penetrated to the farthest points at which maize could be grown. It is clear, moreover, that maize did not spread alone but was accompanied in its diffusion by a whole series of other traits that had either originated in the Southeast or first become acclimated there.

The worker in any of the surrounding regions finds evidences not merely of diffusion but of actual migrations coming into his particular area from the Southeast, but until the history of that region is better known, it is impossible for him to tell when such migrants left the Southeast, what part of it they came from, what their cultural or racial affiliations may have been, or how they are linked to other cultures marginal to the same area. For years we have been working about the edges, leaving [4] the center, where the solutions to most of our problems lie, almost untouched.

Interest in southeastern problems is not confined to those working in the areas to the north and west. Maize is a tropical plant and must have been brought to the Southeast from some region still farther to the south. Its presence is a proof of some early contact with Mexico, and many other traits of southeastern culture strongly suggest Mexican influence. The great domiciliary mounds of the Southeast, for example, are much like Mexican pyramids, and Mexican and southeastern art have many forms in common. Further work in the Southeast may reveal when these various elements entered the region and by what routes, clearing up certain problems as to the time and place at which these traits were developed in Mexico itself. For example, while Mexican traits are recognizable in both the southeastern and southwestern culture complexes, very few of these traits are the same, suggesting that the diffusions to these two areas came from different centers within Mexico.

However, we outsiders have come here to gain knowledge from experts, so I will end and turn the meeting back to the chairman.

Dr. Wissler:
I am sure we are under great obligations to Professor Linton for this expression of the interest of outsiders in the problems of the Southeast. It seems rather interesting that the living cultures of the South should have

apparently vanished more completely than those of the North, and while we know little about the northern, we seem to know even less about the peoples of the southern states who were here when the whites came. The archaeology of the South looms rather large for that reason. We must depend on an archaeological approach to the problems of the Southeast. We are all intensely interested in the results of studies made by those who are working in this area. It seems to me a misfortune that so many archaeologists in the United States have worked elsewhere.

It is impressive to find, as I did at a recent conference in New Mexico, that there are 40 reputable archaeologists working within the limits of New Mexico and Arizona. The American Anthropological Association, at its last winter meeting, began a classification of southwestern cultures. I sincerely hope the Southeast will soon come into its own and make as rapid strides as the Southwest, for its problems are just as important.

We are fortunate in having with us Dr. Swanton, who has for many years made this region his field, and is perhaps the most competent person to speak on its problems. He will speak on "The Southeastern Indians of History."

[5] *Dr. Swanton:*
Some part of the section which is the subject of this paper has been known to Europeans almost from the time of Columbus. Indeed, had that famous navigator not altered his course on October 7, 1492, he would probably have reached the northernmost of the Bahama Islands, learned of the great continent beyond, and become the discoverer of North America. As it happened, he was turned southward, and any American ethnologist will tell you that he did not reach ethnological North America on any of his trips.

It is probable that Ponce de León, the reputed discoverer of the peninsula, had at least one predecessor, but Ponce de León had the honor of naming it, and thus acquired immortality, though one, indeed, somewhat different from that of which he was in search. The documents of his expedition, however, throw little light on the natives of the land, except to indicate that they objected to European exploitation. In 1519 Alonso Alvarez de Pinedo coasted the northern shore of the Gulf and careened his vessels in what was probably Mobile Bay or River. Mention of 40 villages there suggests that the displacement of population on the Gulf Coast, of which we have archaeological evidence, was relatively late.

In the years 1520–1526 came the Ayllon expeditions, important to us as yielding our first body of information regarding the Siouan tribes of South Carolina. This was recorded by Peter Martyr from the lips of an Indian named Francisco of Chicaora carried away from some place near Winyaw Bay in 1521. It has been much garbled by later writers and translators, and the original contained misunderstandings and exaggerations; but it is a valuable body of material nonetheless, one which unfortunately escaped the usually careful eye of Mooney.

The Narvaez expedition is important for the knowledge of the cruder tribes of the Texas coast obtained by Cabeza de Vaca. It also preserves a unique reference to the use of slings by Indians near Pensacola. In 1539–1543 De Soto's army passed through portions of almost every southern state below Virginia and Kentucky, and the chronicles of that expedition are of vast importance to us as furnishing a cross section of aboriginal conditions in the interior about 150 years before it was again visited. As particular attention is to be given to this expedition later, we will pass on.

The impressive attempt to plant a colony on the Gulf coast near Pensacola, in 1559–1560, which was led by Don Tristan de Luna, yields only one item of importance—knowledge of a tribe called Napochies, which seems to have been living not far from the site of the famous Moundville. From a place name appearing in the narratives of the expedition there is some reason to think that this tribe was related to the Choctaw.

[6] In 1562 came the first attempted French settlement by a small body of Huguenots near the present Beaufort, South Carolina. In the spring, they built a small vessel in which they set sail for France, a few finally reaching it, though the greater number died of hardships on the way. As usual, the French were excellent observers and recorders, and their narrative gives us our earliest view of the Cusabo Indians, the easternmost tribe of the great Muskhogean family.

In 1564 came a more elaborate attempt at Huguenot colonization under Renaud de Laudonniére, established on the lower course of St. Johns River in the country of the Timucua. As is well known, this colony was destroyed by the Spaniards under Pedro Menendez in 1565; but it is an interesting commentary on the culture and interests of French and Spaniards that, except in the matter of language and social organization, we learn more of the lives, beliefs, and customs of the Florida Indians from this one French expedition than from all subsequent Spanish writers down to the cession of Florida to England, and its final sale to the United States. The writings of Laudonniére and Le Moyne, and the illustrations of the latter, in spite of their obvious defects, will seemingly always constitute our main sources of information regarding the Timucua people.

Soon after the Spanish conquest of Florida we get some scanty light on the peoples of Beaufort or Santa Elena and the inland tribes as far as the Tennessee and the Coosa. Our information regarding these last comes through reports by the Spanish captain Juan Pardo and his superior, Governor Vandera of Santa Elena.

From this time on, the Spaniards continued to maintain posts and mission stations somewhat irregularly from the mouth of the St. Johns along the coast to Santa Elena and even somewhat beyond. The documents of this period yield small amounts of information as to the situation relative to the Indians, and it is hoped that more still repose in the great manuscript depositories of Spain. It is probable, however, that few South Carolina and Georgia boys and girls are aware of the long period of colonial history that preceded the settlement of Charleston in 1670.

Although Jesuit missionaries were first to begin work both in Florida and South Carolina, they were soon displaced by the Franciscans. In 1597 the missions that these monks had established along the entire extent of the Georgia coast were destroyed in a native uprising, and part of the Indians fled inland. Missionary work was pushed with vigor among the remainder, however, but more particularly among the Timucua from Cumberland Sound south, and in the early part of the seventeenth century practically all of them became Christians.

Attempts to convert the Calusa and other Indians of south Florida were, however, abortive and were finally abandoned. In another [7] direction, however, they were attended with greater success, i.e., in the Apalachee province, which lay about the present Tallahassee. In 1633, in response to repeated solicitation by the Indians, two monks entered the country, and it became nominally Christian after a great native uprising in 1647. Apalachee were also involved for a time in the great Timucua rebellion of 1656. This latter seems to have marked the beginning of the end of the Timucua, who declined steadily in numbers. Some of their missions were turned over to Indians from the coast of Georgia and to Yamasee who had sought refuge from the English. Later, however, these Yamasee withdrew from the missions and went over to the colonists of South Carolina, being settled at the southernmost extremity of that colony.

In 1703–1704, the Apalachee were attacked by Colonel Moore of South Carolina at the head of 50 volunteers and a thousand Creek allies. As a result, part were carried off to the Savannah and settled there, and part fled to Pensacola, which had been founded in 1698, and to the French post at Mobile. In 1715 the Spanish Indians were increased by a new insurrection on the part of the Yamasee, and for a time there were a number of Indian towns about St. Augustine. These, however, were rapidly driven to the west or decimated.

The last of the Timucua settled on Tomoco Creek and died out there, the Apalachee fled to Mobile and then to Louisiana, and the Yamasee withdrew inland and finally formed a band of Seminole. The last of the south Florida Indians on the eastern coast of the peninsula probably went to Cuba in 1763, when the colony passed into the possession of the English, but those on the west coast, remnants of the famous Calusa tribe, held their ground until the very end of the Seminole War, 1840–1841, when they either united with the Seminole Indians or crossed to Havana.

The period of Spanish occupancy is almost barren of ethnological information, except that the Timucua language and something of the social organization and beliefs are preserved in the religious books gotten out by the missionaries.

Spanish occupancy of Texas presents a curious parallel to their settlement of Florida, since in both cases it was provoked by a French attempt at colonization. In the case of Texas, however, this French colonization was purely accidental, due to La Salle's failure to locate the mouth of the Mississippi in 1685. His colony, as we know, settled on Matagorda Bay, was beset by misfortunes and finally destroyed by the Indians.

Before this event, however, La Salle made a heroic attempt to reach the Mississippi overland, in the course of which he was murdered by some of his companions in the country of the Hasinai. Some of the [8] survivors continued on to the Mississippi, and the narrative of Henri Joutel is one of our best, as it is our earliest, extensive account of the Caddo Indians. This abortive attempt did, however, stimulate Spanish activities, and in 1689 Alonso de Leon visited the site of the abandoned fort. The year following he went as far as the Caddo towns in eastern Texas, where his clerical companion, Father Massanet, founded the first Texas mission. Soon after, missionary work here was abandoned, but it was soon renewed and extended also to the Coahuilteco tribes of southwest Texas. Among these, near the present San Antonio, the most important Texas missions were built, including that which was to become famous as the Alamo. The missions reached their most flourishing condition about the middle of the eighteenth century but declined after that date along with the numbers of the Indians, and in 1812 they were suppressed by the Spanish government.

After having been so unfortunate at the eastern and western extremities of the Gulf, the French had better success in the intervening territory along the Mississippi. In 1673 Marquette descended from Canada as far as the mouth of the Arkansas River. In 1682 La Salle reached the Gulf, but, as we have seen, his attempt to reach the river from France by sea ended disastrously.

In 1699, however, the Sieur d'Iberville established the first permanent settlement of the colony of Louisiana in what is now Biloxi Bay, and soon afterwards (1702) the eastern seat of French power was moved to Mobile Bay, and to the present Mobile in 1710, while the western seat was established permanently at New Orleans in 1718. In the meantime, Canadian voyageurs, explorers, and missionaries had been descending the Mississippi; and French power, rather tardily followed by French colonization, worked inland to central Alabama on the east, to Natchitoches on the west, and in increasing volume up the Mississippi River. Missionary work was attempted among the natives by the Jesuits and other orders, but even the most persistent of the missionaries, Father Davion, who had established himself among the Tunica, finally abandoned the field in despair.

In 1729 a body blow was given to the prosperity of the French by the Natchez uprising, in which about 200 Frenchmen were killed. The war which followed and the subsequent disastrous Chickasaw expeditions both held back the colony and reduced the numbers and the importance of the Indians under French suzerainty. An early war with the Chitimacha had already decimated that tribe. Most of the smaller tribes now sink into obscurity, and we hear little of any except the Choctaw, whose size and position between the colony and the Indians in the English interest rendered them of cardinal importance, an importance of which they were fully cognizant. The Creeks were also retained in part in support of the French, but could never be relied on as a body, and even among the Choctaw English influence brought about for a time a bitter civil war. This condition was bequeathed to Spain in 1763, along with Louisiana, and was only brought to an end by the annexation of Louisiana to the American Union in 1803.

[9] French contact with the Indians has yielded us the two important works of Le Page Du Pratz and Dumont de Montigny, besides the writings of Penicaut and the missionaries Le Petit and Charlevoix on the

Natchez, Bossu on the Choctaw and Alabama, Joutel on the Caddo, and an important anonymous publication on the Choctaw recently printed under editorship of the present writer.

English activities in the Southeast really began with the attempted Raleigh Colony 1585–1587. Abortive from every other point of view, it has supplied us in the White drawings with one of the best series of pictures of the American Indians of this early period, and, in the Hariot narrative, with one of the best accounts of their economic life. The Virginia Colony was established only on the margin of the Southeast, but it had a powerful influence in the area, as evidenced by the fact that the Creek name for the white Americans is "Watcina," or Virginians. The narratives of Smith, Strachey, and Beverley, along with that of Hariot, give us most of our knowledge of the tidewater Algonquians, and Virginia traders and explorers were the first Englishmen to penetrate the Piedmont country of the Carolinas and the southern Appalachians. Our earliest information of importance regarding the Siouan tribes of the East also comes from Virginia travelers, such as Batts and Fallan, and John Lederer; but our greatest authority, John Lawson, set out from the capital of the southern English colony second to be founded, South Carolina, and from its very inception in 1670 this new colony began to exert a strong influence on all of the southeastern tribes as far as the Mississippi.

Held back temporarily by the Tuscarora wars of 1711–1713 and the Yamasee war of 1715, English influences emanating from South Carolina soon became dominant among the Chickasaw and all of the Creeks except those close to the French Fort Toulouse at the junction of the Coosa and Tallapoosa rivers, enlisting also a powerful faction among the Choctaw, the Natchez, and other lower Mississippi tribes. It was partly responsible for the Natchez uprising. In 1733 English influences were still further strengthened by the founding of Georgia and by the cession of Florida to Great Britain in 1763.

In 1776 the authority of the new American Union succeeded that of England and in time, as we know, displaced the Spaniards and French. The later history of the Southeast is one of steadily extending white settlement, steadily increasing friction with the Indian tribes, and the inevitable but sad story of Indian removal, containing chapters to which we Americans can hardly "point with pride." The small republics into which the emigrated tribes organized themselves constituted interesting experiments in Indian self-government on European patterns, but, as we know, these now are things of the past.

[10] In the old territory of the Gulf and the lower Atlantic province, small bands of Indians have continued to carry on an obscure and struggling existence until the present time, and they, more than their western relatives, have proved prolific subjects of scientific study. Opportunities for such study, however, are rapidly disappearing and will soon be ended. The remnant peoples include the Cherokee of North Carolina, the Catawba of South Carolina, the Seminole of Florida, the Choctaw of Mississippi, the Koasati of Louisiana, and the Alabama of Texas, besides a number of small but important fragments of peoples in Louisiana and some mixed-blood bands in Virginia.

Later-English and early American contact have left us with some few descriptions of great value of the Indians in addition to those already mentioned, such as the narratives of Adair, Romans, Bartram, Timberlake, Hawkins, Swan, Stiggins, and Hitchcock, but they are all too few. Noteworthy among more recent studies are those of Gatschet and J. O. Dorsey on language, and Mooney, Halbert, Speck, and Olbrechts on general ethnology.

Summarizing the ethnological history of the Southeast since first white contact, we may say that the sixteenth century was largely taken up with exploration and discovery, while the seventeenth down to the Yamassee and Natchez wars was occupied with the story of contacts between the whites and the small tribes, and with the breakup of the latter; and history from that time on has been concerned with the great Indian nations, the Creeks, Seminole, Choctaw, Chickasaw, and Cherokee.

(Culture of the tribes of the Southeast) . . . We now turn to a consideration of the Indians themselves. All of you are sufficiently familiar with Indians in general to make it unnecessary for me to mention those physical characters which are measurably true for all Indians, and the cultural characters which were most widely spread. What we want to consider here is how the Indians of the Southeast differed from others and among themselves.

Ethnologists are wont to classify people in three ways: on the basis of their physical characters, their languages, and their general culture. The first of these, while theoretically the most fundamental, is practically the most difficult to handle, especially in the case of a people as homogeneous as the aborigines of America. Applying the criterion of head form, the character most widely used, we may say in general that there seems to have been a broad-headed people extending east and west north of the Gulf of Mexico and upward along the Mississippi as far as the Great

Lakes, though not always in a continuous belt, while to the north or rather northeast of them we come upon long-headed people. In Florida the two types were very much mixed, and the same is true for considerable areas elsewhere. Broad-headed areas are found farther south in Mexico, and it is particularly interesting to note that the famous Maya Indians [11] were broad-headed, though this fact does not prove that the southeastern Indians were related to them. It will, however, have some interest for us later on.

Hitherto the most satisfactory method of classifying Indian tribes has been on the basis of language, though we must remember that this is a cultural, not a biological, feature, as the same language may be adopted by people otherwise distinct, while biologically related tribes may acquire the use of unrelated languages.

The accompanying map (Appendix, Figure 1) shows all of the important tribes of the section and the linguistic stocks to which they belong. There are slight differences between them because they were intended to apply to somewhat different periods, and they represent somewhat different stages in our advance in knowledge of the region. The map was prepared especially for this conference.

Beginning on the outskirts of the area in which we are interested, we find that the great Algonquian family, occupying a huge extent of territory north of the Great Lakes and as far westward in Canada as the Rocky Mountains, extended southward in two sections until some of the tribes belonging to it were able to play a part in southern history. One tongue of Algonquians ran along the Atlantic Seaboard as far as Pamlico Sound, and we shall have occasion to consider in our treatment of the area the Algonquian groups between this point and the Potomac River, including the famous Powhatan Confederacy, and the Weapemeoc, Chowanoc, and Pamlico of North Carolina. The other tongue extended along both shores of Lake Michigan, through Indiana and Illinois into Kentucky, and later reached Tennessee and Arkansas. We are only concerned here with the Algonquian "Southerners," or Shawnee, for that is what the name means. At a very early date they were found on Cumberland River, to which they had recently moved from the Ohio, and later on, portions of them found temporary homes in Alabama, Georgia, and South Carolina. A part of the Illinois known as Initchigami lived for a time in northeastern Arkansas, but without exerting any appreciable influence on the culture of the region.

Between the two tongues of Algonquians which have been mentioned lay the Iroquoian peoples, who derive their name from the Five Nations of western New York, the Iroquois proper, and included other historic tribes such as the Huron, Neutrals, Erie, and Susquehanna. Our interest, however, is in two detached Iroquoian groups, the one including the Tuscarora, Meherrin, and Nottoway, who lived just inland from the Algonquians of North Carolina and Virginia, and the powerful and populous Cherokee Nation, whose historic seats have always been in the southern extension of the Appalachian Mountains.

[12] In the territory intervening between these three Iroquoian groups of people lay a number of tribes constituting the eastern division of the Siouan family. It derives its name from the Dakota or Sioux of the upper Mississippi, the most prominent member of the western division, which also included other well-known tribes like the Crow, Omaha, Osage, and Winnebago, the latter separated somewhat territorially from the rest. The eastern Siouans were divided into two linguistic groups, a northern covering the Piedmont and mountain sections of Virginia and West Virginia, and including at least one tribe in southern Ohio, the Mosopelea. In the latter part of the seventeenth century, this last descended the Ohio and Mississippi rivers and finally settled on the lower Yazoo, where the French knew them as Offagoula.

There was also a Siouan tribe on the Gulf Coast about Pascagoula River and Biloxi Bay, which seems to represent an earlier but not remote movement from the Ohio. The southern division of the eastern Siouans covered most of the Piedmont country of the two Carolinas and extended to the ocean between Cape Fear and Charleston Harbor. The largest and best-known tribe of this division was the Catawba, but the names of several others are perpetuated in this section, such as the Santee, Congaree, Wateree, Cheraw, Pedee, Waccamaw, Winyaw, Eno, and Shakori.

On the Great Plains beyond the western Siouans was an interesting family represented by the Pawnee, Arikara, Wichita, and Kichai, and including also a group of peoples living in or near the woodlands in northwestern Louisiana, southwestern Arkansas, and northeastern Texas, and giving their name Caddo to the entire stock. It is believed that this last played an important part in the pre-history of the region.

The greater part of the remaining territory in the Southeast was occupied by a linguistic family which took its name, Muskhogean, from that of the dominant people of the Creek Confederacy. By a strange anomaly, this name appears to have been given them by the Algonquian Shawnee. The Muskogee proper seem originally to have consisted of several distinct bands, to which were added a group of tribes speaking dialects of the Hitchiti language, and formerly resident in southern and southeastern Georgia, one of them becoming particularly well-known as the Yamasee. The Alabama, Tuskegee, and Koasati were subordinate tribes connected with the Choctaw in language but politically with the upper Creeks. To these must be added two former Florida tribelets, the Tawasa and Osotci. At a very late period, the Yuchi, to be mentioned presently, joined this confederation and at about the same time so did a portion of the Shawnee. The Seminole were only a late offshoot of the Creeks, not antedating the eighteenth century.

In what is now southeastern Mississippi and southwestern Alabama were the Choctaw, and in northern Mississippi the Chickasaw, where [13] they lived from the earliest period of white contact until removed to what is now Oklahoma. In northwestern Florida, between Aucilla and Apalachicola rivers were the Apalachee, who spoke a language related to Choctaw; on the Georgia coast a confederation of peoples which seems to have been mainly Muskogee; and in South Carolina between the Savannah River and Charleston a smaller confederation probably related to the Yamasee, called Cusabo. The Mobile and Tohome lived near the junction of Alabama and Tombigbee rivers, the Pascagoula on the river of that name, the Acolapissa on Pearl River, the Bayogoula, Mugulasha, Houma, and Okelousa on or near the Mississippi below Red River, and the Chakchiuma, Ibitoupa, and Taposa on the upper Yazoo. These last all spoke languages closely akin to Choctaw and Chickasaw. A related but highly specialized group of tribes included the Natchez, who lived along St. Catherine Creek near the city which bears their name, the Taensa of Lake St. Joseph in northeastern Louisiana, and the Avoyel near the present Marksville.

The tribes of southern Florida—the Calusa, Ais, Guacata, Jeaga, and Takesta—are thought to have spoken dialects of Muskhogean. The Timucua Indians in the northern part of the peninsula have been placed in a distinct linguistic family, but their actual status was evidently something like that of the Natchez, their allies.

On the lower Yazoo and in the neighboring country to the west and south were several small tribes, the Tunica, Yazoo, Koroa, Tiou, and Grigra, which were formerly placed in a separate stock but have more recently been united with two other linguistic groups, the Chitimacha of Grand Lake and Bayou Teche, and the Atakapa between Vermilion Bayou and Galveston, the latter including also the Akokisa of Trinity River, and the Bidai, Deadoses, and Patiri of central Texas.

In Texas, beyond the Atakapa, were two small stocks, the Karankawa along the coast and the Tonkawa inland, and beyond these again the Coahuilteco of southern Texas and northeastern Mexico.

There remains still to be noted a tribe, or group of tribes, closely associated with the Savannah River in historic times, from which they moved to the Chattahoochee in the first half of the eighteenth century to unite with the Lower Creeks. These are generally known as Yuchi, bands of whom were also located at various times on the Tennessee River above Muscle Shoals, on the lower course of the Hiwassee and neighboring parts of the upper Tennessee, in west Florida, and in other places.

The culture of all these people was basically similar, owing in large measure, no doubt, to lack of geographical barriers of consequence such as deserts, lofty mountains, and wide rivers. The southern Appalachians, and at an earlier time the Ozark and Ouachita plateaus, were the homes of people rather than boundaries between people. The only natural [14] boundary of any consequence was the Mississippi River, which in general divided the Muskhogean peoples proper from those of other stocks. La Salle found the tribes on one side usually hostile to those on the other. Yet even here there were numerous exceptions. At an early date the Quapaw had settlements on both sides and so did the Tunica and Koroa. The Taensa on the west side of the great stream were closely related to the Natchez on the east side, and lower down there were Choctaw-speaking people on both banks.

On the other hand, the assumption that geographical barriers are necessary in order to bring about differences between peoples receives something of a setback in view of the situation in the territory of what is now Louisiana. Here, with no natural barriers other than bayous, which were rather means of intercommunication than the reverse, we find six languages were spoken either totally unrelated to one another or so widely separated as to be mutually unintelligible, and a similar complexity extended westward through Texas into Mexico. In linguistic complication this region is surpassed only by California and Oregon.

Do not conclude, however, that topography exerted no influence at all on the distribution of the aboriginal population and the course of their history. To prove this I submit the reproduction of a map issued by the U.S. Geological Survey which shows the physical divisions of the section.

(See Appendix, Figure 2.) On this I have entered Roman numerals indicating the location of Indian tribes at the end of the seventeenth century in the order of their numerical importance.

It will be seen at once that the premier position is held by the one tribe (the Cherokee) that occupied the eastern mountain *massif,* on and near the southern expansion of the Blue Ridge. Second place is shared by two tribes on the coastal plain, and the third by five on the coastal plain and one in the Piedmont country; while the fourth group includes three people on the coast of Florida, Louisiana, and Texas, and two on or near the Mississippi.

Speaking generally, we may say that the physical areas rank in this order: (1) the southern Appalachians, (2) the inland section of the coastal plain, (3) the coast itself, (4) the Piedmont Plateau. By rank I mean rank in size of tribes, not in population per square mile, which yields a different result. Note particularly how many tribes are scattered along the fall line between the Piedmont Plateau and the coastal plain, where so many important cities have since been built. With the aborigines the attraction was food, just as with us it is power.

There is not sufficient time for a detailed account of cultural differences in the Gulf area, only a few broad outlines.

[15] From southwestern Louisiana and extending around the northwestern angle of the Gulf of Mexico, almost to Panuco, lived an enormous number of small tribes or bands, showing, as has been said, great linguistic diversity, addicted but slightly to horticulture, living on wild fruits, seeds, and roots, along the coast on alligators and by fishing, and inland by hunting. This is the one spot in North America north of Mexico where cannibalism seems to have been widely prevalent.

Although exhibiting great diversities, the rest of the Southeast may be treated as a unit. Early historical notices and the surviving Indian remains seem to indicate that at a not-remote period the most dense population was along the coast, due to the abundant supplies of fish and shellfish, and for similar reasons along the Mississippi. This rule still held in Virginia, North Carolina, the southern part of South Carolina, Georgia, much of Florida, part of Louisiana, and Texas, down into the historic period; but the introduction of horticulture from the south, including the raising of corn, beans, and pumpkins, had brought about a revolution in places, so that many of the tribes, though still relying to some extent on river fisheries, had abandoned parts of the Gulf Coast. The ground was cultivated in large measure by community labors, and this perhaps paved the way for those native states of which the so-called five civilized tribes were the last representatives.

Lack of two things—a knowledge of the use of fertilizer and a domestic animal suited for farmwork—prevented southeastern culture from becoming anything more than seminomadic. An annual abandonment of the towns for flesh food was thus rendered necessary, and in all but a few favored spots, periodic abandonment of land which had run out, or from which suitable firewood had been gleaned. These economic factors set certain limits on the cultural development of the Southeast, which must always be kept in mind.

Roughly speaking, the economic lives of these Indians resolved themselves into a summer horticultural and fishing season and a winter hunting season. They had to return to their towns in time to plant the fields, after which some Indians continued to remain about the towns to keep watch over them, but others dispersed in small parties to live on fish, shellfish, small game animals, berries, roots, and so on. The early corn also served to carry them over until July or August, when the new flour corn was ready to eat, the so-called green corn ceremony was held, and there was for a time abundance of food. From then on until October or November the products of the fields—corn, beans, and pumpkins—supplemented by such game as could be found near home, by fish taken in traps or by poisoning, by sturgeon in the northeast, and by wild roots and berries, rendered life comparatively easy. During this time most of the ceremonials, particularly those of a social nature, took place. Afterwards the people scattered to various parts to hunt, and during this [16] time much of the manufacturing was done—baskets, textiles, wooden and horn objects, pipes, and other articles being produced for home consumption or for trade. Those tribes that lived near enough to the sea to benefit by the spring run of herring broke off hunting and established themselves near their fish weirs until it was again time to plant.

Corn was the main support of the population, and beans, pumpkins, and sometimes sunflowers were planted along with it. Tobacco was also raised. Among the central tribes, at least, crops were raised partly in small private fields cultivated by the old women and in large town fields worked by the men and women of the community at the same time, though family plots were distinguished. Surplus food was stored away in granaries raised on posts. A considerable variety of dishes were made from these, the corn being reduced to flour in wooden mortars, and the field crops supplemented by nuts and oil extracted from certain kinds of nuts and acorns.

Wild fruits were eaten fresh or dried, and a kind of bread made principally of persimmons was a staple article of diet throughout the section and is repeatedly mentioned by the chroniclers of the De Soto expedition under the name Ameixas. Explorers speak of numerous roots used as food, but the principal were ground nuts (*Apio tuberosa*), and the kantak, or kunti—roots of several species of smilax. In southern Florida the name was transferred to two species of *Zamia*, out of which flour was made by a rather complicated process.

The most important game animal in historic times was the deer, but persistent mention of the bison seems to indicate that it was of much greater relative importance at a comparatively recent period. The bear was hunted more for its fat than its flesh. Among small game we hear most often of squirrels and rabbits, and the turkey was naturally the principal game bird, though many sorts of wild ducks and geese were eaten, and thousands of wild pigeons by those who lived near one of the famous pigeon roosts. Herring and, in the northeast, sturgeon are the principal kinds of fish specifically mentioned, because they happened to appear at times of food scarcity. Alligators, crawfish, shellfish, and practically all things edible were levied upon by the natives of the section, but some Indians are said to have been prejudiced against the opossum and the wolf. Dogs were eaten only ceremonially and by a few tribes, including the Natchez.

Deer were usually stalked by single hunters who made use of the prepared head of an animal of the same kind, but, at least in the northeast, deer drives were also used. Bears were sought out in their dens, driven out by means of fire and shot when they tried to escape. They were sometimes allowed to breed in certain tabooed areas or bear parks. Small game was left largely to the boys, who often used blowguns against [17] them, but rabbits and probably other animals were trapped. Fish were caught by means of hooks, shot with arrows, speared (often with the help of fire at night), taken in nets, in fish weirs made of stakes along the coast or in stone weirs on the inland rivers. In dry seasons pools left along stream courses were dragged with crails or the fish stupefied by means of buckeye, devil's shoestring, and other plants.

Clothing was mainly of deer, bear, and sometimes bison skin, and consisted of a breechclout for the men and a short skirt for the women, a shirt or cloak, leggings for protection against bushes and briars, moccasins, used mainly in traveling, and heavy robes to be donned only in the most severe weather. The cloaks were sometimes made of bird feathers worked in beautiful patterns, and a type of garment common in the central part of the area and worn more particularly by women was woven of mulberry bark, grasses, or other vegetable materials on a down-weaving loom. Women employed these both as skirts and cloaks. In Florida they substituted Spanish moss, and the men of southern Florida wore breechclouts made of grass. Women, except when in mourning, quite uniformly wore their hair long, but the style differed greatly among men, some tribes allowing it to keep its full length, some cutting off the hair on one side, some shaving all but a roach, and some all but a scalplock. Ornaments were worn in profusion by those of both sexes who could afford them, paint was a *sine qua non,* and the intricacy and skill of the tattooings in this region were frequent subjects of comment. Garters, belts, and headbands were woven of bison or opossum hair and ornamented with beads. Shell, bone, and copper beads, copper plates, copper wire, and (in Florida) dyed fish bladders were all used as ear ornaments. Bracelets, rings, armbands, and gorgets of copper and shell, hair ornaments of bison hair and copper, were in use in various parts of the country, with nasal ornaments in certain sections.

Early travelers were very much struck by the looseness of the relations between the sexes before marriage and the severity with which delinquents were punished afterward. The Creeks scourged both offenders equally, the Siouan peoples only the man, and the Chickasaw only the woman; a double standard of morality which—in the case last mentioned—some Europeans might regard as a sign of civilization.

The Choctaw, as is well known, separated the bones of their dead from the flesh and preserved only the former, at first in a mortuary house but when that became overfilled in a mound constructed for the purpose. This custom of preserving the bones was widely spread but often was confined to the chiefs. We find it on the lower Mississippi, in Florida, on the Savannah, among the Biloxi, and in Virginia, but in the last-mentioned region the bones were put back into the skin and the rest of the space filled with sand, while the flesh was also preserved in a basket. The Creeks and Chickasaw, however, buried in the earth, often [18] under the floor of the house itself. Santee burial seems to have been similar but on the tops of mounds, and in general the common people except the Choctaw and some related tribes seem to have been buried in the earth. I know of no reference to urn burial in the literature, but the Creeks sealed the bodies of stillborn children inside of hollow trees.

The social organization of the tribes of this area differed widely. The Natchez constituted a theocratic absolutism, the ruling caste claiming descent on the female side from the solar culture hero. Timucua rule also tended to hereditary absolutism as nearly as we can judge. The Chitimacha, unlike the Natchez, had true endogamous castes.

The power of Algonquian and Siouan chiefs varied greatly, some being very feeble, while Powhatan had built up a kind of Indian empire and exacted tribute somewhat after the manner of an Old World sovereign. The Creeks were controlled much of the time by a kind of military aristocracy, but the hereditary privileged positions were heavily diluted by accessions from the lower classes, who had reached them through merit. It is a curious fact that two of the most powerful tribes, the two which were most populous, namely the Cherokee and Choctaws, appear to have been ruled mainly by chiefs who had attained their positions through personal merit. Perhaps the Choctaw were the most democratic of all tribes in the section.

Except for the doubtful cases of the Chitimacha and Biloxi, tribes having totemic clans occupied a coterminous area of the Creeks, being about in the center geographically and culturally. The Timucua totemic system was almost as well developed, and so was that of the eastern Caddo. Among the western Caddo, clans seem to have been endogamous as well as exogamous, and they varied in social standing. The Cherokee had 7 fixed clans, though these were said traditionally to have been reduced from 14. Among some of the Siouan tribes four lineages were recognized, descended from as many female ancestors. As we cannot translate the names of these women, we do not know whether the "lineages" were totemic divisions or not. Among the Algonquian tribes, succession to chieftainship descended in the female line, but there were no totemic clans, though these existed among the neighboring Tuscarora and Susquehanna. The Choctaw system appears to have grown up independently of that of the Creeks. They comprised two exogamous sections, subdivided into a great number of cantons, or bands, with local names. The Chickasaw organization was originally the same but had superposed on it the totemic clans of the Creeks. Dual divisions, or moieties, were in evidence, not only among the Choctaw but the Chickasaw and Creeks as well. There is no evidence of such groupings elsewhere. The Creeks also had a dual division of towns.

[19] In aboriginal times the popular religious interest was probably in a great number of supernatural beings animating animals, plants, and other objects in nature, and the activities of these beings were intimately tied up with disease and medical practice. The tribal cults, however, give clear evidence that there was, alongside of this, recognition of a supreme being associated with the sun or sky and represented on earth in the form of fire. But since, in course of time, contact with ordinary things polluted this fire, it was necessary to renew it periodically; and this was one of the principal objects of the Creek busk ceremonial. The Natchez, however, maintained a perpetual fire, and the Cherokee are said to have done so at a former period. The renewal of the fire was, at the same time, the signal for the renewal of other things both tangible and intangible, and for a general pardon for all offenses except murder.

Among the Natchez, Taensa, and apparently the Caddo, the solar cult was intimately bound up with the governmental and social organization. Most tribes regarded the sun as male but a few, including the Cherokee, Yuchi, Shawnee, and perhaps Chitimacha, considered it female. There are traces also of a worship of the Corn Mother and certain other spirits of a general character associated particularly with the busk. This busk was always held when the flour corn of the new crop was first fit to use, and practically every tribe in the Southeast had some special ceremony connected with this event, while the Creeks and Natchez, at least, seem to have had an extended series of ceremonies lasting all summer.

Men conversant of the sacred mysteries were either "self-made," like the Creek kethlas, or prophets, the rainmakers, and so on, or members of certain native schools of doctors. These last embraced the more expert healers and also the regular priesthood such as the "firemaker," who might be described as the high priest of the town. The great ceremonial season began about April and extended to October, but the principal ceremony, which might be called the ceremony of first fruits, came when the flour corn of the new year was ready to eat, usually in July or August.

It will not be profitable, nor is it feasible, to go further into details regarding the culture of the Gulf tribes, but certain general facts may be mentioned.

(1) West of the Caddo country of northeastern Texas and the Chitimacha of southern Louisiana, and extending far into Mexico, almost

as far on the Gulf Coast as Tampico, was an area occupied by an enormous number of very small bands possessed of a very crude culture.

(2) Southern Florida, due in large measure to its semitropical climate, exhibits certain cultural peculiarities, including absence of corn and resort to foods supplied by nature, particularly the "white kunti," [20] or *Zamia,* mentioned above, and a considerable supplantation of skin garments by clothing made from vegetable substances.

(3) In historic times, over much of the territory included in our southeastern states, and apparently over the whole of it at a relatively recent pre-historic period, the population was heaviest on the coast, but in the southern part of the territory corn raising in the river valleys had proved so much more satisfactory than dependence on a fish diet that the population had shifted inland. Along the northeastern margin, in the Tidewater country of Virginia and Maryland and in part of the Piedmont area back of it, commerce had begun to play an important part in the lives of the people; a shell currency, roanoke, had come into use; and money had made some progress in moderating the infliction of corporal punishment or the death penalty in offenses of all kinds. This area, indeed, seems to show a sporadic tendency to converge on a pattern like that of the North Pacific littoral.

(4) It is evident that the culture of the central region had been markedly modified by influences and probably invasions emanating from the northwest. Whether these began outside of the Gulf area or within it, the fact seems evident. As far as we may judge, the movement, both of population and of culture, was down the Mississippi and eastward across country, in considerable measure following the course of the Tennessee River.

The above is a very rough and fragmentary review of the Indian tribes of the Southeast and their culture as known to Europeans of the eighteenth century. My next task will be to suggest a reconstruction of their still earlier history, but before that is attempted, other contributors to the conference will inform us regarding the traces of these Indians which have been recovered by the spade of the archaeologist, and on which such a reconstruction must in part be based.

The chairman, Dr. Wissler, next called on Mr. Stirling to speak on "The Pre-Historic Southern Indians."

Mr. Stirling:
It is significant that any attempt to develop a systematic procedure for archaeology in the Southeast must begin with our knowledge of the locations and movements of tribes in historic times. While early records are far from being as complete as we would like them, we nevertheless know as much about the early history of the Southeast as of any area of comparable size in the United States. Owing largely to the researches of Dr. Swanton, most of this historical information has now been synthesized [21] and made easy of access to the student. (See Appendix, Figure 7)

The method of procedure of the archaeologist should be, of course, to work from the known to the unknown. There exist many early aboriginal sites which can be definitely located and the dates of occupancy of which are known. Some of these sites have more than one period of occupancy by one or several tribes. The first problem in developing the archaeology of the given locality is to isolate the known historic cultures leaving a residue of unknown pre-historic, should such exist. Both vertical and horizontal stratigraphy can usually be applied.

In most of the culture area which we have under consideration agriculture was practiced and pottery was made. Because of its imperishable nature and variety and flexibility of form, pottery will probably always be the most instructive medium with which the archaeologist has to work. Pottery of the Southeast is remarkably homogeneous in style when the size of the area is taken into consideration. Stamped, incised, rouletted, and banded wares are common; painted pottery being much less characteristic of the area as a whole. Certain shapes are likewise characteristic of regions within the area, such as bottles, effigies, and lobed and noded ware. Of the surface decorations, scroll designs are probably the most common.

From our knowledge of the pottery used by the historic tribes, many significant hints are offered regarding pre-historic movements of peoples. It is, perhaps, quite significant that early Caddo ware closely resembles much of the pre-historic ware farther to the east, as at Moundville, Ala., Etowah, Ga., and the northern Gulf Coast of Florida. Dr. Swanton has suggested the interesting possibilities of Tunica influence. In other cases, we find interesting parallels in widely separated pre-historic sites, such as the resemblances between the so-called Hopewell pottery of the Ohio and that of Louisiana sites, such as at Marksville (a recent discovery made by Mr. Setzler). The stamped pottery of the Georgia coast and of Etowah is strikingly similar to the stamped ware on the northwest Florida Gulf Coast, where it merges with the Caddolike pottery of the west.

In developing archaeological areas, certain factors which are not material in nature must be taken into consideration. For example, we must look for associations within the different linguistic groups which we find in the Southeast and must also consider their social and political organization, studies about which frequently give us information concerning secondary visible traits of material culture, as, for instance, the square grounds of the Creeks, the council houses of northern Florida, mortuary temples, etc.

[22] The difficulty of defining a general culture area is obvious. A culture area after all is an arbitrary and artificial device whereby a certain region characterized by distinctive traits is set apart for purposes of consideration. We should not let this spoil our perspective on the interrelationship of cultures as a flow rather than as a series of static jumps. It is only to be expected that certain traits characterizing any region are likely to merge into marginal areas until the problem arises as to where we must stop and at which point we are to draw the limits of the area which we have under consideration. It is quite possible, however, to recognize in the Southeast a general area which may be definitely contrasted with other areas of similar extent, as, for example, the Southwest. Local developments which may be assigned to certain areas can be recognized as definitely characteristic of limited areas within the general region and in some instances can be applied to known tribes.

The pioneer archaeological work of Clarence B. Moore contributed a great deal of information concerning the horizontal distribution of characteristic types of artifacts for the Southeast, and at the beginning of the present century Dr. William H. Holmes produced a general synthesis of pottery types for the eastern United States, recognizing five principal areas. With vastly more information at hand, it is now possible to segregate these areas to a much more accurate degree.

The more remote sources of cultural influences in the Southeast today can only be speculated on. It appears safe to assume that the general agricultural pottery-making culture prevalent in the area is southern in origin. Most archaeologists recognize an affinity with Mexican culture and a rather surprising lack of affinity with the Southwest. Similarly, in spite of the proximity of Florida to the Bahamas and Cuba, we find definite ties with Antillean culture possibly lacking completely. Algonquian and Iroquoian influences from the north are prominent, particularly as would be expected in the northern part of the area. Influences directly attributable to the typical culture centers of the Southeast carry us as far west as eastern Texas, Arkansas, and Missouri and north up the Missouri to North Dakota, up the Mississippi to Wisconsin and Minnesota, and north and east up the Ohio to western Pennsylvania. On the East Coast, these influences are felt as far north as Virginia.

Within the area of typical southern culture, we have tentatively outlined 13 archaeological areas characterized by recognizable traits. It is not, of course, possible to actually outline culture areas on a geographical basis. Sites representing the same culture can be found at such widely separated points as Arkansas, Moundville, and Crystal River, Fla., or as Mr. Setzler has recently shown, between Marksville, La., and Hopewell, Ohio. These distributions have their own significance.

[23] If we draw a line across the Florida peninsula from a point south of Tampa Bay to Cape Canaveral, we find the area south of it to be a low subtropical region characterized by a lack of agriculture. A rather large littoral population is indicated. Although pottery was made through this region, it is mostly of a crude, heavy, sand-tempered variety, characterized by very simple decorative motifs. Lake Okeechobee might well be considered the center of this area. In aboriginal times a large population, presumably living mainly on wild vegetable products from the lake, constructed large geometric earthworks along the line of the Everglades, where they maintained large communities. These sites are characterized by a rather high mound constructed at one end of a rectangular court surrounded by embankments and a large semicircular embankment enclosing the mound. Historical information on these interior peoples is almost entirely lacking.

The northern Florida coast of the St. John's River area again has a characteristic archaeological culture. Large shell mounds with occasional effigy mounds, rather crude, heavy pottery of a type not found outside the area, copper, and check-stamped designs on pottery are conspicuous characteristics of the region, which shows some Georgia influences.

The northwest Florida coast constitutes another distinctive area, the characteristic artifacts of which are pre-historic. The pottery is a well made banded ware similar to the Caddo pottery of the west in its general appearance and characterized by negative designs. Mortuary pottery is typically "killed." Along with this pottery we find stamped ware almost identical in type with that centering in Georgia. We apparently have here a convergence of two different cultures from the north.

From northern Florida up to and including southern and western North Carolina, we have another area characterized by a specialized development in stamped ware. Concentric circles and rectangles, as well as complex geometric designs, are the characteristic units employed in stamped-pottery decoration. It is possible that this is likewise the point of origin of the widely distributed checked-stamped pottery. Along the Georgia coast we find a center for cord-marked, paddled pottery, which extends west to the Mississippi. This latter type of design was used by the historic Cherokee. Urn burials, cremations, and dog burials are another outstanding feature of this south Atlantic area, which suggests Iroquoian influences.

In the mountainous area extending from southern West Virginia to and including southeastern Tennessee and southwestern North Carolina, we find the Cherokee area, a region which is not particularly homogeneous in culture, as it tends to borrow cultural traits from its marginal areas.

[24] Southern Kentucky, western Tennessee, northern Alabama, and eastern Mississippi constitute what Moorehead calls the Cumberland-Tennessee area, one of the most interesting and complex of all of the subculture areas under discussion. Within this region we can find representative artifacts illustrating practically all of the types in the area. It was a region of a large aboriginal population, and archaeological remains are very abundant throughout.

In the region between the upper waters of the Pearl River and Tombigbee River of southern Mississippi, we find the early culture of the Choctaw Indians distributed among the more conspicuous archaeological sites.

From the delta of the Mississippi to the Apalachicola River we have another Gulf littoral culture which is an extension of and merges into the northwest Florida littoral culture, but which has certain traits which set it apart.

From the delta of the Mississippi to the mouth of the Ohio River, we have a region extending along both sides of the Mississippi which might be called the effigy-pot area.

The region of southwestern Arkansas, northwestern Louisiana, and eastern Texas is characterized by archaeological remains of the Caddoan Indians. Because of the striking similarities of this Caddo ware with much of the characteristic ware in the southeastern area generally, it becomes one of the most significant of all of the subculture areas on account of the known affiliations of this type of pottery.

Certain of the large mound groups, such as Moundville, Ala., Etowah, Ga., and Macon, Ga., might possibly be set apart as type areas of their own. Mr. Collins has pointed out certain suggestive comparisons between Moundville and the Natchez.

In eastern Kentucky we have another distinctive area made known principally through the work of Webb and Funkhouser, which has been designated as a pre-Algonquian area, which may indicate along with the Ozark bluff dwellers a very early substratum of culture.

The boundaries are not fixed sharply for any of the subareas above outlined, but they each offer certain groups of traits which make them instructive for any regional study of pre-history. The fact that there is so much overlapping of traits makes it profitable to consider at this time some of the general distributions found within the area as a whole. We might consider first the subject of the mounds. Roughly, mounds might be classed under three different headings: domiciliary mounds, burial mounds, and refuse heaps. The functions of these did not always remain distinct. Burials are sometimes found in refuse heaps or in domiciliary mounds. Refuse heaps were frequently flattened and altered so as to serve as domiciliary mounds, particularly in the case of many of the Florida shell mounds. The material of which mounds were constructed depended principally on the region. In Florida, artificial mounds were generally constructed of sand. On the coast, the shell mounds, of course, were more or less accidental deposits of refuse. In the north, burial mounds and domiciliary mounds were typically constructed of earth. Rocks were rarely utilized, and masonry is, of course, entirely absent.

In Florida and the adjacent areas, the dwelling sites were shell heaps, more or less artificially constructed, or in some cases sand or earth. The burials were interred in separate sand mounds. In the lower Mississippi Valley the dwelling sites were raised earth mounds. Burials were sometimes in these and sometimes in adjacent fields. North and east of Florida the distinction between burial sites and mounds becomes quite vague. Burials were apparently in parts of the dwelling mounds. The large mound groups so characteristic of the upper Mississippi extend as far south as Alabama and Georgia, where they are found at Moundville, Ala., Etowah, and Macon. Effigy mounds have about the same range as the large mound groups. Outside of Florida and the Gulf Coast, the domiciliary mounds are generally rectangular and flat-topped, often with a ramp. Burials were not characteristic in these larger mounds. These groups of exceptionally large rectangular mounds extend as far north as Cahokia.

Copper had a universal, though rather limited, use throughout the area. Its employment for utilitarian purposes, such as for axes, decreased from the upper Mississippi Valley, where many are found, to Florida, where none occur. By the time copper reached the southern part of this territory it became so valuable that it was only used for ornamentation, usually as a coating over wood, stone, or some other substance. For this purpose, it was pounded out into sheet form. Used by itself, it served ornamental purposes in the form of plates or discs of sheet copper with designs embossed on them, or in some cases, as along the Tennessee River or northward, reel-shaped ornaments were cut out. Designs were cut out at Moundville. One of the most common uses of copper was for ear plugs. In the Moundville region, triangular pendants were cut out of sheet copper. Another common use of sheet copper is in the manufacture of tubular beads, which are found throughout much of the southern area.

The type of ear plug most commonly used was the disc type. This material is rather ambiguous as used by Moore and probably includes several types. One variety found in Florida consisted of two perforated discs tied to the ear by a cord. In Arkansas, Tennessee, and Florida, they seem to have consisted of discs with an upper rim which hooked into the hole in the ear. These were constructed frequently of other materials that were copper coated. Spool-shaped ear plugs of copper so characteristic of the upper Mississippi Valley occur as far south as peninsular Florida but are [26] rather uncommon in the southern part of the region. Pin-shaped ear plugs of shell or pottery characterize the middle Mississippi and Tennessee regions. Another type of ear plug consisted of two pieces of wood hollowed out and filled with pebbles. The outside was coated with copper. These are the so-called pod-shaped ear plugs of Moore. Their area of distribution is within that of the pin-shaped types.

Chunky stones, or discoidal stones, and pottery discs cover the entire area with the exception of peninsular Florida. In Arkansas the sherd discs often have central holes, although both perforated and unperforated types occur together.

Circular shell gorgets are one of the characteristic items of the area. These can be divided into six classes on the basis of design. Those with cross designs are found in Illinois, Missouri, Arkansas, and Florida. The type of scalloped-shell disc is very abundant in Tennessee and is also found in Arkansas. These peripheral scalloped patterns resemble either cogs or the petals of a flower. The third type of design is a conventionalized serpent found principally in Tennessee and Georgia. Ornaments with incised faces on them come from Arkansas, Tennessee, and Virginia. From Moundville, Ala., and Etowah, Ga., come shells incised with human figures. Another interesting design is a four-sided figure with star and crested woodpeckers. In southern Alabama, Georgia, and Florida, these shell ornaments are much more crude, and decorations consist of rude incisions and holes cut through the shell. It is obvious that this region is marginal and degenerate as far as this type of artifact is concerned. There is an interesting association between shell gorgets of the best type and the embossed figures in sheet copper. Suggestive of the incised shell discs are the stone discs or paint pallets which occur at Moundville, Ala., Etowah, Ga., and are also found in southern Illinois.

Tobacco pipes constitute another interesting case of distribution. The curved-base mound type of pipe is found in the upper Mississippi Valley region and extends into Virginia. In the Mississippi Valley, a type of platform pipe was developed by forming projections from the bowl of the pipe. This type may be found across the central part of the region to and including Georgia. A specialized middle Mississippi Valley development is a type of elbow pipe resting on a flat base. In the Caddo region the pipes were made long and slim with a stem projecting for a distance beyond the bowl. In several sites along the northwest coast of Florida, Moore mentions finding what he terms monitor pipes. The only one that he illustrates is strongly reminiscent of the Ohio mound type. Large stone effigy pipes are found in all the region except Florida and adjacent areas. This type of pipe reaches its greatest development in the region of Tennessee and extends over into Arkansas. Biconical pipes are found generally in the region south of the Ohio River.

[27] What McGuire calls the southern mound type—a pipe shaped in a similar way to modern pipes—has a rather narrow stem and large bowl. This is to be found in Tennessee, Georgia, and the Carolinas. The block-shaped pipe is found to be centered around the middle Mississippi Valley region. These pipes properly fall under McGuire's classification of biconical pipes. Tubular pipes are listed by McGuire as being found in Kentucky, Virginia, Tennessee, Georgia, and the Carolinas. A specimen from Crystal River, Fla., may be one of these, but there is some doubt as to

whether or not it is a pipe. Discoidal pipes are found in the general region of the states adjoining the junction of the Ohio with the Mississippi River.

It might be well, on account of the importance of pottery, to go into a little more detail as to some of the general features of southern pottery. Shell tempering is more or less universal in the area. Sand tempering was not so generally used, although it is quite characteristic of the Georgia coast and the northwest coast of Florida. Fiber-tempered ware is found throughout Florida, where it is a characteristic development. Various pastes are used baked into gray, yellow, brown, and red wares. Of the color types mentioned by Moore, only his black ware seems to indicate much significance as a cultural trait. Through polishing and coming into contact with carbon while being baked, the pottery takes a shiny black finish. This ware was quite characteristic of various parts of Arkansas and is found also in Alabama, northwestern Florida, and Georgia.

The use of knobs for decorating pottery has had a fairly definite distribution extending from Arkansas through southern Tennessee and into Alabama. Scalloped and notched rims are found in the middle Mississippi region and along the northwest coast of Florida. More important are the methods of decoration more generally utilized. Incising is universal throughout the area. Punctate decorations are also found all over the Southeast. The two combined, however, are much more important in Florida and the Gulf Coast than they are in the middle Mississippi region. They were little used on the Georgia coast where stamped decorations predominated. Trailed designs reached their peak in the middle Mississippi region, where they were used for making scrolls, spirals, concentric circles, etc. Two distinct methods of manufacture took place here. The Natchez used a single instrument, whereas the Choctaws formed their trail designs with combs.

A feature of northwest Florida pottery is the development of a negative type of design formed by incising and tattooing with the open areas frequently broken by incised lines terminating in a small circle or triangle, a feature which may have a dubious Antillean connection. The untouched portions of the surface of the pot form the design. This style also extends up into the middle Mississippi region, although it does not have the importance here that it possesses in Florida and the Gulf Coast.

[28] Stamped ware is a feature of Florida and southern Alabama and Georgia. The most common and most widely distributed type is the small check stamp. What Moore terms the complicated stamp is most charac-

teristic of the Georgia coast and is also quite common on the northwest Florida coast. These stamps are in curvilinear and rectilinear designs of many varieties. Cord-wrapped paddle marking centers about the Georgia coast but occurs somewhat generally through the Gulf area.

Colored pottery has an interesting, although not very extensive distribution throughout the region. We find red slip being applied to pottery in the middle Mississippi Valley, the Gulf Coast, and Florida. It reaches its greatest development in the middle Mississippi region. In this area, much of the pottery is found with a red slip over the entire vessel. The occurrence of this ware in Florida is sporadic and usually consists of a single specimen here and there or a few sherds. It occurs most frequently in the northwest coast region, where it seems to have taken on the special function of being used exclusively for mortuary ware. The same applies to designs in red, which are very rare in Florida, and where this occurs it is usually on mortuary ware. These designs in Florida, as a rule, consist of bands. In the middle Mississippi region we find a greater use of color and the frequent use of polychrome designs. An interesting specialization in southern Arkansas is the use of red and white pigments and incised lines on black-ware pottery. A few examples reminiscent of this technique were found at Point Washington on the northwest Florida coast where kaolin and a pink substance, probably hematite, have been rubbed into incised lines. A unique type of pottery decoration is the use of green paint. It is uncertain just what this green paint indicated. It was not baked on the pottery, according to Harrington, but was apparently rubbed on after baking or perhaps applied just before being deposited in the graves. Moore also discovered this in certain parts of Alabama, and one instance of it was found in Kentucky and another in Louisiana.

The use of low relief as a means of decoration is fairly extensively used in the Arkansas region. In Florida a crude type of this relief decoration occurs in mortuary pottery and consists principally of bird heads and animal designs in crude low relief.

Certain design elements are likewise intriguing in discussing connections between different areas. One of the most important design elements in the middle Mississippi region consists of scrolls. These are made by both trailing and painting and also occur in negative designs. The same type of design has spread to the northwest Florida coast. Concentric circles show a more limited occurrence within the same region as the scrolls. They are also frequently utilized as one of the more common of the complicated stamp patterns of Georgia and Florida. The swastika is a fairly common design element in the middle Mississippi [29] region and extends as far east as Alabama. Step designs are found in the Arkansas region and more rarely in the southern Alabama, Mississippi, and Gulf Coast regions. Definite serpent designs are found in several places throughout the region, but probably a more conventionalized form is much more general, although not so easily recognized. Incised feather-serpent designs are found at Moundville, Ala., the Hollywood mound in Georgia, and from Arkansas. There is a possibility that certain of the scroll decorations may have been a conventionalization of serpent designs. Another design characteristic of the middle Mississippi region is the star design on pottery. The human hand or hand-and-eye design is very prominent at Moundville, Ala. This motif is also found near Apalachicola, Fla., Crystal River, Fla., and at Naples, Ill., indicating a rather wide distribution of a highly specialized item.

In the northern Florida region, a special type of pottery vessel containing from two to five separate compartments was found. The only one of these four types that really forms a definite type is the five-compartment vessel, which occurs in the northwestern part of Florida. These are formed of four compartments of equal size encircling a central compartment elevated above the other four. All of the compartment vessels are relatively shallow. Flat bases on pottery vessels, while widely distributed, are nevertheless localized to individual sites here and there. The only area where they occur very frequently is in northern Louisiana.

A specialization of the northwestern Florida coast is the construction of small five-pointed dishes. Similar to these and in the same general area are vessels with flaring four-pointed rims. The use of flaring rims, however, extends farther and includes Alabama and Georgia, and there is one example from Arkansas. Vessels constructed with four lobes, or swellings, in the body of the pot are found occasionally throughout the area. On the northwest Florida coast we find occasionally three-lobed vessels. Quadrilateral and trilateral vessels are confined to the northwest Florida coast. In Florida, vessels when so equipped have four legs. In the middle Mississippi region they have three legs which take on a variety of shapes.

Bottle-shaped vessels are limited to the middle Mississippi region and adjacent areas, extending as far eastward as the Etowah region in Georgia. The teapot type of vessel is characteristic of Arkansas and adjacent regions. The specialized form of bowl with two cream-pitcher spouts occurs along parts of the Tennessee River. These are usually decorated with knobs.

Animal-effigy bowls are found possibly in the central Mississippi region and in Florida. These two areas have effigies of different types, however. Those of the Mississippi region, in addition to being of much [30] better workmanship, are more numerous and take the forms of turtles, frogs, and various animals, including human beings. In Florida we find mostly crude bird effigies, and these as a rule are mortuary vessels. In northwest Florida, we find well made human faces in relief on the sides of pottery vessels not of effigy shape. In Arkansas is an interesting development of the human-head effigy vase. The type of bowl having a bird head projecting on one side and a tail on the other shows the same distribution as the effigy vessels. These extend also into Florida, where the bowls are shallower and wider than they are in the north.

Toy or diminutive vessels have a wide though sparse distribution. These may have been made by children, while others may have been a type of mortuary ware.

The commonest method of constructing pottery for suspension was the making of loop handles on it. This method was universal throughout the area but rare in peninsular Florida. The use of holes in the rim was restricted to the Mississippi region across to Florida. Handles formed by projections outward from the rim are scattered throughout the region. The same situation holds true for knobs projecting outwards to serve as handles. Instances of the distribution of artifacts of this kind could be prolonged more or less indefinitely.

Before concluding my topic, I should like to mention one other important item which should be considered by the archaeologist. This is the matter of skeletal material found in the mounds and cemeteries. All too frequently, investigators have not had the patience to remove poorly preserved skeletal material on account of its fragile nature and thus have destroyed in many places one of the most valuable of our indications concerning the former inhabitants of the region, namely, the remains of these inhabitants themselves. In addition to normal physical variations within the area, there are interesting occurrences of special types of head deformation which occur both historically and pre-historically and which in the future may give us especially valuable information in bringing about cultural ties.

The problem of ancient man in the Southeast is one which had perhaps best be avoided at this time. As in all parts of the country, there are occasional suggestive finds indicating the association of human artifacts or human remains with those of extinct animal forms. Such finds have been reported most frequently from Florida and Tennessee, but as yet the evidence does not seem sufficiently convincing to enable us to accept any of these without considerable reserve.

It is perhaps in the cave regions of the Ozarks and of Kentucky that we may hope to find the most ancient of our human remains, finds in dry caves being particularly valuable because of the fact that organic [31] materials are frequently preserved, whereas in the relatively humid climate of the Southeast, this is not possible in the open sites.

Insomuch as Dr. Swanton is taking up the subject of key historical sites, I have limited my discussion principally to general distribution areas and to problems of pre-history, which, of course, can never be entirely divorced from the problems of the historical Indians. This general synthesis is meant to be merely suggestive. It is perhaps yet too early to make any attempt at a final defining of pre-historic culture areas for the Southeast, such as has been done so well in the southwestern part of the United States, but a definite beginning has been made.

Afternoon session, Mr. Brannon, chairman

Mr. Brannon called on Mr. Dellinger to report on recent archaeological work in the state of Arkansas as the first in a series of reports of recent work in the southern states.

Mr. Dellinger:
Since the main outlines of archaeology in the eastern and southern divisions of my state (Arkansas) are fairly well known, I intend to confine my remarks to the recent work in the bluff shelters in the northwestern part of the state. The problems raised here may possibly assist some of my friends in understanding the perishable materials which once existed in their respective areas.

The region under consideration lies in the northwest part of Arkansas and southwestern Missouri in what is usually spoken of as the Ozark uplift. Here the topography is of course quite hilly and sculptured out into deep valleys and steep hillsides. The surface of these hills is composed of Boone limestone, one of the Pennsylvania series. This overlays a soft shale. Due to weathering, a number of shelters or caverns have been formed, ranging anywhere from several hundred yards in length and a hundred feet in depth to 15 or 20 feet by 10 or 12 feet in depth. In the ashes and dry dust beneath these shelters we found a very favorable place for the preservation of all the objects used in the everyday life of the bluff dwellers.

In the back of the shelters we find even the beds of leaves and bluestem grass (*Andropogon furcatus*) where these people slept and under which they buried their dead. In the burials we have found babies still resting on their cradles, having been disturbed only by the gnawing of [32] wood rats. Their cradles were made of the stems of ironweed bound together with bark. In some cases all the trappings of the burials are still intact, showing the buckskin moccasins and leggings and the feather-down blankets. In other instances the babies were carried on cradles made of split cane wickerwork. In these burials certain smaller wickerwork objects seem to have served as diaper boards. When the infant was larger, the mother seems to have placed him in a carriage of bluestem grass, which was slung over her shoulder, so that the baby could accompany his mother as she went about her agricultural tasks.

The clothing of the Ozark bluff dweller consisted of woven sandals of rattlesnake-head grass (*Eryngium yuccifolium*), a belt of the fibers of Indian hemp (*Apocynum canabinum*) around the waist, and attached to this belt a loincloth made of a bunch of twisted bluestem grass. Their clothing in the winter consisted of buckskin moccasins with puckered toes and a drawstring around the ankle. These moccasins resemble very closely the drawstring tobacco pouches of the present time. During inclement weather, heavy overshoes of canary grass (*Phalaris*) were worn over the moccasins. Fragments of buckskin leggings and shirts have also been found. Over this, in very inclement weather, a cloak of feather-down cloth was worn.

Many awls and bone and wooden needles for the manufacture of these objects have been recovered. Some of the awls have the ends wrapped to protect the hands. The women made a great variety of baskets out of splints made from the whip cane (*Arundinaria tecta*). These baskets differ greatly in size and shape, ranging from an inch or two in diameter to large hampers holding more than a bushel. The baskets are made of quartered cane. Usually the outer bark of the quarter was removed for the finer baskets, and the coarser ones were made out of the remainder. These baskets were used as sieves and for other such utilitarian purposes as storage and winnowing. Some of the better ones have designs put on with some red juice, perhaps from the pokeberry. Many coiled baskets resembling those from the Southwest, and also some with stitches similar to those used by the Pawnees on their gambling baskets, were present. A great many bags made from Indian hemp and rattlesnake-head grass have been found to contain seeds for planting and for food. Most of these baskets have a sort of drawstring arrangement at the end.

The fragments of bags and baskets were often used to line storage pits, in which were placed the various agricultural products used for food. I am indebted to Dr. Melvin R. Gilmore of the University of Michigan for the identification of these specimens. The foods ordinarily found are of the following cultivated varieties: flint and dent corn, summer and winter squash, pumpkins, sunflowers (*Helianthus annuus*), beans (*Phaseolus vulgaris*), large bottle gourds and small egg gourds, seeds of the giant ragweed (*Ambrosia trifida*), *Chenopodium nuttaliae,* and stores of *Iva ciliata.* The smaller seeds were quite often stored in gourds. [33] However, they are occasionally found in bags. In addition to these cultivated varieties, we find great quantities of hickory nuts, walnuts, acorns, chinquapins, persimmon, and other native seed. The Indian women used either the shell hoe or the crooked digging stick to cultivate their crops.

The animal food of these people consisted of practically all the native animals found here today, including even the grasshopper. The deer and turkey seem to have composed the greater part of this food, however. A great quantity of fish bones and skeletons of the species found in the streams of this region have been located in the camp sites. Occasionally a bone fishhook or piece of fishnet or even the remains of a cane fish basket are found. Since we have no evidence of the presence of the bow and arrow in the lower layers, it seems that the larger animals must have been taken with the aid of spears, the foreshafts of which have been found in practically every site. Sometimes the fragments of an atlatl of the primitive type discovered by Harrington have been found. Several coils of rope about three-quarters of an inch in diameter with a stick attached to one end and looped at the other have been located. These may have been used as snares for game. The presence of nets resembling bird snares would indicate that small game was probably taken in nets.

These foods may have been eaten in wooden bowls or in the shells of the highland terrapin (*Testudo triungis*). Many fragments of wooden and bark bowls have been secured. The polished shells of the terrapin occur in almost every site. Many of the coiled baskets seem to have been badly charred, which together with the presence of paddle-shaped sticks, suggests that the grains may have been parched by dropping hot rocks on them in the baskets. Mortars and grinding stones are present in each shelter. Hominy holes, as described by Webb in Kentucky, are very rare.

The discovery of reed flutes indicates that these people had some time left after they had provided for their ordinary maintenance. Their esthetic sense was also taken care of by the practice of adorning their bodies with shell beads, gorgets, and also beads made from seeds of the Ozark gromwell. The presence of quantities of red ocher suggests that they probably painted their bodies, as well as the backs of their shelters. Many of the walls show crude pictographs representing men, deer, turtles, lizards, snakes, and beavers. One sandstone shelter contains a great number of petroglyphs. These are quite unusual. Most of them are angular in outline. Unfortunately, the shelters with pictographs have proved a strong drawing card for seekers of Spanish gold. There are many legends among the hillfolk of fabulous wealth buried beneath the shelters by De Soto and his followers. A divining rod and a week's "grub stake" are sufficient to start gold fever at any time.

[34] The Ozark bluff dweller was buried at the back of the shelter behind the fallen rocks in places similar to the beds, if not directly beneath them. The burials are of people of medium height with long, narrow heads. As a rule, the men are somewhat stockily built. The common burial custom was to place the body in the flexed, or knee–chest, position and enclose it in a bag of feather-down cloth. This is then tied firmly at the opening and usually about the middle and ends with bark and strong cords. The burial is placed in a pit lined with bluestem grass. As a rule, no food is placed with the body. However, in one baby burial we found 11 ears of corn, about a quart of sunflower seed, and a few acorns and chinquapins. In a burial of a dog about a quart of dried beans in the shells was recovered. These, however, are exceptions rather than the rule. In one region, the burials were made beneath cane baskets. In every instance, just in front of the left shoulder was placed a corn cob with a cord looped around it. Spear points, knives, corn, and fragments of gourd were placed with these bodies. The presence of partially burned human bones in many of the burials suggests cremation. However, the habit of placing ashes over the body may indicate that the burning may have been accidental. The ashes may have contained sufficient live coals to have ignited the grave furniture, thus burning the body.

At present this culture seems to be the oldest in the South. The entire absence of tobacco, pipes, pottery, and the bow and arrow tends to substantiate that statement. We have located only one celt, and that, we are inclined to think, was an intrusion from a later culture. No polished axes have been found in any of these sites. Chipped flint work is very scarce, consisting largely of crude willow-leaf shaped knives and dart points with square bases. The presence of the sunflower, flint corn, coiled baskets, and woven moccasins rather suggests a southwestern influence. Dent corn and cane basketry, however, are characteristic of the southern Indian. In the top layer of some of the sites in the southern part of the Ozarks we have found pottery cooking vessels resembling the urn-shaped vessel of the Caddo to the south. However, it is not believed that these belonged to the bluff dwellers.

Mr. Brannon next called on Dr. Jones to report on recent work in Arkansas. (Dr. Jones stated that he had been assisted in the preparation of his report by Mr. DeJarnette.)

Dr. Jones:
For the past five years, the Alabama Museum of Natural History has taken a very definite interest in the celebrated aboriginal site at Moundville. Funds were raised largely by private donation for the purchase of the 175 acres of land comprising the mound tract, and the unencumbered title to the property is now vested in the museum.

[35] The type locality and center of the Moundville culture is located on a high, essentially level plain, known as the University Terrace, between the town of Moundville and the Warrior River, in adjacent parts of Tuscaloosa and Hale counties. A number of outlying sites in the vicinity have also been definitely assigned to the Moundville culture. The results outlined in this paper are based on the extensive researches of Clarence B. Moore, of the Academy of Natural Sciences of Philadelphia, and those of the last four years carried out by the staff of the Alabama Museum of Natural History. Moore spent several field seasons in Alabama, two of which were at Moundville, in 1905 and 1906. The work of the Museum has been at four sites on the Tennessee River, two on the Chattahoochee, five on the Warrior and one in the Mobile delta. The authors wish to acknowledge the splendid cooperation of the National Research Council, without which this year's fieldwork could not have been carried out.

The Moundville culture was highly specialized, particularly in the design and finish of objects of clay, stone, shell, bone, and copper. Any sites properly referable to this culture should show positive evidence of these arts. The culture is characterized by medium to large mounds of the truncated-pyramid type. The bases are normally square, but sometimes elongate. There are 34 mounds in the central group at Moundville, 18 of which form a hollow square. The highest is B, 58 1/2 feet high, and the longest structure is the plateau just north of B, which is about two acres in extent. Some of the smaller structures have suffered considerable damage from erosion and cultivation, while six have been entirely obliterated. Truncated pyramids of the Moundville type occur at Florence, outside of the limits of the culture; at Hobbs Island in the Tennessee River; and at Bottle Creek, in the heart of the Mobile delta. All three of these sites were investigated during this year, resulting in the tentative conclusion that the Tennessee River sites belong with the Middle Mississippi or Tennessee-Cumberland cultures, whereas an entirely new designation will probably have to be made for Bottle Creek.

Mounds. The mounds belong to the domiciliary or ceremonial class and were never used for burials, except occasional superficial interments. They were constructed of clay, which was brought in from adjoining areas and not taken from the plateau. Mound C showed two periods of construction, and it is likely that the rest of the larger ones will show two or more periods.

Burials. The method of burial was in the flesh, in pits just large enough to accommodate the body, and the vast majority of the 1,600 skeletons removed by the museum were fully extended on the back. The depth at which burials were encountered ranged from a few inches to approximately 7 feet below the present surface. In some places pits were so numerous that individual ones were difficult to determine, while the site northeast of N carried five different levels of interments. Occasionally bundle burials were encountered, and frequently aboriginal disturbances [36] were noted. It is not certain that the bundle burials were due to aboriginal disturbances, but they were so rare that they do not require special consideration. Moore reported occasional cremations. It is the opinion of the authors that this was an exception to the rule. Apparently no deformation of the skull was practiced by this people.

Objects associated with burials. Evidently, objects constituting the personal property of the individual were deposited with the remains. Such objects were usually placed back of or near the head. The list is a large one but might be summarized as follows: clay water bottles, pots, and bowls;

pipes, discs, and discoidals of clay and stone; beads of shell, clay, and copper; copper and stone ornaments; ceremonial axes of copper and stone; stone celts; and other implements.

Smoking. The comparative scarcity of pipes would imply that smoking was not very common among the Moundville people. These are of two distinct classes: small ones for individual use and large, elaborately carved ones for tribal or ceremonial purposes. The smaller types may be of clay or stone, while the latter are invariably of stone.

Pottery. It is not so much the manufacture of pottery, although a remarkable percentage of it is of excellent ware, but rather the intricate and delicate incised designs which give to the Moundville culture such a distinct and interesting pottery type. As a rule, the incised lines are so delicately executed that one can scarcely feel them with the fingers. Water bottles carry the most elaborate engravings and, in fact, undecorated bottles are rare. Bowls are frequently adorned with splendid designs, while pots are normally plain, except for handles and occasional effigy shapes. The pottery is shell-tempered and of several types of ware.

Stone objects. In this division, we have the well-known "rattlesnake" disc, beautifully fashioned from fine-grained sandstone. Stone discs are characteristic of the Moundville culture, some 60 having been recovered by the museum and by Moore. Other objects of stone include ceremonial axes, celts, pipes, discoidals, pendants, arrowheads, hammer stones, and other implements.

Copper. Objects, mostly for personal adornment, were occasionally wrought from native copper. Among the lot, ear plugs consisting of discs of copper-coated wood were the most widely used. Copper pendants, gorgets, and beads were sometimes found.

Paint. The use of paint must have been common, judging from the amount of it found in the burial pits. The usual pigments were red (ground hematite and ocher); yellow (ground limonite and ocher); white (lead carbonate made from galena); black (graphite); and green (glauconite and one small lump of unknown material).

[37] *Relative age.* The sites of the Moundville culture must be considered pre-historic, for no objects showing European contact have ever been found at any of them, even on the surface. Camp debris often extends to depths of 4 feet or more, and disturbed soil sometimes reaches depths of 6 to 10 feet. The building of the mounds required an immense amount of time and labor. The presence of both extinct and living species of shells is

of some significance. The bones range from well-preserved skeletons to mere lines in the soil, even the crowns of the teeth having completely disappeared in some instances. As yet, the authors are unprepared to assign a definite date to the Moundville culture, dismissing this question with the statement that it appears to be the oldest in Alabama.

Mr. Collins was next called on to report on the present condition of archaeological research in Mississippi.

Mr. Collins:
The state of Mississippi was in early historic times the center of a native population conservatively estimated to have numbered between 25,000 and 30,000 individuals. Included within the present boundaries of the state was the greater part of the territory of the Choctaw, Chickasaw, and Natchez, the three largest groups, with the exception of the Creek Confederacy, of the tribes comprising the great Muskhogean family. The Chickasaw, whose territory extended into western Tennessee, had their principal settlements in Pontotoc and Chickasaw counties, in the northern part of the state. The Natchez, who figured so prominently in the early history of the territory, lived in Adams County, on the Mississippi. The most numerous of all the Mississippi tribes were the Choctaw, who occupied the southern half of the state and adjacent parts of Louisiana and Alabama. Three smaller and less important Muskhogean tribes were the Chakchiuma, Taposa, and Ibitoupa. There were also representatives of the Siouan and Tunican stocks. The Siouan groups were the Biloxi, a small tribe on the Gulf Coast, and another small group, the Ofo, on the Yazee. The Tunican group consisted of six small tribes in western Mississippi, the Tunica proper, Yazoo, and Koroa on the lower Yazoo River, and the Tioux, Grigra, and another village of the Koroa farther down the Mississippi in the Natchez country.

Our knowledge of the ethnology of the Mississippi Indians is based almost entirely on the work of Dr. John R. Swanton, whose careful researches have thrown much light on the linguistic and cultural affinities of the Muskhogean and other southern stocks. There yet remains the task of determining the limits of the various groups in pre-historic times—their relations one to another and to other southeastern groups—an undertaking [38] that as yet has been hardly begun. Although Mississippi is rich in aboriginal remains and a considerable number of these have been investigated, it cannot be said that the work done has clarified to any great extent the archaeological problems involved. The early investigators, in accordance with the unfortunate tendency of the time, too often proceeded on the assumption that the accumulation of specimens was an end in itself rather than a means toward the elucidation of archaeological problems.

The most important immediate problem of Mississippi archaeology, as of the Southeast in general, is to establish a basis for a chronology of pre-historic sites. From the fragmentary nature of the available evidence, this will have to be for the most part a disjointed and patchwork chronology, far less perfect and comprehensive than that which has been worked out in other areas, notably in the Southwest, where ruins of all periods are well preserved and where at times even such perishable materials as basketry, textiles, and wood are found; and where in addition there still exist native tribes whose customs, social structure, and economic activities continue along much the same lines as those of their pre-historic ancestors. The task of working out a chronology for southeastern archaeology will be much more difficult, and there is therefore all the more reason for painstaking examination and study of such aboriginal remains as are still available. The obvious beginning toward such a study, as toward any other, is to start with the known and work back toward the unknown to determine wherever possible the nature of the remains left behind by the historic Indians, most of whom have long since disappeared or been removed to reservations. Practically, this means locating exactly from historical sources the sites of old Indian villages and collecting what may be available for comparison with similar material from sites of unknown age.

Anyone who has examined an abandoned Indian village site of the historic period soon realizes that he is faced with a paucity of material, usually for the reason that such sites have been under cultivation for many years. As a rule the surface will show only a scattering of potsherds and flint implements and rejectage. Fortunately, however, potsherds are of decided value as chronological determinants and, if present in sufficient quantities to show the entire pottery range of the site, are of far more significance than a number of complete vessels which might not happen to show such a range. In fact, the obliterating effect of white civilization has reached such a point that at many aboriginal sites potsherds are the only really useful material that the archaeologist is able to salvage. The lowly potsherd thus seems destined to bear much of the weight of the chronology that we all hope may sometime be established for southern archaeology.

In 1925, by utilizing the pioneer work of Henry S. Halbert, I was able to locate and make collections from certain historic Choctaw [39] village sites in eastern Mississippi. The result was the determination of the historic Choctaw type of pottery, on the basis of which comparison with pottery from sites of unknown age is now possible. A few years later similar work was undertaken for the Mississippi Department of Archives and History by Moreau B. Chambers and James A. Ford, who were able to locate certain historic Natchez and Tunica sites in western Mississippi. In a forthcoming paper by James A. Ford the potsherds from these historic sites, as well as those from neighboring sites of unknown age, are to be described.

Choctaw pottery bears a decoration of straight or curved bands of finely incised lines applied for the most part to the upper part of the vessel. The bands, formed usually by five or six fine lines, were produced by trailing a comblike implement across the surface. No entire vessels have been found, but the shapes of the sherds indicate a low, rounded bowl as the prevalent form. Aside from a very few sherds bearing crude and nondescript designs, this banded decoration is the only one found at the historic Choctaw village sites examined. It appears to have been a strictly local type; I know of only one instance in which it has been found elsewhere: a single sherd from a historic Natchez village site on St. Catherine's Creek.

The Natchez pottery, as determined by collections made at this historic Natchez site, is characterized by a somewhat similar general style, consisting of a scroll or meander decoration sometimes combined with cross-hatched areas. It should be noted, however, that the lines forming the scrolls are more deeply incised and fewer in number than those on Choctaw vessels; in addition, they were made freehand and singly instead of with a comblike implement. Another fact at once apparent is that this Natchez ware was by no means a local type, as was the Choctaw. On the contrary, it is a well-known southern type, variants of it having been found by Moore at a number of sites, not only on the lower Mississippi in the vicinity of Natchez and Vicksburg but even as far away as the upper Ouachita and Red River valleys in Arkansas and Louisiana. Here it occurs both singly and in combination with the more elaborate Ouachita and Red River patterns. To the east it has been found by members of the Alabama Anthropological Society on the lower Tallapoosa River and in Baldwin County on the Gulf Coast; its easternmost extent seems to have been northwestern Florida where, as in southern Alabama, it is associated with the rim decoration of lines and of bird and other effigy heads characteristic of that part of the Gulf Coast.

It should be noted that the Natchez pottery type occurs also at Moundville. At the historic Natchez site, Chambers also found a piece of limestone pipe of the general Moundville type. Five of these flat-based Moundville pipes, showing winged serpents, an eagle, and a crouching human figure, were found some years ago at the great Seltzertown mound group a [40] short distance from Natchez. These facts are significant as indicating a possible relationship between Moundville and the later Natchez.

The third known historic pottery type in Mississippi is that of the Tunica, a determination based on collections made at the old Tunica site of Fort St. Peter on the lower Yazoo River by Ford and Chambers in 1929. The characteristic feature is not a surface but a rim decoration. The rim is somewhat enlarged, and along it runs a row of indentations or scallops with a single encircling line either on or just inside the rim. Again, as in the case of the Natchez, the Tunica ware was not a local type. This simple rim decoration extended at least into Tennessee and doubtless will be found to have occurred elsewhere.

With the historic types of a region determined, there naturally follows a comparison with materials from neighboring sites of unknown age. Thus in the Natchez and Tunica territories Ford and Chambers have found sites where other and presumably earlier pottery types occurred. The most striking of these is a decoration formed of parallel lines drawn at such an angle as to have a distinct overhanging appearance. This type predominates at certain pre-historic sites in western Mississippi where historic Natchez ware is absent, but a linkage is furnished by the finding of a few such sherds at the historic Tunica site.

In eastern Mississippi there is at least one instance of a still more striking difference between the historic ware—in this case Choctaw—and the pre-historic. A mound in Clarke County in the center of the Choctaw territory on partial excavation yielded pottery of an entirely different type, bearing cord and textile impressions and curved, stamped designs of the Georgia type.

These examples of cultural differences that have a suggestive chronological value are taken from two limited areas in eastern and western Mississippi and are mentioned merely as examples of a method, which, if followed out, promises to aid materially in the solution of our archaeological problems.

Certain other outstanding ceramic types that occur in the state might be mentioned, such as the cord-marked ware of the upper Tombigbee; the red-and-white painted ware occurring in the Yazoo Valley and along the Mississippi, a type which extends well into Arkansas; the grotesque animal heads attached to vessel rims such as are more common in eastern Arkansas but which have been found in northern Mississippi in association with typical Moundville art; the pottery of the Gulf Coast, including check-stamped ware and other vessels with outflaring rims and lines and punctate decoration, types which extend in an unbroken line along the coast from Florida to western Louisiana but rarely penetrate beyond the Gulf littoral; and finally in the western part of the state a few examples [41] of what Setzler has recently shown to be Hopewell pottery, thus carrying our southern ceramic types considerably beyond their generally recognized boundaries and enlarging the scope of the problems involved. Such, very briefly, are the more important pottery types native to Mississippi. These, and others not mentioned, must be studied further and their ranges in and outside the state determined. In the work that should follow we must seize upon every clue, no matter how small, that throws light on their respective chronological positions.

Other classes of archaeological materials are of course not to be neglected: pipes, the many forms of stone and bone implements, and copper and shell ornaments. Valuable evidence is also to be obtained from house remains. Although but few such remains have been reported in the Southeast, there is reason to believe that careful searching will reveal them at many places, particularly at village sites. From historical evidence we know that both circular and rectangular houses occurred in Mississippi. During the past summer, Chambers found rectangular house remains in the Yazoo Valley, and two years ago Ford, Chambers, and I excavated at a pre-historic village site in Yazoo County and found the postholes of three circular dwellings. One of these had been a large and complex structure, with three concentric rows of posts of different sizes set in trenches ranging from 1 to 2 1/2 feet in depth. Such a structure would seem to conform more to the council houses of the Creeks and Cherokees than to any known Mississippi type.

We need to know much more about the nature of our Mississippi houses: the distribution of the circular and rectangular types, and the cultural features associated with each, whether the two occur together, as in the case of the Chickasaw, who used a rectangular house in the summer and a circular semisubterranean one in winter; and whether the Choctaw likewise had a circular winter dwelling in addition to the rectangular type that has been reported for them.

Those engaged in archaeological work in the Southeast are in a position to furnish data of another kind—skeletal material—which will be essential if we are to have anything approaching an adequate reconstruction of our pre-historic cultures. For no matter how complete the reconstruction may be in other regards, the story is only half told as long as the physical types of the people themselves remain unknown. Of the many sins of omission and commission that may rightfully be charged to the early workers in this field, none is more grievous than their failure to preserve skeletal material. As a general rule, only bones in an exceptional state of preservation were saved; others, which with a little care and patience could have been saved, are now lost as completely as if they had been wantonly destroyed.

[42] In attempting to determine the physical types of the Mississippi aborigines, we are unfortunately handicapped at the start by a lack of information on the historic tribes. Physically the southeastern Indians are often spoken of as falling into two groups: one, broad and high-headed, represented by the Choctaw, Alabama, and Natchez, and the other lower and more oblong-headed, represented by the Creeks and Chickasaw. Yet measurements taken some 40 years ago on the living, though undoubtedly to some extent mixed bloods, showed the Creeks to have been closer to the Choctaw, with a cephalic index of above 81, than to the Chickasaw, who had an index of just below 80. Measurements on a larger series of Mississippi Choctaw, also to some extent mixed bloods, which I made in 1925 and '26, revealed a somewhat lower cephalic index, 80.6, closer to that recorded earlier for the Chickasaw. With anthropometric data on the living yielding such unsatisfactory results, the need is at once apparent for painstaking work in excavating and preserving skeletal material from both historic and pre-historic sites.

This brief and wholly inadequate survey of Mississippi archaeology serves to emphasize the fact that as yet the gaps far outnumber the links in our chain of evidence. I think it right that we realize fully the difficulties and uncertainties inherent in southeastern archaeology, provided that realization does not lead us to despair of ever reaching a solution. The task of establishing a chronology for southeastern archaeology is not a hopeless one, although a vast amount of patience, skill, and hard work must be expended before the result is accomplished.

The chairman next called on Mr. Walker to speak of conditions in Louisiana.

Mr. Walker:
Louisiana occupies a strategic position in regard to the pre-historic cultures of both the Mississippi Valley and the Gulf Coast. It is now generally conceded by students of southeastern ethnology that the original home of many of the tribes found historically in that region is to be sought west of the Mississippi River, and it is most probable that some trace of these earlier migrations will be found within the borders of the present state of Louisiana. In attempting to clarify this problem and demonstrate the relationship of the historic to the pre-historic inhabitants, the archaeologist must work in close cooperation with the ethnologist. We are particularly fortunate in this respect in Louisiana where, as in so many of the other southern states, the untiring research of Dr. Swanton has provided us with about all that is known of the historic Indians.

[43] It is futile to attempt a classification of pre-historic mound cultures in the lower Mississippi Valley until we know more definitely whether or not they have any connection with the principal historic tribes found there: the Arkansas, Tunica, Yazoo, Koroa, Natchez, Taensa, Avoyel, Houma, and Chitimacha. Some of these Indians we know were builders of mounds, but just which ones, and through what stages of development they may have passed, are problems requiring further attention.

There are roughly speaking three geographical pre-historic culture areas in the state—the mounds and shell heaps of the Gulf Coast, the groups of flat-topped quadrilateral mounds of the lower Mississippi Valley, and the sharply conical and truncated mounds of the Red River valley. Each of these areas possesses traits that are distinctive, as well as others held in common, so that no hard and fast lines can be drawn between them. For example, the shell heaps contain the check-stamped pottery characteristic of the whole Gulf Coast as far as Florida, yet in this same section of Louisiana similar pottery is also found in the quadrilateral mounds more typical of the lower Mississippi Valley province. The Atakapa and Chitimacha were the principal historic tribes in this coastal region, and both of them are known to have been mound builders, the former for dwelling sites for their chiefs and the latter for burial places for the same class. Yet as no one so far has investigated the village sites definitely attributable to either of these people, we cannot say for certain just what types of pottery and artifacts they possessed.

In regard to Caddo sites we have been more fortunate. A year ago last summer an Indian burial ground was accidentally discovered near Natchitoches which yielded elaborately engraved and incised highly polished pottery associated with European trade objects such as glass beads and articles of brass and iron. This discovery I have made the subject of a brief report now awaiting publication. Its significance lies in the fact that we have been able to identify this site as the probable one occupied by the Natchitoches Indian village visited by Henri de Tontí in 1690, where he stated that he found the Natchitoches, Ouasita, and Capiché tribes living together. The Natchitoches pottery is almost identical with that beautifully decorated ware found by Moore on the Ouachita River at Glendora and Keno Place. But it should be noted that at neither the Natchitoches nor Ouachita sites were the burials in mounds. This pottery is possibly only one variant of a more general Caddoan ceramic type which we may expect to find in adjacent parts of Texas and Arkansas, but it gives us what we hope will be the key to the major archaeological problem in northwestern Louisiana, the temporal and areal extent of Caddoan culture.

The only other major archaeological province lies in the northeastern part of the state in the "delta" land of the Mississippi Valley and extends westward into the hill country along the tributary streams. The pottery typical of this section I prefer to call Natchesan, meaning Natchezlike, rather than definitely Natchez, because although it bears a striking [44] resemblance to the ware from that historic site, the exact nature of that relationship is only imperfectly understood. This type of pottery occurs both in mounds and in low burial sites, as is the case at Glendora and Keno Place, where the Caddoan pottery also is present. Unfortunately, however, the data from this site, which would tell us whether or not the two were contemporaneous, is not detailed enough to permit a positive statement on this point, but my guess is that the Natchesan type is the earlier, based on the fact that only the Ouachita tribe, a branch of the Caddoan stock, is mentioned in the earliest historic accounts dealing with this region. An interesting feature of both types of pottery is their similar use of the scroll design, which appears more tightly coiled and better executed in the Caddoan ware than in the Natchesan, suggesting a definite interrelation between them.

Subtypes and variants of the general Natchesan pattern may be expected in subsequent identification of the pottery of the Taensa, Koroa, Avoyel, and Houma. A suggestion of Tunica rim types has been found at the Larto Lake mounds and probably will be discovered elsewhere in that portion of the state.

In order to differentiate the truly pre-historic or at least nonhistoric sites, we are compelled to utilize the geographical terminology of the older classification: lower Mississippi Valley and Gulf Coast culture areas. The shell heaps and mounds along the coast of Louisiana contain typical Gulf Coast pottery with stamped, punctate, and line-and-dot decoration. This ceramic type is not, however, limited to the littoral region, as it also occurs in sites as far north as Red River and perhaps beyond. It has been found associated with burials exhibiting the fronto-occipital form of head deformation. But this physical characteristic likewise is present in sites where there is a very different kind of pottery. Another ceramic feature more common to Florida is the manufacture of mortuary pottery with intentional "killing," circular holes left purposely in the bottom of the vessels. In Louisiana this variety of pottery has been reported so far only from sites near the southern end of Catahoula Lake.

After eliminating the Caddoan and Natchesan ceramic types we find certain other kinds in the lower Mississippi Valley area. Chief among these is an elaborately decorated ware employing the deep grooves, filled spaces, and double-headed bird figures regarded as characteristic of a certain well-known Ohio mound culture. This has been found in the Marksville and Jonesville groups and is regarded by Setzler as similar to the Hopewell pottery. It probably has a much wider distribution in the state. Mr. Setzler, in a forthcoming article in the *American Anthropologist,* will describe this variety in more detail. The ceramic similarity is all the more remarkable because of the presence in Louisiana of certain other concomitants of the northern culture, such as groups of [45] mounds surrounded by earthen embankments and ditches, burials on prepared floors at the bottom of mounds, and the presence of galena, copper objects, figurines, platform pipes, and woven mats of the one-over-one-under pattern. Thus new possibilities are opened for tracing the course of mound cultures generally regarded as upper Mississippi in origin.

Cord-marked, painted, and effigy wares are also found in northeastern Louisiana, as in the adjacent sections of Arkansas and Mississippi. The so-called clapboard, or overlap decoration, which Collins reports from the pre-historic Deasonville, Mississippi, site, is also present in Louisiana at Jonesville, Marksville, and Larto Lake. Apparently the east and west sides of the Mississippi River possess closely correlated sites, which is not surprising considering that we are here probably dealing with people who depended on canoes for their principal means of transportation. A great river could in no sense be considered a barrier between such tribes.

Regarding the many other kinds of archaeological specimens usually given so much space in both public and private museums, little can be said as to their place in the cultural patterns outlined above, for the reason that too little attention has been given by their collectors to the data necessary to establish such associations. It is not permissible to assume that all such objects are the products of the same makers merely because they happen to be found in the same mound or village site. Unless the method of excavation employed is such that the exact vertical as well as horizontal position of all the artifacts can be determined, it is useless to attempt further classification based on typology alone. I will, therefore, merely call attention to some of the prevailing forms and mention a few which may be unique.

In the northwestern province, large numbers of beautifully worked flint artifacts are found, ranging from small notched and stemmed arrow points to large leaf-shaped spear blades and knives. They occur principally as surface or subsurface finds over a wide area and are presumably the work of the Caddo tribes who last inhabited the region. Farther down Red River, however, centering in Natchitoches Parish, curious tiny flint points with curved barbs or with barbs only on one side are found which closely resemble certain scales from the garpike. The significance of this is now apparent in the light of Du Pratz's description of the manufacture of war arrows of the Natchez, which he says were ordinarily armed with scales of the garfish fixed in place by means of fish glue. These tiny flint points were very likely specialized forms designed with much the same intent. Double-notched projectile points, that is, points with a second pair of notches along the blade, are also characteristic of this section. Undoubtedly they represent another specialized form, but for what specific use is unknown. Another feature that so far as known is unique is the use of petrified palm wood for the manufacture of chipped [46] and polished artifacts such as knives, projectile points, and celts.

Smoking pipes are the only other class of artifacts that exhibit forms of unusual interest. Elbow pipes of clay are present in the upper Ouachita Valley and its tributaries. Here also are found animal effigy pipes of stone, principally sandstone. Platform, or monitor, pipes occur at certain sites in Avoyelles Parish as has been mentioned. Perhaps most interesting are the human effigy forms in which the figure is carved facing the smoker, holding the bowl toward him. These have been discovered at several places in the Red River valley.

Increasing interest is being shown in the construction of mounds as well as in the nature of their contents. The most surprising addition to our knowledge in this direction is the discovery of great sheets of cane placed in the base of one of the largest mounds in the state, the great mound at Jonesville, formerly called Troyville. William Dunbar's description in 1804 states that the mound stood 80 feet high, a steep truncated cone rising from a double-terraced base covering nearly an acre of ground, barely wide enough at the summit for a few men to stand on top. It suffered many changes of height since that time, until the final stage of destruction took place during the summer of 1931, at which time it was cut down nearly to street level and the dirt carried off to build an approach to the new bridge being built across Black River. During this demolition large areas of cane were disclosed not far below the top of the mound as it then stood. Since that was the first time such a large mound had ever been so thoroughly leveled to the ground, it offered a wonderful opportunity to see what lay below it and to study more carefully the cane exposed in the bottom. We have just finished a season of intensive excavation at this site, the results of which will be published at a later date. It will suffice to say that we found the cane sheets to consist of split pieces of common swamp cane laid with the utmost care side by side and recrossed at right angles, but not interwoven, by other pieces, everywhere oriented approximately toward the four cardinal points. These sheets were compressed into layers varying from 1 inch to nearly 2 feet in thickness, which we found present over an area some 150 feet long by 125 feet wide, nowhere reaching to the outer edge of the mound. They presented a very uneven appearance, in some places horizontal, in others nearly vertical, and elsewhere sloping at various degrees or extending in wavy lines. Even in this bottom 5 feet of the mound we could distinguish two periods of construction where an earlier mound had been incorporated into the later great mound. How many more building periods may have been present in the 75 feet which had towered over this base is a matter of conjecture.

Turning for a few minutes to present problems and future plans, it does not seem likely that the tree-ring method of chronology applied so successfully in the Southwest will work as well in the Mississippi [47] Valley. The rainfall in Louisiana is too heavy and lasts too long to permit much seasonal differentiation in the rings, and the type of timber available does not lend itself readily to such a dendrochronology. Aside from the actual observation of stratified horizons in a mound or village site, the only other approach to this problem would seem to be the possible determination of the successive changes in the river channels, particularly of the Mississippi, and the alluvial deposits left around the base of mounds in their vicinity. It is my personal opinion that if we only knew the history of such large mound groups as now appear to be some distance away from the larger rivers, we would find that at the time the mounds were built, the rivers ran much nearer them than they do today. If it were possible to determine the length of time required for this shift in the channel, we could perhaps establish the period of the building and occupation of the mounds. It is clear, for instance, that absolute identification of the sites visited by De Soto along the Mississippi depends on knowing just where the river ran 400 years ago. The clue may lie in the study of the formations of the larger cutoffs or oxbow lakes so plentiful in the lower Mississippi Valley. For the solution of this problem, we look toward the geologists and geographers.

Unfortunately, destruction of the great mound at Jonesville is not the only occurrence of this kind in the state. Numerous other examples could be cited where mounds have been destroyed in the work of cutting roads or building levees and where shell mounds have served to furnish the material for roadbeds. Naturally these public works cannot be stopped, but we are attempting to enlist the cooperation of all agencies likely to disturb any pre-historic sites. We are asking them to report promptly to the nearest scientific institution any accidental discoveries of this kind.

In common with most other parts of the mound area, we have here the problem of the treasure hunter, the man who is certain that gold or silver coins are to be found hidden in every Indian mound. Romantic tales of pirates' or robbers' loot are rife in Louisiana and are responsible for much of the vandalism that has occurred. The commercial relic hunter is in a little more excusable position, as he can at least claim that he is preserving the contents of the mounds for future study. But both of these groups should be made to see that the mounds of the state are her most precious pre-historic possessions and that their proper study and preservation is of the utmost importance. I sincerely hope that the responsible citizens of Louisiana will take action to stop this destruction of ancient remains before it is too late, otherwise the next generation will see few of them left in the state.

The imperative need is for a detailed archaeological survey, parish by parish, listing all aboriginal sites known, whether still in [48] existence or not. The archaeologists within the state are naturally the ones to undertake this task, and they can perhaps enlist the aid of interested amateurs by asking all collectors to furnish those making the survey with photographs and descriptions of specimens in their collections together with all available data as to the places and circumstances of discovery. Emphasis should be placed more on obtaining as much information as possible about the sites visited than on merely carrying off the relics found in them. Dr. Kniffen of the Louisiana State University has made a good start by making a survey of the east Florida parishes, the report of which is now about ready for the press.

The other part of the program for Louisiana is the search for village sites occupied by known historic tribes. We have yet to locate those of the Acolapissa, Bayogoula, Houma, Tunica, Quinipissa, Chitimacha, Avoyel, Washa, Okelousa, Atakapa, Opelousa, Koroa, Taensa, and the tribes of the Caddo Confederacy other than the Natchitoches. This means careful attention to all sites where trade objects of European origin are found associated with Indian pottery.

Of the protohistoric period in the state we are completely in the dark archaeologically. The clue to this phase is the identification of sites visited by the Spaniards in 1542 and by the French in 1682. Special investigations should be made of all relics purporting to date back to either of these periods of exploration. Sites known to contain only pre-historic material should not, of course, be neglected, as there is much work to be done in determining the relationships of the northern and southern mound cultures. But it is more important to establish first the succession of historic and proto-historic cultures before attempting to say positively just what cultures belonged strictly to pre-historic times. In this program we hope for close cooperation from our scientific colleagues in the adjoining states, for the problems of Louisiana's aboriginal cultures are also the problems of eastern Texas, Arkansas, and Mississippi.

The chairman next called on Mr. Peacock, who stated that he would attempt to outline the archaeological tasks awaiting Tennessee.

Mr. Peacock:
Among the first observers of aboriginal man in the South were James Adair, William Bartram, and Haywood. Later came a pioneer piece of work by Squier and Davis, but very little data was given on Tennessee, as

only the Stone Fort on Duck River was reported. In 1867 investigations at Citico (Chattanooga) and elsewhere in the eastern part of the state were made by M. C. Read for the Smithsonian.

[49] One of the pioneers in our field was Gen. Gates P. Thruston. But doubtless the best informed archaeologist in Tennessee was the late W. E. Myer. Contemporaneously, Clarence B. Moore of the Academy of Natural Sciences of Philadelphia spent two seasons in the state. The Tennessee River valley was given his major attention. His report is interesting in description of specimens, but very little was attempted in culture study.

The past several years have seen the work of our late state archaeologist P. E. Cox, and the East Tennessee Archaeological Society.

The cultures of Tennessee might be grouped geographically as follows: (1) west Tennessee, (2) stone grave—Nashville, (3) rockshelter—Cumberland plateau, (4) Tennessee River valley. We now find several interesting cultures and questions to study. Some sites to be mentioned are well known but have not been sufficiently studied to exhaust their possibilities.

Mound Bottom: In Cheatham County on Big Harpeth River there is a site on the Taylor farm known as Mound Bottom. Here is a series of conical and pyramidal mounds. Lying between the mounds and bend of the river is a stone-grave cemetery. In 1926 Mr. P. E. Cox and the writer made a brief examination of this site. Tests of the mounds disclosed no burials. An examination of about 20 stone graves showed that apparently burials were made of disarticulated remains, as the graves were of an average length of about 4 feet. There was a paucity of artifacts. No evidence was found of a village site sufficiently large to have supported the apparent number of people necessary for so much construction.

Across the river there is a bluff rising several hundred feet. In the rock near the highest point there is cut a double-winged mace similar to those represented in embossed copper plates found at Etowah by the Smithsonian in 1885. This is cut full size with lines averaging about a quarter of an inch and is on a horizontal plane. It is reasonable to assume that the man leaving this record came from the village site below.

Rockshelters: The Cumberland plateau is our greatest field for the study of rockshelters. Very little data have been secured on them. Mr. P. E. Cox made an examination of some sites on Obey River in 1925.

At the request of the owner, the East Tennessee Archaeological Society made a survey in April 1932 on the J. F. Robbins place in Pickett County. This shelter is about 150 feet above the Obey River and near the top of the bluff overlooking the stream. It has an exposure N 15 W. The opening is 105 feet long and 35 feet at its greatest depth, [50] gradually tapering off toward each end. Clearance varied from 22 feet at the outer edge to 18 inches at the back. Over the entire floor surface there are ash and midden deposits which vary from 6 inches to 3 feet in depth. A large section, in time past, fell from the overhang. This piece is 6 feet thick by 21 feet at its greatest width and 48 feet long.

Until early last March, apparently no one had ever observed that the shelter was once occupied by man. At this time, fate decreed that a dog should den a coon under this fallen rock. A farm tenant secured his hoe and in an attempt to get the coon, drew out a mass of ashes and bones. Further scraping brought forth cane matting and coarse cloth. The hunters made a fire out of the matting and cloth to warm themselves but saved the cranium and some other skeletal material of an adult female. Pot hunters later disturbed most of the accessible floor area.

When examined late in April, we found that the fallen block protects the deepest deposit of ashes and midden. When practical to remove this rock, a virgin area awaits the archaeologist. Practically no information is available as to the relations of this shelter to others and to valley cultures.

Duck River flint: Several years ago Mr. Cox, at the request of Dr. Moorehead, attempted unsuccessfully to locate the source of supply of the flint from which the famous Duck River flint swords were made. There are still real possibilities that this material came from west Tennessee.

Tennessee River valley: The engraved-shell culture of the Tennessee River valley has produced the well-known rattle design, and although much collecting has been done, there is still a field for study here.

Within this area there are also several local pottery cultures which are intriguing and wait for proper classification and orientation in relation to others. One of these is to be found on the Connor place 15 miles above Chattanooga on the Tennessee River. Here surface finds of potsherds with human effigies are quite prolific. Certain types found here have not occurred on any other sites noted. There is much to be accomplished in a study of the time levels and relationships of cultures. It is noticeable that adjacent village sites are different, one being agricultural and the other being apparently of an unsettled living condition.

Citico-Chattanooga: This is the Citico mound and village site explored by M. C. Read in 1867 and Clarence B. Moore in 1915–1916. Mr. Moore uncovered 106 burials in the village site but encountered none when he examined the mound. Quoting from his report *Aboriginal Sites* [51] *on*

Tennessee River, the mound "was domiciliary and not a burial mound, as an excavation 12 feet square sunk by us to a depth of 12 feet from the center of the summit-plateau encountered no interments or signs of interments. Evidence of former digging was found in places but no trace of skeletal remains was present in the material."

In June 1916 a highway was constructed through the site, and most of the mound was destroyed by a steam shovel. Even under such unfavorable conditions, an important burial was discovered and some of the artifacts were recovered. A string of 12 or 15 shell beads about the size of hen eggs and a copper crown were recovered. These were later acquired by W. E. Myer. In the village, workmen secured two snakes made of embossed copper. These copper specimens raise the question of a possible connection with the Etowah culture. This culture was reported at Castalian Springs by Myer, and there is maybe a further link at some intermediate site. From Hopewell to Etowah is a long jump, and every bit of intervening evidence of this culture is of great importance.

It has been generally accepted that De Soto crossed from Georgia to the Mississippi River without touching the Tennessee River valley. In 1927 the writer secured from the Citico village site a surface specimen indicating some contact with De Soto's party. This specimen, an effigy head in clay, depicts a horse in armor with a double bit in its mouth. As the women were the potters and usually traveled very short distances from the villages, some scouting party from De Soto's army must have come this far. The reproduction is most accurate and leaves little doubt but that the maker saw the original. With all its past explorations, Citico still offers a most interesting study, particularly in the light of the copper culture.

McKenzie site: In 1930, with assistance of the Smithsonian, the East Tennessee Archaeological Society explored the mounds on the McKenzie place seven miles above Chattanooga. A group of three small conical mounds was completely examined. In mound A four adults were buried superimposed. This group died possibly of disease or warfare. Influences indicating European contacts were entirely absent. Artifacts were very few. No excavations were made in the village site. A surface find of interest was a large nutting stone with 51 pits. Three or four gossiping squaws could have easily worked around it at one time.

Evans site: The society [the East Tennessee Archaeological Society] has recently been working on a mound on the Evans place in Sequatchie County. No pottery has been found, and so far all burials indicate craftsmen in shell and leather. This work is expected to be completed this coming spring.

Stone graves: In the past week, one of our group has reported the discovery of stone graves. This type of burial is rare in the Tennessee River valley. A careful survey of these is planned in the near future.

[52] *State survey:* Tennessee has witnessed many years' activity in exploration of its aboriginal remains and still offers much to the archaeological field. As has been the case in most states, considerable digging has always preceded recognition of the fact that a careful survey of the entire state is not only desirable but essential for the best and most complete results. No effort, so far as the writer knows, has ever been made to develop such a survey in Tennessee. We recognize the fact that Myer, through his own efforts, more closely met this need than any other individual or organization.

Through the march of our modern civilization, various pre-historic sites are being obliterated. This is particularly true in the valley of the Tennessee River, where in the past two years the United States government has completed a survey which makes possible the development of some 17 dams for power and flood prevention on the Tennessee and its larger tributaries. In most cases the completion of these dams will obliterate pre-historic sites. The dam at Hales Bar permanently inundated an important site on the Bennett place. Another case is the completion of Calderwood Dam on Little Tennessee River, which destroyed a mound and site.

As economic conditions readjust themselves and further development of these projects becomes necessary, there will be other losses to science in the Tennessee River valley. These conditions make it essential that particularly those of Tennessee who are interested in preservation and study of our pre-historic remains should lend every aid toward the crystallization of a statewide interest in this field.

Our first need is to educate the public so that they may appreciate the fact that it is desirable to preserve and permit scientific study of these remains. Coupled with this appreciation should also come an awakening that those who are using the archaeological field to traffic and barter are detrimental to the real interests of every scientific effort that is being made.

It has been the privilege of the East Tennessee Archaeological Society through personal contact with owners to preserve several remains that may be studied by others in the future. We believe that the majority of property owners where such sites are found will be reasonable and have a pride in assisting in a statewide movement to further scientific study.

Regardless of how long it takes to inaugurate such a survey, our group at Chattanooga intends to carry on with surveys in our vicinity, giving as much time to this as excavating.

[53] The next speaker called on by the chairman was Mr. Pearce, who spoke on the significance of the east Texas archaeological field.

Mr. Pearce:
[Ed. note: The superscript numbers in Pearce's paper are keyed to bibliographical sources; these can be found following the paper.]

For the outstanding characters of this field with the distinctive differences of the subareas, one is referred to the last number of the *American Anthropologist,* No. 4, 1932.[1] After an allusion to some returns from last summer's fieldwork there will follow a discussion of relations to outlying fields and to the remoter fields of Mexico and the far Southwest.

The field as treated in this paper comprises a block in the northeast corner of the state, approximately 150 miles square. It is heavily wooded and well watered.

Archaeological evidence shows that the early inhabitants lived in villages, had extensive agriculture, and possessed a culture closely assimilated to that of the south Mississippi Valley.

During the summer of 1932 a University of Texas expedition into this region explored two burial sites in which evidence of white contacts were found, thus establishing relations of some of the returns from fieldwork with the historical tribes. Much of the material from north of the Sabine we now know was Caddo, and we can identify some of the material in the Neches Valley as Assanai. There is considerable tendency toward unity within the river valleys and toward noticeable diversity as one goes from one valley to another.

Along Red River, heads were artificially deformed by binding in a way which forced the forehead downward and gave the skull an elongated cylindrical form. At the Sanders place all male heads seem to have been so treated. Two bands were used, one passing over the forehead, the other over the top of the head, the two coming together in the region of the occiput.

To take up now relations with outlying fields, we will deal first with the coastal region to the south. Burial of pottery with the dead has not been observed south of Nacogdoches, but pottery in the form of sherds abounds to the coast, though it changes its character radically before the coast is reached. In fact, the east Texas elements and influence can be traced in adjacent fields in diminishing intensity in all directions to the south and west. On the coast the elements in common with east Texas proper are:

[54] 1. an abundance of pottery in the form of sherds,
2. beads of the conch shell identical in form with those to the north; also some crude shell pendants, but no fine incised gorgets have been found in this region by the University of Texas,
3. small arrow points and small scrapers, though the arrow points differ considerably in form from those of east Texas.

The differences of the two regions are striking, though the margins between them are vague except in the practices of burying property with the dead and of building earthen mounds.

There is on the coast little or no evidence of agriculture immediately adjacent to the sea; there are no traces of polished stone or of mud-and-thatch houses.

Though the pottery on the coast is abundant and often of fine quality, the methods of decoration are so different from those of east Texas that one cannot suppose there was much influence of either region on the other in this connection. Sherds on the coast are so small that it is hard to make out the forms of vessels, but a slip of hematite pigment was often applied for decoration as in the north.

Designs are few and were always cut into the soft unburnt clay and in straight lines. Hatching in triangular figures is found, as are some notches on edges and a few dotted lines. In the middle coast, asphalt was often applied for waterproofing and in lines for decoration. Pottery in this region is sometimes decorated on the inside, a practice virtually unknown in northeast Texas. Tempering is with sand or not at all, rarely with shell.

Mr. E. A. Anderson of Brownsville, a very capable amateur archaeologist who has made a study of the coast about Brownsville, says in a letter that the higher cultures of Mexico never extended farther north on the coast than southern Tamaulipas, i.e., Tampico.

Below Falfurias, pottery is scarcer than to the north and exhibits pronounced Haustec characters, particularly in the use of painted figures. There are evident differences in the pottery and in other culture elements in different sections of the coast.

One large earthen mound exists on the coast. It is on the banks of the Guadalupe, about 8 miles below Victoria, and is about 18 feet high and 500 feet in length by 150 feet in width. This was trenched along the short diameter by the University of Texas during last summer. We encountered ten burials at varying depths, all but one of bundle [55] character. Nothing was found with the burials except a few crude shell pendants. Broken and splintered animal bones were common. Some 35 crude flint artifacts and 2 small pieces of sherd were found. All bone materials were heavily mineralized. Some of the limb bones indicate men of large size, a character of the Karankawa. The contents of this mound show close affiliation with the coast practices generally and are markedly at variance with returns from mounds to the north. Bundle burials are common around Galveston Bay. Moore mentions them as "bunched" burials and encountered them frequently in the valley of the Atchafalaya and along the lower Mississippi.[2] He found them as we did on the Texas coast, without accompanying artifacts.

The relations of northeast Texas with Arkansas are striking. Comparisons are based on the work of Harrington[3], Moore[2], and Dellinger[4], and the collections of the Lemly Bros. at Hope, Ark. Going northward from Red River, one encounters rapidly increasing differences as follows: increasing specimens of fine stone work, effigy clay vessels, especially those involving the human head and form (except in the most rudimentary form, these are not found in Texas), vessels with teapot spouts and with handle spouts (these are not found in Texas), a diminution of carved pottery (not found, says Moore, north of the Arkansas), the appearance of vessels with legs (unknown in Texas except with the ring base). Only one toy bottle from Red River, in the University of Texas collections, has legs. Elements in common with Arkansas are the ring base with four legs, double forms (one above the other), nipplelike corners or projections, and carved scroll designs cut through a red slip, giving a relief effect; forms of bottles and bowls remain largely the same, notably the cone-shaped neck on bottles, nodes on bowls—especially when intended for use as drums or tambourines—and conch-shell cups (some decorated). Common elements become conspicuously less frequent as one passes through Arkansas. Holmes[5] and Moore make this river the dividing line between the middle and lower Mississippi cultures. The pottery of the middle Mississippi is markedly inferior to that of the lower Mississippi.

With Louisiana the relations should be close and probably are, but the writer has not seen returns or printed accounts of returns from western

Louisiana in sufficient quantity to be sure of conclusions. What he has seen in collections and in the accounts of Moore[2], Cyrus Thomas[6], and Holmes[5] indicates practical identity for some distance to the east of the state line. Carving designs on burnt pottery continues to the Mississippi.

Odd to say, affiliations in the forms and decoration of pottery are rather close with the northern Gulf Coast of Florida, much closer in fact, than with the coast of Texas. This observation is based on the accounts of Moore[3], Holmes[5], and Vaillant.[5]

[56] Common elements with the Florida coast may be listed as follows:

1. incision of designs,
2. stippled areas,
3. complicated scrolls,
4. cylindrical bottle necks,
5. pipe bowls and stems of much the same form,
6. scalloped margins,
7. identical profile for large bowls,
8. complicated designs in parallel curved lines,
9. double vessels,
10. olla-like vessels (one with complicated hooked or swastika design laid over the top of the vessel in a kind of mantle is very like one from Morris Co., Texas),
11. featherweight, highly porous wares,
12. burial of pottery with the dead, in contrast to the Texas coast practice.

This comparison applies only to the better types of Florida pottery, and it must be added that the corresponding types are, in east Texas, far superior. Nonetheless it is possible to avoid the inference of a common source for at least some of these elements.

To the west, the east Texas influence extends up the streams in diminishing form as one gets farther away from this culture center. It is observable in the burnt-rock mounds of central Texas in the upper level in the form of occasional sherds and of characteristic fine arrow points, in occasional boat stones, stone plaques, polished axes, and shell gorgets, and probably also in the increased attention to agriculture in this upper level as compared with the two lower levels. The pottery of the prairies and plains—and there are at least traces of pottery over nearly the whole of Texas—affiliates with east Texas rather than with the Pueblo area. Pueblo influence is confined to the extreme western margins of the state and is not very pronounced even there. A type of laurel-leaf blade shaped to a drill at one end is found all over Texas and in Arkansas to the Mississippi. It belongs to the upper level of central Texas middens.

In the Panhandle, Holden has just discovered important remains of a high type of village life involving extensive pottery manufacture of a type that shows at least one marked trait in common with the eastern lower Mississippi and lower Appalachian areas; viz., designs made, seemingly, by rolling some form of woven fabric, wrapped about a stick or mallet, over the soft surface of the pot.[8] This discovery seems to the writer to be highly significant and, taken with the fact that the scattered sherds from the Panhandle down the Red and Brazos rivers increase in number as those streams approach the east Texas area and have [57] an east Texas character throughout, would seem to indicate that cultural practices of the prairies and plains were dominated from the east instead of from the Pueblo area. This the writer is convinced was the case.

Finally, relations with the Pueblo area and with south Mexico must have brief mention. In a recent monograph on "Some Resemblances in the Ceramics of Central and North America," Dr. George C. Vaillant deplores the lack of knowledge of the archaeology of Texas, the need for which his article, and many another, forcibly brings out.[7]

With the Pueblo area, the common factors are the same as between the Pueblo and the lower Mississippi in general, to which the east Texas area undoubtedly belongs. Of the two areas, the Pueblo, a high culture center, seems to be younger, and the peoples of the forests seem to have been the more powerful and aggressive and to have penetrated the prairies and maintained closer relations there with the warlike tribes of those parts.

As for common traits with Central America, there are some, but only such as are common with the mound-builder area in general. There is no evidence on the Texas coast that a large migration movement from south Mexico ever passed into the Mississippi Valley over this route. The coastal evidence of relations with distant Mexico may be summed up as follows.

Pottery exists along the whole coast, some of it high grade, but fishing will lead to pottery making as well as to agriculture because both result in a relatively sedentary life. Again, that the pottery of the coast was largely a local development is indicated by the fact that it varies in quantity and character so sharply in different parts.

One sherd of Haustec type has been found as far north as Corpus Christi, and it is not uncommon about Brownsville but, again, it does not extend north above Corpus Christi. One large mound mentioned above, of typical mound-builder type exists near the mouth of the Guadalupe, but its contents are affiliated closely to local coastal traits. Earthen mounds are found below Tampico, but the one mentioned is the only one the writer knows of on the Texas coast.

There are deep refuse deposits at certain places on the coast, indicating great age and evidencing pottery and a high type of stone chipping, but again they are not identical with like materials in east Texas.

[58] Other negative evidence lies in the total absence from the coast area of the symbols, pottery forms, and polished stone implements that are sometimes strikingly similar, if not identical, in Central America and the Mississippi Valley.

No stratification has yet been found in the middens or mounds of the area that would indicate definitely the presence there of higher culture in early times. Nonetheless, the writer believes that a slow trickle of Mexican influence passed in early times up the Texas coast into the mound-builder area. This may have included some migration that moved over the whole route. This assumption would account for the Guadalupe mound and for certain fine specimens of pottery on the coast. Possibly even domestic plants passed up that line but if so, they did not linger in the coast area.

This coast was not an inviting route for the migration of even primitive peoples of a relatively high culture for the following reasons: (1) the land is low and swampy, subject to floods, infested with mosquitoes, snakes, and alligators, and often impenetrable thorn brush comes right to the water's edge; and (2) the nomadic warlike prairie tribes held the hinterland and often penetrated to the sea. The degraded Karankawa of historical times represent the effects of the coast on culture, and they were held strictly to the sea by the Tonkawa, Comanche, and other prairie tribes.

The writer's opinion is, therefore, that most of the common elements of mound builder with Mayan culture passed by way of the West Indies and Florida. Some of it may have come by way of northern South America and the Little Antilles, as suggested by urn burials and other factors mentioned in the accounts of Nordenskiöld[9] and Gower.[10]

[59]

Bibliography

1. Pearce, J. E., "Archaeology of East Texas," *American Anthropologist,* Number 4, 1932.
2. Moore, Clarence, "Antiquities of the Ouachita Valley."
   " " "Antiquities of the St. Francis, White, and Black Rivers."
   " " "Some Aboriginal Sites on the Mississippi River."
   " " "Some Aboriginal Sites on the Red River." Reprints from the *Journal of the Academy of Natural Science,* Phil.
3. Harrington, M. R., "Certain Caddo Sites in Arkansas," *Indian Notes, Heye Museum of the American Indian.*
4. Dellinger, S. C., Collections at University of Arkansas.
5. Holmes, W. H., "Aboriginal Pottery of the Eastern United States," *20th Annual Report of the Bureau of American Ethnology.*
   " " "Pre-historic Textile Fabrics of U.S. Derived from Impression on Pottery," *4th Annual Report of the Bureau of American Ethnology.*
   " " "Ancient Pottery of Mississippi Valley," *4th Annual Report of the Bureau of American Ethnology.*
6. Thomas, Cyrus, "Mound Exploration of the Bureau of Ethnology," *12th Annual Report of the Bureau of American Ethnology.*
7. Vaillant, George C., "Some Resemblances in the Ceramics of Central and North America," *The Medallion,* Gila Pueblo, Globe, Ariz.
8. Holden, W. C., "Recent Archaeological Discoveries in the Panhandle," *Southwestern Political Science Quarterly,* Vol. 13, No. 3.
9. Nordenskiöld, E., Various Articles in the *Journal des Americanistes.* "Origin of the Indian Civilizations in South America," Vol. 9.
10. Gower, Charlotte D., "The Northern and Southern Affiliations of Antillean Culture," *Memoirs of American Anthropological Association,* 1927.

[60]   [Dinner and evening session, Dr. Wissler presiding.]

Only one speaker had been scheduled for the evening, and Dr. Wissler called on Dr. Swanton, without formalities, to speak on the subject "The Relation of the Southeast to General Culture Problems of American Pre-History."

Dr. Swanton:
In my earlier paper I attempted to give an account of the kind of people who inhabited the territory in which we are interested at the close of the seventeenth century and of the general character of their culture. Later speakers have treated of the distribution of the aboriginal remains found in the same region. We now have to inquire what relation existed between the two, and what light may be thrown by the use of both sets of data on the past history of the section and its connection with surrounding areas. Some such relations earlier contributors to the conference have already established.

Our quest is made particularly difficult on account of the absolute failure of practically all of our early authorities to describe native artifacts clearly, particularly those relatively imperishable objects with which we are chiefly concerned. In spite of a few luckily preserved textiles and articles of skin, we are, as you know, very nearly limited to pottery, pipes, objects of shell such as gorgets, and certain other artifacts of bone, stone, and copper. To these must of course be added the testimony of the shapes and varieties of earthworks, the manner in which they are built, and the testimony of the position and condition of the objects contained in mounds, graves, on village sites, and so on, when carefully explored. These last data may be called semiperishable, their utility depending on the care and thoroughness with which the investigator conducts his work. That, however, is your business; my own duty now is to show to what extent our knowledge of the living obtained through documents or from the Indians themselves may help to interpret your findings and add interest to them.

The tribal maps given in my first paper (Appendix, Figures 1 and 2) indicate the distribution of Indian tribes in the eastern part of the present United States about 1650–1700. Before attempting to relate these to the archaeological areas which have also been indicated, it will be well to consider some earlier historical data preserved mainly in the chronicles of the expedition of Hernando De Soto. As this famous explorer passed through the entire region of the Gulf states from east to west between the years

1539 and 1543, the narratives of his expedition throw a momentary flash of light on the entire region and enable us to note whether any displacements of tribes had taken place between the date of his expedition and the period of English and French colonization over a hundred years later. [61] Spanish occupancy of Florida and the Georgia coast was of course continuous from a period only a few years subsequent to that of De Soto's expedition, but the knowledge gained of any Indian tribes to the west or northwest of the Florida peninsula is surprisingly slight. Most of them are lost to view for almost 150 years.

The accompanying map (Appendix, Figure 3)—for the preparation of which, along with most of the others used in this paper, I am indebted to Mr. Frank M. Setzler of the U.S. National Museum—shows the probable route of the Spanish explorer, laid down in accordance with the latest and best information available to me. This has been made possible by the labors of many men, but particularly those of Col. John R. Fordyce of Hot Springs, Ark., who contributes a special paper on one part of De Soto's route, and Mr. J. Y. Brame of Montgomery, Ala., who has traced the section of it through Alabama and parts of Georgia with singular care and success. My map will be found to differ somewhat from that of Colonel Fordyce, but only in details. There is no opportunity at this time for an extended discussion and defense of the route as here laid down. I must put that off until a later occasion and beg permission to be a little dogmatic. Such differences as may exist between my allocation of the route and other theories will have little bearing on our present problem.

It is generally admitted that De Soto landed at Tampa Bay. From that estuary until his army crossed Aucilla River and entered the territory of the Apalachee Indians, he was among tribes speaking dialects of what is known as the Timucua tongue, given in our *Handbook of American Indians* as the Timuquanan linguistic family. This connection is proved by the names, some of which may be translated in whole or in part, and by the frequent occurrence of the phonetic *r* for which Timucua was noted, a sound entirely wanting in the speech of the Muskhogean tribes proper.

The Apalachee are known, through fragments of their language which have come down to us, to have belonged to the Muskhogean stock, and to have spoken a dialect rather closely related to Choctaw. De Soto and his army spent the winter of 1539–1540 at an Apalachee town near the present capital of Florida, Tallahassee, which has the significant meaning, "Old Town." Between this point and Savannah, the place names are all in Muskhogean dialects, and we are able to identify several of the tribes. The first two were related to the Apalachee or Hitchiti. Above them, on Flint River, De Soto seems to have found a body of Creeks a little southeast of their later center on the Chattahoochee. On the Ocmulgee he came to a province later identified as belonging to the Yamasee, and only a little west of the country occupied by them in 1700. Higher up on the same river, he found Hitchiti-speaking people just where later explorers report them before the Yamassee war. I feel there is little doubt that the Indians of the famous province of Cofitachequi were Creeks, ancestors of one [62] of the Lower Creek tribes, and probably either the Kasihta or Coweta.

North of these Creeks, however, as far as the Blue Ridge, and in the territory of the later Underhill Cherokee, lived tribes related to the Catawba. This, then, indicates a movement of peoples, Cherokee entering and Catawba moving out. Guasili seems to be a corruption of Hiwassee, plus a locative ending, and this again appears to be good Cherokee, so that Guasili may have been a Cherokee settlement; but it is the only one that can be so identified because Canasauga, where the Spaniards left the mountains, bears a Creek name and evidently belonged to a Muskhogean tribe. The Chiaha, encountered next, were connected either with the Yamassee or the Hitchiti, the languages of which appear to have been identical. After them comes the Costehe, the later Koasati, again Muskhogean, and finally the "Tali," at the bend of Tennessee River, an enigmatic tribe which I formerly strove to identify with the Talikwa or Tellico Cherokee but am now inclined to regard as Creeks, perhaps the Talladigi or Abihka. When the Virginians entered this country, this part of Tennessee River was still mainly occupied by Muskhogeans, though the Cherokee were pushing farther and farther southward.

While De Soto was on this river, he heard of a province across the mountains to the north called Chisca, and this is the first appearance in history of the Yuchi. Two Spaniards were commissioned to pay the country a visit, but after crossing several lofty ridges, they gave up the attempt and rejoined their comrades. The Indians had informed De Soto that there was gold or copper in this province, a suggestion which fell on fertile soil, and so after they had crossed the Mississippi, they made one more attempt to reach it, but without success.

On his way from Tennessee River to the Coosa, De Soto met a band of Tuskegee, a tribe later represented among the Cherokee and among the Creeks when their town was located at the junction of the rivers Coosa and Tallapoosa. Most of the Upper Creeks he found in practically the same region as that later occupied by them—the Coosa at their historic seat on Coosa River and the Hothliwahali approximately where they continued to live until their removal to Oklahoma. The Tawasa were also about where they were in Hawkins's time (1799), but meanwhile they had moved to the neighborhood of the Apalachicola, thence to Mobile and back to their old homes. The Talisi would seem to have been a Creek tribe living considerably west of any point occupied by Creeks in the eighteenth century, and the Mobile were living farther inland than when the French encountered them at the end of the seventeenth, but not very much farther, while the Choctaw (Pafallaya) and Chickasaw were close to their historic seats. The same seems to have been true of the Chakchiuma.

The greatest changes of all we find west of the Mississippi, in Arkansas and Louisiana, and particularly the former state. The Pacaha have hitherto been identified by most students with the Quapaw, but I think this [63] has been due merely to a confusion of names, and I regard the former as a part of the Tunica. "Tunica oldfields" is almost opposite the sites occupied by these Pacaha in 1541. The Casqui or Casquin, their neighbors and rivals, were without doubt the Kaskinampo of the seventeenth century, and I have elsewhere given reasons for supposing that they were related to the Koasati and ultimately united with them. Quiguate, reported to have been "the largest town in Florida," Col. Fordyce has placed tentatively at the Menard mounds, and with this identification I concur. At any rate, from this place on we begin to recognize place names belonging to a new language which I feel sure is Natchez or some dialect of Natchez.

De Soto now turned westward into the hill country of Arkansas, the Ouachita plateau. The first tribe encountered by him, the Coligoa or Coligua, may have been the later Koroa, kindred of the Tunica; the name of the province of Palisema also looks Tunican. Tanico may be a form of the word Tunica itself, but the identification is highly speculative. However, there can be no doubt that the Tula Indians, who differed entirely in language from those met before, were part of the Caddo. At this point, however, De Soto returned to the southeast and quickly passed into territories which seem clearly to have borne Natchez names. These Natchez place names follow us to Utiangue, where the Spaniards spent the winter of 1541–1542, and to Guachoya on the Mississippi, where De Soto died. And when Moscoso, his successor, started west in hopes of reaching Mexico, Natchez names continue through the salt country but change suddenly when we reach the provinces of Maye and Naguatex, of which terms the former is probably and the latter certainly in the Caddo language.

From this time until the expedition reached the river Daycao we are given names plainly recognizable as Caddo, many of them tribal names in existence when the French and Spaniards entered the country at the end of the seventeenth century. On the western side of the river Daycao, which I take to have been the Trinity, lived a rude people whose language was wholly different from Caddo, and in them we plainly discern either the Tonkawa or Badia. Here, however, the expedition stopped and returned to the Mississippi by the route it had pursued going out. Not finding sufficient corn in Guachoya, which had been ransacked before, headquarters were established in another town with a Natchez name, Aminoya. A town higher up the Mississippi was known as Tagoanate, and one lower down, under the greatest lord in all that land, was known as Quigualtam. These again seem to be Natchez names, and I hold the last to have been the Natchez tribe itself, or a part of the people later so called. After spending the winter of 1542–1543 at Aminoya, the surviving Spaniards built a number of small boats in which they floated down the Mississippi River to its mouth, then making their way along the coast of the Gulf to the Spanish settlement at Panuco.

On reviewing the information furnished by the chroniclers of De Soto, we find that in the subsequent century and a half there were [64] surprisingly few striking changes. The Timucua occupied the very same territory except where they had been supplanted by Spaniards. The boundaries of the Muskhogean stock are also shown to have varied little, except that a part of the Lower Creeks later established on the Ocmulgee and Chattahoochee rivers were then about the present site of Augusta, Georgia. The Yuchi had not yet invaded the Gulf states, though a part of them may have been near Muscle Shoals, and some were apparently on or near the Cumberland. The Cherokee had barely reached the Hiwassee, and the northwestern part of South Carolina—later the seat of the Underhill towns—was inhabited by Siouan peoples related to the Catawba. The Upper Creeks occupied substantially the same region, though one Muskogee tribe seems to have been a little farther to the southwest.

The change in location of the Mobile tribe was evidently due to the Spaniards themselves and was caused by the losses these Indians had suffered at the battle of Mabila. The rest of the Muskhogeans, as well as the Caddo, occupied substantially the same territories in De Soto's time and later, but along the Mississippi River very great displacements of population occurred in the intervening period, the nature of which we cannot completely elucidate. It seems certain that a Muskhogean tribe related to the Koasati was at that time west of the Mississippi River, near the present Helena, and there are also indications that tribes of the Tunican stock were their neighbors both on the Mississippi and inland about the present Hot Springs. From the lower course of Arkansas River southward as far as the Red, however, we get rather clear evidence of peoples speaking dialects related to Natchez, and such names continue through the salt provinces of northwestern Louisiana, though this fact does not prove that those territories were occupied by Natchez Indians.

As we shall have occasion to mention presently, one closely related group of Muskhogean tribes, including the Casqui, who were just mentioned, the Koasati, Tuskegee, and Alabama, seem to have been farther toward the northwest in 1541 than in 1700, and this is in line with Muskhogean traditions which indicate a general movement from the northwest. Such a movement was probably responsible for the cultural differences noted in my first paper between the tribes in the central part of the area and those east, west, and south. In order to make this point somewhat clearer, I subjoin a chart showing the distribution of 17 cultural features. (Appendix, Figure 4). *A* indicates a character, found among certain tribes, and *B* an alternative character found in others. It will be noted that the *A*s are most numerous among the peripheral tribes and the *B*s among the central tribes. As the evidence is fragmentary, I will not give the names of the factors themselves but will merely enter this as a suggestion. It will be noted that the *B* factors are most strongly represented among the Creeks, who are believed to have entered the region in very late times, but the De Soto narratives show us that in the early sixteenth century part of the Creeks did not conform to this pattern. I am therefore inclined to attribute the [65] standardization of Creek culture to a Creek tribe which arrived from the northwest at a late period, and this might conceivably have been the Abihka or Talladigi, to whom Hawkins attributes the origin of the laws against adultery.

Let us now pass on to our principal undertaking, an attempt to relate various groups of tribes to the areas established by archaeologists. As a preparation for this undertaking, I suggest that you compare the tribal and stock maps with the map of archaeological areas (Appendix, Figures 1 and 7).

In the map shown in Figure 5, I have brought together all of the data available to me, from historical, traditional, cultural, and linguistic sources, regarding the probable movements of the tribes of the Southeast in the early historic period, just before and just after the appearance of

Europeans. The identity of each tribe is indicated by one or more initial letters, the key to which is given on the right-hand margin of the map. The linguistic family to which each belonged is shown by the type of line used to mark its migration, and the key to these appears below. Let us take up these stocks in order.

An examination of the lines used to mark migrations of Algonquian tribes shows that many of them radiate from the region of the Great Lakes, even the Delaware of the Atlantic Coast, indicating that as the part of the continent from which they had come. Here is, in fact, a culture which has been generally called Algonquian, and of which McKern believes the culture of the effigy mounds is an offshoot. The Algonquians, however, do not much concern us.

Students of the Iroquois proper regard them as intruders from the west and probably from the southwest. Cherokee tradition, as here reproduced, points to an entrance into the southern Appalachians via the Upper Holston and New rivers—in other words, down the great war trail, the Warriors' Path—and this, extended backward, would carry us along the Kanawha to the upper Ohio. From this area, then, the Iroquoian trails diverge, somewhere south of the Great Lakes and the Algonquian territories. Hereabouts should be found the key to the proto-Iroquoian culture; and I leave archaeologists to determine whether this was the Hopewell or Adena culture, or some other. But I think that I can at least indicate that it was not the culture called after Fort Ancient.

Caddoan peoples have moved from the east and south, but we cannot trace them far in any direction. Sibley states that the Caddo proper supposed they had formerly lived higher up Red River, but another early writer tells us they claimed to have originated in the neighborhood of Hot Springs, Arkansas. The Wichita told Gatschet that they had formerly lived on the Arkansas, and La Harpe found them on the Canadian midway between [66] the Arkansas and their later seats on Red River. Some Pawnee claimed to have come from the east and some from the south, while the Arikara were a late offshoot toward the north. The idea that they once had their homes along the Rio Grande is due, I believe, to a confusion between the Caddoan Shuman tribe and the Suma Indians, and I think that their contacts with Mexico were late rather than early. This agrees with the evidence collected by Dr. Strong, who considers that the Caddoans should be classed among the eastern stocks rather than those of the western Plains and beyond. Whether they had a more remote connection with the Iroquoians, as their language might indicate, or with the

Siouans and part of the Muskhogeans, which may be supported on physical grounds, must be left for later investigators to determine.

One of the most remarkable results is attained by studying the Siouan Indians. Early and consistent Quapaw and Osage legends point to the Ohio above its junction with the Wabash as their early home, while the Kansas, Omaha, and Ponca claimed community of origin with them, and all claimed an earlier residence in the east. Probably the circumstantial narrative given by Dorsey is over-rationalized, but I think there is no doubt it embodies an essential truth. What little we know of the Hidatsa, Crow, and Dakota Indians again points to the east, while Mandan tradition take them down into South Dakota, and tradition and archaeology seem to carry them back into Iowa, with Dr. Strong suggesting for them a connection with the Mill Creek culture. So far as the Hidatsa and Dakota are concerned, more rigid scientific proof is furnished by the really remarkable resemblances among the languages of those tribes: Tutelo in the far east and Biloxi in the far south. This is shown in the accompanying table, on which I have entered, roughly, the number of closest resemblances between six Siouan dialects. It is defective in that it does not include any languages of the Dhegiha or Winnebago groups, but it is sufficiently striking as it stands. Catawba is known to have been wholly distinct.

Eastern Affinities of the Coahuiltecan Languages

Coahuilteco	Languages Spoken Eastward
ta-, me	*ta-*, I (uncompleted action) (Natchez)
na-, I	*ni-*, me (Natchez, Timucua), I (Tunica)
ha-, *ya-*, you	*-ya*, your (Timucua), *-he, he-* (Tunica)
mai-, *ma-*, you (subj.)	*ba-*, you (Natchez)
u-, *wa-*, he; *a-*, him	*oke*, he (Timucua); *ū-* (Tunica); *ha* (Atakapa)
[67] *ka*, (plural)	*ka*, (plural) (Timucua); *-ka* (Chitimacha)
-ho, (pl. with pronouns)	*ho*, (plural) (Hitchiti, Alabama, etc.)
-tci, (pl. w. demons.)	*-tc*, (plural) (H, A, etc.)
oh, ohua, no; *aham*, not	*-ko*, (neg. suf.) (Creek); *-ha* (Tu, A); *ka* (Chit)
apa, (reflexive)	*apc-*, (Ch); *ap*, (medial) (Koasati)
-ma, to (instr.)	*-ma*, in, etc. (Timucua)
-m, -n, (obj. suffix)	*-n, -ⁿ , -m*, (ind. obj.) (Cr, Choc, A, K, H)
-t, (obj. suffix)	*-t*, (nominal conn.) (Cr, Choc, A, K, H)

-kue, (loc. suffix)	*-k^b, -gūc* (N); *-ki, -nki* (Chit); *-kua* (Tim)
e, (interrogation)	*-he* (Tim); *-ho* (H); a (At)
-mi, to possess	*hami* (Tim); immi (Choc)
kawa, to love	*goho* (N); ka (Chit); *kū (At); kuc,* to want (N)
tco, tcu, to take	*tcū* (Tu); *icūl, icaū* (At); *isi* (Cr, A, H, K)
a (pl.), *o* (sing.), to be	*a* (sing.), *ōn* (pl.) (Tu); *a* (def.), *o* (indef.) (A)
tagu, woman	*tek* (Choc); *tägi* (H); taiye (A); *ti-* (f.p.) (T)
ah, to give	*a-i* (Chit); *yūi, yūa* (Tim)
t'il, day	*yil* (At); equela, day, *ela,* sun (Tim); ila (O)

Three small tribes—the Iowa, Oto, and Missouri—were late separatists from the Winnebago. We cannot trace the latter further, but their linguistic connections with the Dakota, Tutelo, and other tribes are clear. Turning now to the south, we find that one Siouan tribe encountered by the French on Yazoo River may be traced historically to southern Ohio, and there are linguistic and cartographical grounds for tracing another—the Biloxi—to the same region, while Tutelo traditions all point westward. Now note that the homeland of five tribes—Quapaw, Osage, Ofo, Biloxi, and Tutelo—proves to be in the region of the Fort Ancient culture and [68] that the movements of several others were distinctly away from that region. It is not necessary to suppose that the Fort Ancient culture represents the one shared in that identical form by all ancient Siouan people, for it is probable, indeed almost certain, that several of the so-called cultures or phases of the upper Mississippi were due to them. This appears to be indicated for the Upper Mississippi phase, so called, for the Mill Creek phase, and for the bluff culture of Illinois. Confirmation of such an association of people and culture seems to be strongly indicated by the recent movement of a Siouan people from Ohio, the Ofo, and the presence of European objects in Fort Ancient culture remains at Madisonville.

The Catawba and their allies were probably differentiated earliest from the remaining Siouans. We seem to trace them into the Appalachian Mountains, where they were displaced by the Cherokee, and perhaps even back through them to Kentucky. There exists a possibility that the eastern area of stamped ware was associated with them. From this branch of the Siouan family comes our earliest ethnological information of magnitude from any tribe in North America—that supplied to Peter Martyr by Francisco of Chicora—or, as there is little doubt we should call him, Francisco the Shakori.

The migration legends of all Muskhogean tribes, with one or two trifling exceptions, point to a general movement from the west, and particularly the northwest. Legends preserved by some Titchiti among the Lower Creeks indicate a former location on the Gulf Coast, and this agrees with Spanish references to a Sabacola or Sawokli province west of Apalachicola River, the Sawokli having later united with the Hitchiti. Some early migration legends of the Alabama Indians also speak of a northward movement, but this is explained historically by the former residence of a part of that tribe near Mobile and later settlement of what is now Montgomery. The Natchez legend, as reported by Du Pratz, points to the southwest, but De la Vente, the missionary, understood that they had lived farther northwest. Possibly this movement was only local, as there seems evidence for a relatively long occupancy of the region where they were found in historic times. As we have seen, there seems to be linguistic evidence that they formerly extended as far north as the Arkansas and as far west as the salt domes of northern Louisiana. You have already learned of the identification as historic Natchez pottery of a beautiful type with glossy surface and ornamented with artistic scrolls. The area over which these pots are found agrees closely with that which seems to have been occupied by Natchesan people in De Soto's day.

Regarding the Choctaw, Chickasaw, and Chakchiuma, all we know is that traditions pointed westward for their homeland—in later times specifically designated the Rocky Mountains—but one can put little reliance on this latter assertion. Shortly before the French ascended Mississippi River, part of the Chakchiuma seem to have broken away from the main body of the tribe, then apparently living on Yalobusha River, and moved down to a point opposite the mouth of Red River, where they became known as Houma. The rest of the nation later united with the Choctaw and Chickasaw.

There are several versions of the Muskogee legend, all of which bring the people from the west, except in the case of the Tukabahchee, who believed that they had come from the north. They, however, are said not to have constituted a constituent part of the Muskogee nation until relatively late. These legends for the most part speak of crossing some great river or other body of water, often identified with the Mississippi, and like those of the Choctaw and Chickasaw also carry them back to the Rocky

Mountains or even beyond. The most circumstantial of these legends, the one related to Governor Oglethorpe by Chekilli, traces their course from a stream called "Coloose-hutche" west of Coosa River, across the Coosa, and by readily identifiable landmarks across the Tallapoosa to the Chattahoochee where, according to tradition, the Creek Confederacy was formed. As we have seen, they spread eastward to the neighborhood of Augusta, Georgia, and southward to the Georgia coast, later withdrawing again to the Chattahoochee, the Coosa, and the Tallapoosa.

One of the Muskogee tribes, which came to be known as Eufaula, lived for a time on Euharlee Creek, which flows into Etowah River just below the famous group of mounds of that name, and afterward moved by successive stages to Talladega Creek, Alabama, and to the lower course of the Tallapoosa River, where they gave off colonies to the Chattahoochee, and finally to the Red House Hammock north of Tampa Bay, Florida.

Some constituent members of the Creek Confederacy also have interesting histories. The Alabama in De Soto's time were in northern Mississippi, northwest of their historic seats. The Tasqui, whom he found on or near Canoe Creek in northern Alabama, were probably the Tuskegee, and at any rate the Tuskegee were there 25 years later. Here they appear to have split into two bodies, one of which moved up the Tennessee River and united with the Cherokee, while the other descended the Coosa to its mouth and long occupied a town close to the point where the Coosa and Tallapoosa rivers come together, though for a time they moved as far east as the Ocmulgee.

The Koasati, or part of them, were on Pine Island when De Soto passed through the country, and at a much later period Tennessee River was called by their name, River of the Cussatees. While they were still on the Tennessee, however, they were joined by another people mentioned in the De Soto documents, after whom that stream was sometimes called— the Kaskinampo. As I have already said, there is every reason to believe that this tribe was the Casqui or Casquin encountered by De Soto west of the [70] Mississippi in the neighborhood of Helena, Arkansas. Noting this fact and comparing with it the historic drift of the Alabama and Tuskegee, cognate tribes, and the similar drift of the Muskogee, Chickasaw, and Choctaw, we have, it seems to me, a possible clue regarding the identity of the people responsible for the effigy ware of the St. Francis region. It is perhaps not accidental that the only words of the Napochi language spoken by the tribe located historically nearest to Moundville are pure Choctaw. The relations between the Moundville culture (or phase) and that of the Middle Mississippi seem quite evident, and so does much of that of the Tennessee, while we find traces of it in the lower level of Nacoochee, in the Hollywood Mound close to Augusta, Georgia, and at Etowah.

Besides historic and traditional evidence tending to link the Muskhogeans with northeastern Arkansas, there are linguistic and cultural suggestions that at one time the Muskogee proper were nearer the Caddo than were the Chickasaw and also that they were far enough north to have adopted at least one word from Shawnee—Pinwa, or turkey. To this may be added the cultural evidence given in my first paper (page 5). It is possible that most of the effigy region was occupied by the Muskogee and the Alabama group but that the ancestors of the Choctaw and Chickasaw were southeast of them. It appears to me that the eastern part of the Gulf region retains indications of occupancy by many different peoples, and I suspect it will take very careful work to unravel the pre-history of the section.

The only information we have regarding pre-historic movements of the Tunica Indians is the fact that their name has been retained by an aboriginal site in northwestern Mississippi called Tunica Oldfields and slender evidence that the Pacaha Indians found by De Soto a little north of this point, but on the opposite side of the Mississippi, spoke the Tunica language. What their history was back of this period is a mystery, as is how they happened to get so far north of their congeners, the Chitimacha. However, the Chitimacha point to the Natchez country for their own origin, and therefore they may themselves have come down the Mississippi. Possibly these two peoples and the kindred Atakapa were pressed down from the northwest by Caddoan and Muskhogean tribes; the Mississippi archaeologists Messrs. Ford and Chambers have found evidence that the Tunica were predecessors of the Natchez in at least part of this area. The culture of the Tonkawa, Karankawa, and Coahuilteco was rude when white men first came in contact with them, and evidence is lacking that it was ever anything else.

If the Indians of south Florida were, as seems probable, related to the Hitchiti and Choctaw, the question arises how they got by or through the more remotely connected Timucua, but we now know that in De Soto's time there was a Timucua-speaking band of Indians—the Tawasa—living in central Alabama, and it is possible that Timucua occupancy of the peninsula may have been later than has generally been supposed. Since they [71] shared with the Creeks the possession of totemic clans, it is possible that they advanced into Florida in front of the Muskogee immigration, and were preceded by some Muskhogeans of the Choctaw connection, since the Choctaw were entirely without totemic subdivisions. Possibly they were the ancient occupants of St. Andrews and Choctawhatchee bays in west Florida, the extensive remains of which have been explored but not exhausted by Clarence Moore.

I have left for last the consideration of our mysterious wanderers the Yuchi, whom we have seen to occupy so many different points in the Gulf region. These were the Chisca of De Soto and Pardo, whose homes in the middle of the sixteenth century were on the western flanks of the southernmost Appalachian chains. Detached bodies of Yuchi are known to have been on the middle Tennessee, possibly also on the Ohio and Green rivers, while some got as far north as La Salle's fort in Illinois. Almost all references to them mention the fact that they were occupying stockaded towns. De Soto's messengers did not reach their country, but in 1566 Morgano, Pardo's lieutenant, captured two stockaded towns belonging to them. In 1677 the Spaniards and Apalachee found the Florida Yuchi also living in a stockade; and about the same time Needham and Arthur visited a Yuchi stockade on the headwaters of the Tennessee. If the Westo were Yuchi, we may add the stockaded Westo town visited by Henry Woodward of South Carolina somewhere near Augusta, Georgia. Another testimony is as to the high type of Indian represented by them.

Here then we have the problem: a people consisting of a number of distinct bands moving from the north from time to time and occupying widely separated parts of the country, where they usually protected themselves behind stockades. The Yuchi language has been studied, and though it is undoubtedly closer to the Muskhogean-Siouan languages than to any others, it differs widely from even these. A people who remained apart from all others long enough to have developed such a distinct tongue must infallibly have had at one time an equally distinct culture, though the culture would not be as fixed as the languages and may have been lost soon after they left their ancient home. The question, then, for the archaeologist to solve is: What was the ancient culture of the Yuchi? Is it perhaps represented in some culture already investigated by the archaeologist? Could it have been the Hopewell or Adena culture of Ohio, or one of those discovered farther west?

Let us briefly summarize: we find Algonquian peoples in more recent times seeming to radiate from the region of the Great Lakes and presumably associated with the Great Lakes culture of the archaeologists. We find

Iroquoian threads leading to a point near the upper Ohio and those of the Caddoans pointing eastward. Might they be traced to a junction point? Siouan tribes diverged in the main from the Ohio River region as a whole, a number of Siouan tribes being traceable directly thitherward, [72] and one of them at least seeming to be identified directly with what has been called the Fort Ancient culture; thus it is probable that Siouan tribes were responsible for certain of the cultures found farther toward the northwest. The Natchez and their allies appear to have shared a culture found widely extended in the region with which the tribe was always associated, while many of the other Muskhogean peoples seem to have spread from the middle Mississippi, leaving us with the hypothesis that the so-called middle Mississippi culture was created by them. Traditions of the Tunican stock point to the north, and possibly they came in from the northwest, yet they preceded the Natchez in parts of their territory. Certain types of pot rims have been gathered from Tunica sites, but it remains to be seen whether these and other traces will enable us to extend their pre-history still farther into the dim past. Traces of the Timucua will probably be found northwest of the peninsula of Florida, but as yet we lack definite information of any such. Finally, and perhaps most important of all, archaeologists must set before themselves the quest for a culture that may be tied into the mysterious Yuchi.

In conclusion, I am going to venture some general speculations relative to the pre-history of the eastern United States. It appears to me, then, that the Algonquian peoples may have been among the earliest occupants of the section, and that at a remote period they may have extended much farther south and southwest than at present. Possibly they may have once occupied the entire country to the Gulf. This earlier extension is indicated by the work of M. R. Harrington on the upper Tennessee, by the work of McKern in Wisconsin, Strong in Nebraska, and Keyes in Iowa. Later, or in addition, there came waves of people from the west, possibly down the Missouri or around the Gulf from Mexico, but more probably across the southern Plains via Red River. If they came from the north, they may have reached the Black Hills by the Great Northern Trail (see Appendix, Figure 5) and crossed from there to the Platte or Kansas rivers, or possibly the Arkansas, or they may have kept on farther south and worked across by the Red as just suggested. Or again, they may not have moved eastward until they had lived for a long period on or near the Rio Grande.

This latter is, indeed, indicated by the bluff culture of the Ozarks, explored by Bushnell and more extensively by M. R. Harrington, and perhaps connected with the oldest remains found by Webb and Funkhouser in Kentucky and the Black Sand culture of Illinois. These phases seem to belong to an earlier epoch in the history of the Southeast than any of the rest, which are distinctly non-Puebloan. Do they represent an early stage in the evolution of southeastern culture, or a kind of island projected out of the Southwest and having no decisive influence on the cultural evolution of the region as a whole? That the seat of this was so near the seat of the effigy culture is probably not accidental. It is one of the problems for archaeologists to solve.

[73] Testimony to movement of peoples into the Gulf region from the west is yielded by physical anthropology and linguistics. The first shows us that the broadheads of the section find their nearest relatives westward and southward rather than northward. Linguistics indicates that the tongues of the Southeast, from southern Texas to Florida, are kindred in structure and probably had a single origin in the remote past, as nearly as a single origin may be predicated of a language. Proof of this is contained in the accompanying table, showing points of relationship between Coahuilteco, spoken on both sides of the Rio Grande, and languages as far east as Florida. Apart from these, however, and equally independent of the Algonquian tongues are the Iroquoian and Caddoan families, and it is perhaps not accidental that certain points of resemblance have been noted between these two. Did they, at some period in the past, form a connected belt from east to west between the Algonquians to the north and the Gulf tribes proper? Or are we to rely rather on physical characteristics and seek a common origin for the Caddo and the southern Siouans? Did they have something to do with the unidentified cultures of the Middle West? These, again, are problems for archaeologists to solve.

Table Showing Results of a Comparison of 117 Terms in Six Siouan Languages, the Number of Closest Resemblances Being Indicated in Each Case

Biloxi	and	Ofo	38
Biloxi	and	Hidatsa	36
Biloxi	and	Tutelo	34
Biloxi	and	Dakota	33
Tutelo	and	Hidatsa	29
Ofo	and	Dakota	26

Tutelo	and	Dakota	25
Hidatsa	and	Dakota	24
Hidatsa	and	Mandan	20
Dakota	and	Mandan	20
Ofo	and	Hidatsa	17
Biloxi	and	Mandan	13
Tutelo	and	Mandan	13
Ofo	and	Tutelo	12
Ofo	and	Mandan	11

In short, aside from the bluff dwellers of the Ozarks and one or two other very primitive groups, we have evidence of at least three major bodies of people in the eastern United States: the Algonquian about the Great Lakes and northward, with extensions far to the south in historic times and much farther probably in pre-historic times; the Gulf people, later split into a considerable number of groups and showing affinities as far west as the Rio Grande, and how much farther only future investigation [74] can tell; and an intermediate group made up possibly of the Iroquoian and Caddoan stocks. At an early date these last may have driven the Algonquians north. Later the Gulf tribes, represented by Siouans, may have split them in two, to be separated in turn by an Algonquian recoil. What relation, if any, did the Ozark Indians bear to these?

I put these thoughts forth as suggestions, not as dogmas. They are to be examined, revised, completely made over, and in some cases doubtless wholly rejected. But it is important beyond all else for you archaeologists to tie your discoveries into known tribes, after having done which you may trace them back into the mysterious past as far as you will, and your work will have more interest and more meaning for you and for us all.

As a sort of first aid to the pursuit of this enterprise, I subjoin a chart (Appendix, Figure 6) which gives the general location of certain key sites known to have been occupied by historic tribes, which it would be very desirable to link up with specific types of archaeological remains.

Adjournment

After informal discussion, during which Dr. Swanton answered several questions from members of the conference, the meeting adjourned, to meet the following morning at 10:00.

Tuesday, December 20

Morning session, Dr. Guthe, chairman

The third day of the conference was devoted to discussion of methods of pre-historic research, using the three definite divisions of fieldwork, laboratory work, and publication. Formal discussions of these three subjects were given in the morning, with informal round tables in the afternoon. Dr. Guthe, chairman of the morning session, called first on Dr. Cole, who spoke on "Exploration and Excavation."

Dr. Cole:
Something appears to be radically wrong with the popular idea of the archaeologist. He is usually pictured as a bewhiskered old gentleman sitting in an attic studying a pot or bone. Yet here we are, a group of beardless youths, attempting to solve the problems of pre-history. Apparently times have changed, and, perhaps, our attitude toward our materials has likewise changed.

[75] What is the purpose of archaeology? Perhaps the simplest answer is "To make the past live again," as Dr. Guthe has phrased it in one of his reports.

We are no longer satisfied with the mere possession of collections, in learning the development of ceramic art, or of stone chipping. We are not content with a mere classification of cultures or in knowing the successive occupations of an aboriginal site. Interesting as these topics may be, they are, as Goldenweiser says, only scientific gossip, and as valuable as gossip in general.

We now seek to know the total culture of each group we study—not isolated facts. When we know our cultures and plot them on the map we see that they tend to take on geography. As we excavate we can learn the sequence of cultures and thus can view our subject in time and in space.

With such materials at hand, we are in a position to study the dynamics of cultural growth. We can see to what extent a culture is dependent on its environment and to what extent its accommodation to local conditions is governed by its prior history. We can see what happens to objects and crafts which diffuse into an area, and we can see the effects of contacts on people through trade and migration.

But such a study requires the gathering of all the evidence. It means that every possible technique must be employed. Nothing may be discarded as useless until its meaning is fully considered.

The first step in securing such evidence should be the survey. Proceeding from township to township, from county to county, we should place on the map every known aboriginal site of whatever character. Its present condition, owner, and other pertinent information should appear on a survey sheet. All objects in the hands of local collectors should be studied and the frequency of their occurrence noted.

In this manner we can secure an idea of our problems in advance. We can have some idea of the probable cultures we may encounter. We find desirable sites for excavation, and often we learn the exact place from which many objects have been taken.

The survey party should also make careful notes concerning the geology and geography of the region, for such information may often lead to significant results. As an example: a party working under the direction of Dr. Arthur Kelly of the University of Illinois recently uncovered the historic Indian settlement of Kaskaskia located on Plum Island. A survey of the adjacent region showed that in the not-distant past the island had been a part of the mainland and at different times had been subjected to overflow. Excavation of the village site revealed the life of the [76] historic Indians and also of a long period of occupancy antedating the arrival of the French. Not content with this information, the party cut below until it came on an old land surface. There, far below the recent level were found campfires evidently built by man. By studying soil profiles and successive layers of deposits above the fires, geologists were able to assign an approximate date of 4,000 years to the time when man first occupied the site.

The survey then is an important preliminary to more intensive study.

The second step is excavation. Here a great responsibility rests on the archaeologist, for as he excavates he destroys a page of history which can never be rewritten. Whatever part of its story he fails to decipher is gone forever. It thus becomes his duty to secure all the record. He has no justification to record and preserve only those things which interest him. The next investigator may need just those facts which he has passed by. Unless one is willing to make a complete study, he has no right to open an aboriginal site. An object which at the moment seems to be trite and trivial may prove to be the key to important problems.

Before any work is started, test pits should be sunk on all sides of the site. The depth of the upper humus layer should be noted and the extent of surface leeching tested by acid. Careful study of soil conditions should be made and special attention given to evidences of underlying humus layers or other signs of geological changes.

Once the nature of the surrounding territory is known it is time to begin work on the site. If it is a mound it is staked out in squares (5-foot squares are usually most convenient). A trench is started at right angles to the axis of the mound and is carried down at least 2 feet below the base. The face of the trench is now carried forward into the mound itself by cutting thin strips from top to bottom. At the same time, the top is cut back horizontally for a distance of a foot or more. If this procedure is followed it is possible to see successive humus layers as well as to note all evidences of intrusions.

An excellent example of the value of this method is afforded by the work of Mr. George Langford in his excavations near Channahan, Illinois. Here a low mound was built over several bodies. For years it stood undisturbed, and a layer of humus matter gathered on its surface. Later, a second group of Indians added greatly to this mound and buried their dead. Again the mound stood undisturbed until a humus layer formed on its surface. This occurred three times before the historic Indians cut in from the top and interred their dead. As Mr. Langford cut into this mound the successive humus layers stood out as clearly as natural stratification. Intrusive burials broke through these lines and indicated clearly the period to which they belonged.

Every object encountered should be carefully noted and its location and relationships recorded. All indications of intrusions should especially be watched for.

While opening a mound near Lewistown, Illinois, the University of Chicago field party encountered, about 2 feet below the base of the mound, the bones of a fossil musk ox. Close to it and at the same level lay a human skeleton. The ox belonged to the late glacial period. Here apparently was an authentic case of Pleistocene man. But the technique just described—of cutting both horizontally and vertically—revealed a dim line of disturbance which started well up in the mound and extended below the human skeleton. It was evident that the burial belonged to the period of the mound, not to glacial times. Thus a fine newspaper story was utterly ruined, but the truth of the situation was revealed.

Shells encountered in the excavation may tell of wet, humid times; plant and animal life may likewise give evidence of climatic conditions. Even the burial mound may tell us something of the food supply of the builders.

Potsherds may reveal the art of weaving, while their temper, shape, and decoration may tell of the movement and contacts of cultures.

Skeletons may likewise reveal movements of people, while examples of pathology may give us hints of what happened to the earlier settlers in the land. Hundreds of skeletons have been thoughtlessly destroyed by excavators. Yet it is a queer idea that an investigator may be interested in objects and not in the man who made them.

Once a skeleton is encountered, it should be carefully cleaned by means of brush and orangewood stick. No bone should be moved until the whole body is revealed and until its relationship to all neighboring objects is carefully noted. It should then be fully recorded, photographed, and the bones numbered before it is disturbed. Many skeletons are so fragile that they cannot be moved until treated. If damp they should be protected from sunlight and allowed to dry for a few hours. They should then be treated with successive applications of acetone until the bone is thoroughly penetrated, after which ambroid should be applied until the bone is solid.

In opening mounds, we should keep full records and see that they are written up each night. The difference between looting and scientific work depends to a large extent on the completeness of the record.

While not as spectacular as mounds, village sites and refuse heaps often contain the most important data. Objects here are usually [78] broken, but in them the former occupants have often left us an unintentional yet very complete record of their daily life. For example, the lakefront park at Chicago is really a great city dump. Just imagine what a fertile field this will be for the archaeologist of a thousand years hence. What is going into the dump today?—automobiles, incandescent lights, radios, and other objects of our culture. What went in 30 years ago?—oil lamps, horseshoes, wagon wheels, corsets. If we go far enough back we may come to the days of Fort Dearborn and back of that to the Indian.

A village site is best uncovered by a series of trenches much like those used in mound work. A cut is made down to undisturbed soil and the earth is thrown backward as excavation proceeds. Horizontal and vertical cutting should be employed in hopes of revealing successive periods of occupancy. The worker should never come in from the top. He should never be on top of his trench, otherwise lines of stratification will almost certainly be lost.

A village site or refuse heap of considerable depth may indicate a long period of occupancy and afford an opportunity to study cultural change within a given group. In such an excavation we obtain the best and most complete record of food supply, of house types, and the like.

Grave sites occur in many regions and if properly excavated may reveal a story far more complete than the settlements in the open.

We should always be on the watch for evidence of early man. In the glaciated districts, in old fills, and in ancient lakebeds we may hope to find traces of human occupancy. Today we have many hints that man may have reached America in Pleistocene times. Careful study of all commercial excavations and river cuttings may reveal positive proof of this early invasion.

Time has allowed me to touch only briefly on methods and objectives.

We should always keep our aims in mind. We should gather all the evidence, and we should keep a full record. Finally, we should have frequent conferences of the workers in adjacent fields, for in this manner we widen our horizons and perfect our methods.

The chairman next called on Mr. Judd, who spoke on "Laboratory and Museum Work."

[79] *Mr. Judd:*
If a specimen is worth collecting, it is worth preserving. This is our motto at the National Museum, and this is my theme for the present occasion. But let us bear in mind that there will always be an honest difference of opinion as to what specimens are worth collecting in the first place. Men have their hobbies. Some hoard arrowheads; others accumulate stamps, pewter mugs, or white elephants. On a Washington golf course I once secretly cursed an elderly gentleman who followed our foursome for a time and explained his irritating search at every tee by confessing that he was greatly intrigued by the variety of discarded match papers.

The man who cornered the market on cigar-store Indians had to rent a warehouse. Somewhere, someone is probably collecting early American bathtubs. We are not all motivated by the same interests. What one would save, another would destroy. But each of us engaged in museum work, each one of us who feels the urge to start a collection, should frequently ask ourselves this question: What is our purpose in collecting?

Too often we have no definite purpose. Too often in our collecting we merely gratify an instinct to accumulate and hoard personal property. Squirrels and packrats have this same instinct. Acquisitiveness is a well-known human trait and has been ever since our early progenitors first began to throw rocks at their neighbors. As boys we often came home with pockets bulging with marbles, string, dead mice, and what not. As men, many of us go right on collecting oddities of one sort or another and build museums to house them.

Because our personal interests vary, museums differ. Every existing museum had its inception in the idea of some one individual. The number of museums which died with the passing of their respective sponsors surpasses the number of those now functioning. Only an adequately endowed and ably administered museum can be expected to carry on from one generation to the next. Like libraries, museums are adjuncts to our educational system; they preserve, in visual form, a brief index to mankind's intellectual achievements. Museums and libraries are the means through which we may best transmit to the future our cultural heritage from the past.

In general, museums may be divided into three classes: museums of the fine arts, museums of natural history, and historical museums. This conference is concerned solely with institutions of the latter type, namely, the historical museum. For we are dealing with history whether our efforts be directed toward the Indian tribes who inhabited our country prior to the advent of Europeans or toward the period of Spanish, French, and English colonization. Where written history begins, there pre-history ends. Our objective is to lower the barriers between history and pre-history; to push forward into the unknown that dim line which marks the [80] frontier of recorded history.

Libraries guard and protect those rare documents which pertain to the Spanish conquest of the Western hemisphere. The Native American tribes whom the Spaniards displaced left no written language; we are able to gauge their degree of civilization only through study of such fragments of their culture as ethnologists may glean from the narratives of early travelers and from study of the material remains abandoned by the ancestors of those historic tribes. It is perfectly obvious that these latter remains can never tell a complete story, since only the imperishable artifacts will have survived under average conditions. No matter how carefully the excavator may perform his task, he cannot recover what no longer exists. But a trained eye may detect a potsherd or a lump of clay with the imprint of textiles, and bones from a rubbish heap may indicate, in part, the local food supply. Every worked fragment, in itself perhaps of little moment, is a thread for the fabric we seek to reconstruct. A scrap of papyrus means more to an Egyptologist than to one unable to read Egyptian hieroglyphics. Likewise, a stone ax or an earthenware vessel, the data for which have been carefully preserved, is more than a mere curio to the serious student of southeastern pre-history. An old Indian village site is a patchwork puzzle that can be solved, at least in part, by the properly qualified historian.

Now there is a widespread belief that in the interpretation of prehistoric Indian remains, one man's guess is as good as another's. Only those of us who have devoted some years to the subject know the fallacy of this. If I, as a museum worker, were to venture an opinion, I should say that the properly qualified student of ancient civilizations needs, first of all, an abundance of common, ordinary horse sense. He should be familiar with the findings of his predecessors in the same field. He needs, in addition to other, equally diverse subjects, at least a working knowledge of all those natural phenomena which influenced the lives and thoughts of the primitive people he is studying—biological, botanical, and geological. He needs keen perception and the wit to observe and accurately record every significant factor. It is patent that all the varied information packed into the notebooks of such a student cannot be fully digested and utilized in the field. Notebooks and specimens must go to the laboratory for final study prior to publication. And every explorer knows that no matter how diligent he may be in the field, preparation of a report shows several points neglected or even overlooked.

Because it is humanly impossible to conclude a given bit of research at the site of excavation—no project may be regarded as complete until its results are ready for publication—the observer will naturally bring to his laboratory a considerable quantity of fragmentary material which has a direct bearing on his current investigations but which is of little use otherwise. All this rubbish—and rubbish it often is—will [81] naturally receive full consideration during the preparation of a report, but whether or not it is to be preserved thereafter will depend on the combined judgment of the investigator and the museum curator. If discarded, such material rightfully should be returned to its place of origin; if retained, it should be so marked that there can be no possibility of future confusion.

This latter admonition is based on personal experience. Even the most painstaking individual is not infallible. I have made more mistakes than any other man now living, but one of the mistakes I have learned to avoid, as a result of my 20-odd years in the National Museum, is that of setting aside material before it is adequately numbered and described. The costliness of human carelessness or inattention is astounding. Laboratory assistants are not always interested in the work for which they are paid: I have found a Hopi bowl from Arizona catalogued as from Florida just because the gentleman who gave it happened to be living in Florida at the time of presentation; I have found specimens wrongly numbered and wrongly stored; I have found catalogue cards hopelessly involved and confused; I have spent days tracing and correcting the errors made by a wandering mind. And in consequence of all this I have reached the conclusion that no phase of museum work is more important than that which has to do with the preparation of records. These are not made for the present alone; they must be perfectly intelligible to the unknown student of the future. It may be taken as a maxim that the scientific usefulness of any museum specimen varies in direct proportion to the completeness and trustworthiness of the information concerning it.

What is our purpose in collecting? More specifically, what is our purpose in collecting the tools and utensils of Indian tribes no longer living? If museums are primarily adjuncts to our educational system, as they should be, then the materials placed in museums must be regarded as worthwhile only insofar as they are informative and instructive. And this depends both on the manner in which the material is displayed and the completeness of the data relating to it. To the true scientist, these data are of the utmost importance; specimens serve merely to illustrate the data. During my years as a student of pre-history, I have witnessed the passing of the curio cabinet that used to be a feature of every parlor. For the most part, Indian relics are no longer collected merely for the sake of the relics themselves. Curio dealers have thin picking today. And yet exaggerated stories of the prices they received 30 years or more ago keep fresh in some quarters the impression that anything of Indian origin has a real commercial value.

Scarcely a week goes by but what there passes over my desk at least one letter offering Indian relics for sale. The letter may refer to a single specimen found while plowing a field, or to a collection which has cost its owner heavily in time and money. Such material is invariably useless for scientific purposes; I know of no museum today that buys Indian [82] relics. The individual who is now collecting such objects under the belief that he is making an investment which will pay a handsome profit whenever he wishes to cash in on it is doomed to disappointment. I have seen this proved time and again. Except for the fun one gets out of it, there is no longer any reward in collecting Indian artifacts. Neither is there any likelihood that the collector or his heirs will ever recover more than a fraction of the original cost. Only last spring the executors of a certain estate tried in vain to sell for a few thousand dollars a collection which is said to have cost over one million. Museums are not in the market today for Indian artifacts. Such objects are of worth only insofar as they serve to illustrate the unwritten history of the people who made them.

Because Indian relics did formerly have a certain commercial value, the business of making fraudulent antiquities came into being. Frauds are still made in every country where archaeology awakens local interest. But even a lazy man will not waste his energy faking antiquities unless he can profit through sale of them. Four Kentuckians were fined three years ago for using the United States mails to sell alleged Indian relics made on a grindstone run by an old Ford engine out in the woodshed. On various trips to Arizona, I have seen quantities of beautifully chipped arrowheads and blades, of so-called ceremonial obsidian. The purchaser is assured this bright red or green material is extremely rare and precious; he is not told that it comes chiefly from the lanterns that guard railway switches along the Santa Fe and Southern Pacific. I am reliably informed that the cost of replacing this stolen glass is not inconsiderable. It seems incredible that even a schoolboy should mistake red or green glass for obsidian, but the truth is that these fake arrowheads and knives are now found in private collections far beyond the borders of Arizona.

I cite these two examples from the many which have come to my attention merely by way of illustrating the fact that the business of manufacturing fraudulent Indian relics is still alive and that the museum curator must constantly be on guard. There is no law against this practice just as there is no law against boring wormholes in antique furniture. So long as there is any commercial demand for old Indian artifacts there will be an incentive to fake them. Most of these frauds go to individuals, but eventually they will be offered, in good faith, as gifts to museums. The museum curator must therefore be critical not only of the material he accepts but also of the accompanying data, since the person who fakes a relic will not hesitate to tell a plausible story to support it.

Only in recent years have museums come to discriminate as regards the material they accept, even as gifts. New acquisitions must fit in with the recognized policy and purpose of the institution; they must be needed in one of the two categories into which museum collections are [83] invariably divided, namely, that for study and that for exhibition.

It has been my observation that museum exhibits must be simply and attractively arranged if they are to invite attention. The average visitor rarely takes time for more than a casual glance at any case. Crowded shelves tend to confuse and discourage one. "You can lead a horse to water but you cannot make him drink" is an old proverb that applies particularly to museum visitors. Whether or not an exhibit of Indian artifacts will convey to the uninformed any understanding of, or appreciation for, aboriginal arts and industries, must depend very largely on the skill and understanding of the curator. Museums are places of impressions. Museum exhibits, therefore, should be so arranged as to present definite pictures to the lay mind.

The trained investigator, on the other hand, insists on drawing his own conclusions. He will turn to the museum storeroom rather than to the exhibition halls. His chief desire is full and accurate information regarding the objects before him. It is for this reason that I keep stressing the point that the scientific usefulness of any collection depends largely on the data available for each specimen. The locality where it was found must be exactly known; its associations *in situ* and the conditions of finding are equally important. The historical value of any collection is thus dependent on the thoroughness with which the original excavator recorded his observations—the clarity of his report, whether published or unpublished. By this criterion, the average collector of curios makes little or no contribution to human knowledge if he looks on an artifact merely as something to possess and utterly disregards its historical significance. If we may judge the future by the past, most collections now privately owned will eventually be offered as gifts to museums, and many of those gifts will be declined for lack of adequate data.

As a museum curator, I cannot too strongly emphasize this necessity for full and complete information about every specimen. Fifty or even 25 years ago, this did not seem at all important. More recently, however, the study of pre-history has developed into a science, and the foundation stones of every science are accuracy and impartial judgment. We are gradually eliminating from the national collections artifacts which do not measure up to current standards, for the requirements of the future will be even more exact than are those of the present.

At the National Museum we no longer purchase archaeological material; we no longer accept it as a gift unless it will serve either for study or for exhibition. Mere curios have no place in our present program; storage space is too precious to waste on objects that cannot be utilized. Such material as we do accept is recorded with the utmost care and with a view to the permanency of those records. The Smithsonian Institution was created by act of Congress in 1846, with an endowment provided by an Englishman [84] 20 years previously to establish in Washington "an institution for the increase and diffusion of knowledge among men." We are still guided by the broad principle Smithson indicated. Since our current methods are based on the experience of many workers over a period of almost 90 years, perhaps you will be interested in learning just how we handle a new acquisition.

For purposes of illustration we may liken the National Museum to a funnel with three tubes representing, respectively, the Departments of Anthropology, Biology, and Geology. Each of these three major units soon separates into two or more lesser divisions as, to use the Department of Anthropology for an example, into the Divisions of Archaeology, Ethnology, and Physical Anthropology. Thus official matters poured in at the top of our funnel lead directly, and without delay, to the staff member most interested and, presumably, best informed.

Each new acquisition, whether it be 1 specimen or 1,000, is regarded as an accession. An accession number is assigned by our receiving office and, eventually, all the original correspondence and memoranda pertaining to a given transaction are returned to that office and filed under the accession number in fireproof steel cases. These form the official museum records. But lists of the specimens received, with descriptive details, are retained by the department and the division. Thus the permanency of our records is reasonably assured; pertinent data are immediately accessible in the division office having custody of the material itself.

A much simpler system would suffice, of course, for a smaller institution. But the National Museum, now receiving each year over 100,000 specimens of every description—geological, biological, and anthropological—has evolved what I regard as a very practical, and not too complicated, method of keeping track of every single item and being able to put a finger on it with a minimum of time and effort expended.

Let us assume that one of our staff returns from archaeological investigations in Arizona. His collection consists of various artifacts, fragmentary and otherwise: a number of human skeletons, animal and bird bones from old rubbish heaps, and perhaps some specimens of local flora more or less closely related to his studies. This entire collection is given one accession number, but the botanical material is sent to the Division of Botany for identification; the bird and animal bones go to the Divisions of Mammalogy and Ornithology; the skeletons to the Division of Physical Anthropology, and the cultural artifacts from the ruins to the Division of Archaeology. With the collector's field catalogue at hand, each division prepares a descriptive list of the material it receives, and this information is embodied in the report of the investigator. After his researches have been completed, we discard the fragmentary material not desired for permanent preservation and of no use to investigators in other institutions.

[85] In the Division of Archaeology every specimen receives a catalogue number. Objects from a single room or dwelling are covered by numbers in sequence; less significant objects such as arrowheads, bone awls, or miscellaneous beads may be grouped under one number. Each catalogue card carries a full description of the specimen recorded and any pertinent information from the collector's field notes. A history card carries a brief summary of the expedition and the site excavated; a key card, slightly larger than the others and of different color, provides an index to the individual accession and tends to separate it from the next in our card file. Collectors' cards show at once the number of accessions received from any one source and refer directly to each of these in our main catalogue, which is arranged numerically. Both our exhibition and study series are arranged by states rather than by cultural areas since most visitors are more interested in seeing things from their own home state than in noting the distribution of aboriginal civilizations.

As described, all this may seem very complicated. In actual practice, it is surprisingly simple. Our chief concern is to record all available information regarding every specimen at the time it is received. We are trying to anticipate the needs of future students just as we wish our predecessors had anticipated our present requirements. Eventually we shall have cross-references, by culture area and type of specimen, to our main catalogue.

Like many others, the national collections include a considerable number of rare Indian artifacts which are of little historical value today because the information originally furnished was too meager. The exact locality represented, the associations *in situ,* the depth, and other circumstances of finding are among the data essential if any one specimen is to play its full part in "the increase and diffusion of knowledge." Precision is the watchword of the modern archaeologist. Our conviction at the National Museum is that "a specimen worth collecting is worth preserving," and a specimen worth preserving for the future should be accompanied by all the information the present can supply.

Before a collection is catalogued the individual objects should be carefully cleaned, repaired, and restored if necessary. In the field, the investigator rarely has time to do more than keep his notes in order, but he will naturally have at hand such preservatives as may be required for delicate specimens. Partially decayed basketry, textiles, wood, and shell—objects which may disintegrate on exposure—should be treated at once with a celluloid solution or some other substance that protects and strengthens. Any excess of celluloid, together with the dirt it covers, may be removed at leisure in the laboratory. Lack of a little attention in the field has caused the loss of many precious artifacts that might have been saved, just as lack of a little thought and care in the laboratory has brought destruction to many items which can never be replaced.

[86] In characteristic American fashion, we have been prodigal of our resources in pre-history. We have been inclined to save only the whole pieces or those of striking form and color. This is especially true of the amateur collector and the commercial digger, but students of pre-history, including myself, have often missed the important feature while looking for the spectacular.

We have been particularly wasteful with skeletal remains. Broken skeletons have been tossed lightly aside. Now this material, if not too far gone, can be utilized by other investigators if not by ourselves. The School of Dentistry at Columbia University welcomes jaw fragments for its study of aboriginal dentition; other research institutions are tracing age variations and the effect of disease. For these investigators, whose efforts may prove of greater benefit to mankind than our own, we can well afford to go to a bit of trouble. Skeletal remains, like objects of material culture, are distinctly limited in quantity. Neglected at the time of exhumation, they are gone forever.

Treated with celluloid solution or gum arabic before removal, even soft and fragile bones may be recovered for laboratory study. Strips of cotton cloth, or burlap, dipped in flour paste and applied to half-decayed bones will strengthen and support them pending proper attention. Paleontology has reached its present high position among the sciences through patience with fragmentary bones. Archaeologists may profit greatly by adopting field methods of the paleontologist. As a laboratory man, I must plead with the excavator for greater consideration of the skeletal remains he disinters.

Our laboratory work is concerned not only with preservation of the skeletons of pre-historic men but of every vestige of the arts and industries known to those men. Preserving stone artifacts is a relatively simple matter; preserving earthenware vessels, shell ornaments, carved wood, etc., may prove both complex and tedious. The methods we employ at the National Museum vary according to the nature of the specimen and the nature of the disease working to its destruction. Knowledge of the physical and chemical properties of the object to be treated, complete understanding of the changes which have already taken place and of those which will follow if the specimen is left untreated, together with an understanding of the consequences of any treatment—all this is necessary if unhappy results are to be avoided.

It is first necessary to discover and remove the cause of any degenerative changes. If the specimen is damp, it must be dried; if it contains salts of one kind or another, these must be dissolved out; if it is covered with a calcareous deposition, this must be removed in a dilute acid bath. Too much acid may stain the specimen or destroy its surface slip and the applied decoration, in the case of earthenware vessels. Pre-historic pottery from Peru and certain sections of our own Southwest often [87] contains minute alkali crystals which expand in our humid, eastern climate. The flaking which follows—and, if unchecked, it will soon reduce the vessel to a handful of dust—can be stopped by repeated soaking in water and, later, by applications of dilute celluloid solution or paraffin dissolved in gasoline. Paraffin darkens the specimen and is therefore less desirable than common sheet celluloid dissolved in acetone. This latter solution will satisfactorily meet most of the preservative problems arising in any laboratory.

Although our laboratories have from time to time employed various chemicals in the treatment of wood, basketry, bone, shell, etc., we have come of late to depend very largely on Ambroid, a commercial glue manufactured by the Ambroid Company, in Brooklyn, New York. Like celluloid in acetone, Ambroid serves for mending pottery or almost anything else; thinned to a proper degree by the solvent which comes with it, Ambroid may be brushed or sprayed on any specimen liable to decay. Duco, a cement product of the DuPont Company, is equally useful where only small surfaces are involved.

Museum specimens are not only affected by changes in temperature and humidity but many of them are also subject to destruction from moths, beetles, and other insect pests. All substances these insects might feed on must be poisoned and should be examined at regular intervals. When eggs or larvae are present we sprinkle the threatened items liberally with gasoline and leave them for a day or two in a tightly closed box. For poisoning, we still spray or brush the objects with a solution made from one ounce of corrosive sublimate in half a gallon of water and half a gallon of alcohol. Any liquid treatment, however, must be followed by careful brushing in order to restore specimens to their former appearance.

A new and safer poison, ethylene dichloride-carbon tetrachloride, was developed in 1927 by two investigators for the U.S. Department of Agriculture. Mixed three parts of ethylene dichloride by volume to one part of carbon tetrachloride, this poison is more toxic to insects and less harmful to specimens and to man than any other yet devised. Ethylene dichloride has recently replaced the highly explosive carbon disulphide heretofore used in our Division of Birds, thus holding out the hope of a longer and more useful life for all ornithologists.

But I will not anticipate this afternoon's session with details of laboratory methods in the preservation of museum specimens. While the literature on the subject is very meager, those interested will find many helpful suggestions in the latest paper of which I know, that by Mr. D. Leechman in the 1929 annual report of the National Museum of Canada, published in Ottawa, 1931.

[88] Curators and museum preparators grow gray early; theirs is a diversified and responsible task. Theirs is the responsibility of saying what specimens can and should be preserved and what treatment should be applied. Some field men apparently expect the laboratory worker to perform miracles, to correct all errors of judgment made in the field and even to save what, for lack of timely attention, has already passed redemption. So the field often unjustly blames the laboratory while the laboratory blames the field. As a matter of fact, the archeologist at his pre-historic village site and the preparator in his laboratory have a common purpose, namely, to recover and preserve for the future so much as is possible of extinct, aboriginal civilizations. Toward this end we may cooperate to our mutual advantage. I venture the prediction that this initial conference on southern pre-history will shortly prove as highly beneficial to its participants as did that which first brought about a unity of interest and effort among southwestern students with a resultant speedy solution of many problems then existing in the archaeology of the Southwest.

The chairman next called on Dr. Wissler to speak on "Comparative Research and Publication."

Dr. Wissler:
Many persons look on the objective in archaeology as merely the digging up of something. Once the object is found, the matter ends; the object is put into a cabinet to be "gloried over."

Those who preceded me in this program have strived to emphasize the emptiness of such a procedure. A good maxim might be "Don't dig until you have a problem."

In other words, excavation should be considered as merely part of a problem—the first in order of time, but not necessarily the most important. Publication should be regarded as a vital part and not something apart and incidental to be done sometime when one has nothing else to do. By publication I mean preparing an adequate written record of your diggings. This may be printed in the usual manner or filed in an accessible place.

It sometimes seems queer that so much enthusiasm should be given to digging and so little to the study of what is on the surface. Careful, painstaking surface studies are needed as much, or more, than digging; there is no reason for feeling discouraged if you cannot finance a large digging enterprise, for some of the best archaeology can be found without a spade.

[89] But I am to speak of publication. To many, publication means merely the manufacture of the book by the editor and the printer. On this point I have little to offer, nor am I certain that this body cares to discuss that subject. In passing, however, one may venture to remark that the present tendency toward expensively illustrated papers in archaeology may be responsible for some of the difficulties in budgeting institutions. The conservation of space and illustration is an important problem, but the process must begin with the writer, before the manuscript reaches the editor. In any case, we can do little about it here since it is the large institutions that must lead the reform.

Yet if the job of the editor and printer is out of our field, the manufacture of the manuscript is not. The editor can and does do a lot to realize the writer's intent, but he cannot be expected to rewrite the paper. He can go far in trimming your literary tree to look respectable, but he cannot hide a disreputable bush by hanging respectability on its branches.

Nor is the manufacture of a good paper merely a matter of writing; in fact writing is the very last stage of the process.

Those preceding me have emphasized the method for digging, recording, etc., and I suppose the next step is to study the evidence, or whatever you choose to call the objective materials—accumulated, artifacts, photos, notebooks, etc. Like every other academic subject, archaeology has certain more-or-less standardized procedures for such a study. I suppose these may be formulated something like this:

a. Record of the excavation
b. Descriptions of artifacts
c. Comparative statements

In my opinion, everyone who does a bit of digging should prepare a manuscript report covering "a" and "b" and see that a copy is filed in some appropriate institution. If publication is possible, well and good; but when a manuscript is properly prepared and filed, the "digger" has discharged his first duty, and not until he has done this should he feel free to dig again.

I need not comment further on the form and plan suited to such procedures; if you are interested, the subject can doubtless be handled in the round-table conference. In my opinion, inability to print does not excuse the archaeologists in a state from seeing to it that steps are taken to acquire a suitable file for such reports.

[90] We turn to the third task connected with an archaeological project, viz., the comparative statement. As just said, the duty of the "digger" has been discharged when he files a report covering the other two topics in the traditional outline. Proceeding with the third topic is and should be optional.

If, on the other hand, one sets out to do a well-rounded bit of research, then he must meet certain other requirements of tradition. He should inform himself as to the history of archaeology in the vicinity of his operations; he should review the ethnological data as to aboriginal occupation, one problem being to determine whether the site can be associated with a historic tribe. If it should turn out to be a historic site, then one may either gather all the historic data available, attempt to correlate these data with his own findings, or leave the problem to Dr. Swanton.

We return for the moment to the comparative study. In the usual meaning of the term, this implies that you carefully compare your artifacts and site data with those of neighboring sites, states, etc. And so by similarities you attempt to relate your finds with those of other investigators or sites— in other words to the existing scheme of knowledge for the archaeology of the region. Such a study may take you far afield and require that you review in detail materials in distant institutions, something possibly impractical to the resident archaeologist.

You are no doubt getting impatient with this commonplace discussion and wonder why we do not come to the point. In other words, get back to fundamentals. After all, what are the objectives in archaeology?

The usual answer to this question is that we are attempting to recover lost history, particularly the lost history of culture. This has some similarity to the statement that geology seeks to recover the lost history of the earth. Classical archaeology seeks to push back the limits of dated history by excavation but begins to lose interest when inscriptions cease, I suppose on the assumption that where no writing existed there could be no history to get lost.

Our point of view is slightly different. We do not expect writing, but we recognize that knowledge of the past can be had by digging, that if we are careful enough we can read the record in the ground, and that we are still blind to much of the record. Our archaeology is yet scarcely out of the illiterate stage. True, the noting of postholes, traces of weaving, etc., is now commonplace, but it is still a strange language to many. All of us are still blind to many parts of the record. We need, then, some intensive search for new techniques by those especially equipped for such work. In the meantime, our duty is to learn the current language.

[91] Returning to the objectives in archaeology, we see its goal to be the reconstruction of a history of culture. Dr. Swanton has often called attention to the possibility of working backward from sites occupied by tribes when Europeans came on the scene. The southern states offer special opportunities for such approaches. But always the fieldwork of archaeology must aim at discovering more and more traits of cultures.

Wherever one sees an artifact he is moved to ask, How old is it? This universal question may well take the form as to which of two artifacts is the older. There is no escape from this inquiry because a frame of time sequence is the one necessary achievement for the archaeology of any region. If we consider the archaeology of Alabama, we need, first of all, this time sequence. But probably the first step in such an achievement is to differentiate cultures.

What is meant is that we conceive the remains discovered as belonging to a group, village, or tribe living a well-rounded order of human life. The few things one finds serve as indications of the whole. One might so cast the descriptive paper he writes as to reveal the culture of the tribe instead of launching into long descriptions of his artifacts, as if none had ever been found before. As an example one may cite a paper by H. I. Smith, the "Pre-historic Ethnology of an Archaeological Site." A perusal of the outline of that paper will show a surprising amount of reconstruction in culture; practically every specific statement is based on artifacts and excavation data. The ideal of archaeology should be to extract more and more such information from the site. Smith's site was done years ago; with the advance in technique, the "digger" today can scarcely be excused for not giving a much fuller account of the cultures pertaining to the site.

Comparing the data for one site with that of another will ultimately determine the extent of any given culture.

But mere comparative study is hardly enough for the ultimate realization of all these objectives. Dealing as we do with unwritten history, it can scarcely be expected that actual dates can be given in our time sequences. The best substitute is to correlate culture differences with changes in natural phenomena. The correlation of artifacts with geological and biological material has given us the great sequences of the Stone and Bronze ages, but as you all know, similar gross correlations have not worked so well in North America.

Yet it does not follow that no correlations are possible. Ingenuity, the cooperation of other sciences, and wide comparisons have made some progress. In proof of this I need but mention a few examples.

[92] In New Mexico and Arizona the accidental discovery of what is called tree-ring dating as recovered a lot of lost history. Of course, that method is not directly applicable in this area, but the natural phenomenon of climatic change may eventually apply. It has been shown that many travertine deposits in caves are seasonally banded and so subject to age counts. So it is that those who come after us may handle the cave problem much better than we can do it.

But some correlations with natural phenomena can be undertaken now. So far scarcely anything has been done toward relating recent topographical changes to archaeological finds. While it is true that erosion changes are greater in less humid regions, changes are underway—streams are deepening their beds, floods changing the lay of bottomlands, etc. As a suggestion of what may result if one spends a reasonable period of time in the observation and study of a limited area, also without a spade, I refer you to a paper in the *American Anthropologist* of 1931, by Frank Ryan, presenting a method of interpreting the sorting power of water in surface erosion, and thus distinguishing cultures. By noting how artifacts were distributed with respect to surface-drainage courses, this author has been able to set up time differences between them, something like what one finds in a case of stratigraphy. Ryan's method might not work in Alabama, but a similar intensive study of situations will certainly bring additional insight.

The point is not that these suggested methods can be applied directly to any locality but that other relations may be discovered suitable to the place and time. In other words, the state archaeologist can be a student of his locality just as well as a collector of curios.

In short, the state archaeologist may well consider topography, soil, climate, flora, and fauna as having a place in his comparative study. Studies of living tribal cultures have been enriched by considering these facts. For example, if one were to take up a comparative study of archaeology in this state (Alabama) he might first consider the probable pre-historic environment. We are not so far removed from the immediate pre-Columbian period that its geography cannot be reconstructed.

First, considering the South as a whole, can the region be comprehended under well-defined climatic zones?

A recent paper by Hinsdale, University of Michigan, shows what may be done by correlating data on flora and climate. Hinsdale finds the distribution of archaeological remains coincident with certain forest belts. Also, he has correlated the historic Indian population with this same forest distribution. Recently I chanced to see some maps on the snowfall of New England; it turned out that the region of dense Indian population was also the area of least snowfall.

[93] Yet this is not all. It has been said that science may yet ruin mankind, because each new discovery points the way to many new possibilities. We began this discussion by taking note of the possibilities in establishing time sequence by correlating with natural phenomena. If we reflect a moment, it appears that as the last ice age came to an end the climate in these states could scarcely have been what it is now. It is quite possible that even when aboriginal man came into this area, the climate was unfavorable to agriculture. But, you say, this is all guesswork. So far it is. Yet read a recent article in the *American Anthropologist* by Sears and note that by the simple process of exploring samples of certain soils one can tell the flora of the time, and in turn the climate.

So in conclusion we may summarize by noting that the primary duty of the "digger" is to record all the essential facts observed. He may stop there if he likes. But he should engage in sufficient comparative study to put him in the "literate class." Remember that comparative study will go on and on; the data of the digger will be used over and over, but on its quality and integrity will depend the ultimate product. On the other hand, the "digger's" job can rarely be done again. Everything his eye has not been trained to see will be lost forever. And everything his eye saw will be lost if not set down in his report.

Afternoon session, December 20
Round-table discussions, field methods:

Dr. Warren K. Moorehead acted as chairman in the early part of the afternoon of Tuesday, for the round table on fieldwork. Dr. Moorehead expanded his remarks made during the Monday evening session. These were to the effect that we should establish, insofar as possible, the boundaries of our various mound-builder cultures; that he intended to prepare an outline map to be sent to archaeologists specializing in the mound-builder field. He felt that we should adopt the point of view of our legal friends, which is crystallized in the term "preponderance of evidence"; for instance, that we should not claim that Hopewell people were in Louisiana, or Muskhogean in the Iroquois country of northern New York, or Caddoan in Wisconsin merely because a few artifacts have been discovered apparently representing art concepts of these cultures. He advocated caution on the part of observers in both field and museum.

Dr. Moorehead outlined some 14 or 15 units which he employs in measuring the status of mound-builder cultures. Some of these would apply to Ohio, Georgia, or Arkansas. Other units would appear in a given mound-cultural area and be absent in the others. He further advocated more concise or accurate mound-culture nomenclature.

[94] Dr. Strong, Professor Webb, Mr. Dellinger, and others commented at some length on field methods they used.

Laboratory methods:

Dr. Strong acted as chairman for the second period of the afternoon session, during which some time was taken up with a discussion of the best methods of preserving perishable archaeological materials both in the field and later in the laboratory. Among others, the methods given in two books on this subject were recommended ("Preservation of Antiquities," translated from the German of Dr. Friedrich Rathgen, Cambridge, England, 1905; and "Technical Methods in the Preservation of Anthropological Museum Specimens," by D. Leechman, 1929 annual report of the National Museum of Canada, Ottawa, 1931).

It was also suggested that representative series of potsherds from every site be closely studied and analyzed, as well as the unbroken pieces. By such methods, the complete range of a ceramic type may be given, rather than special characteristics based on individually selected pieces. Arrow points and knives from individual sites may also be classified according to any desirable variant of the Thomas Wilson classification, annual report of the U.S. National Museum for 1897, published in 1899 (pp. 887–944). Both the potsherd and chipped-point classification for any well-worked area can thus be briefly and vividly presented in a chart, with the list of the types above and the various sites down the side. When the data are arranged on such a chart, the main cultural groupings based on these two characteristics are readily observable.

It was also suggested that in arranging archaeological types for illustration they be picked out roughly in proportion to their actual occurrence at the site, thus enabling the reader to quickly gain a correct impression of the range of cultural material recovered.

Mr. Brannon discussed the nature of house sites excavated in parts of Alabama. Dr. Cole and Dr. Strong also brought out the fact that complete excavation of house sites, rather than trenching such remains, had yielded the best results in Illinois and Nebraska. It seemed generally to be agreed that only by laying bare the entire floor of an earth lodge or similar aboriginal dwelling could an accurate plan of its arrangement be secured. Since regional comparisons of pottery, artifact, burial, and house types are especially desirable at the present time, more or less uniform methods of preparation of data in the laboratory are essential. Naturally, such considerations brought up the necessity and problems of publication, and Dr. Strong turned the meeting over to Mr. Webb, who presided over the final section of the afternoon session.

[95] *Research and publication:*
In the round-table discussion on research and publication, the necessity of publication was stressed by Dr. Webb, in order that the science of archaeology may profit by the result of work done. Dr. Moorehead and Dr. Wissler cited examples of exploration within their knowledge which had yielded much information and voluminous field notes but which had never been published. Such field notes, no matter how accurately kept, are available to only a very few persons, and thus the value of such explorations are largely lost to the science, since very few of even those active in the field of archaeology know of their existence.

Dr. Moorehead stressed the desirability of writing up field notes before they became "cold" by the passage of time and the intervention of other duties.

Dr. Strong raised the question of how to meet the increased difficulty of publication because of decreased budgets.

Dr. Wissler pointed out that if a decrease in budget made it impossible to publish a complete report on fieldwork, the report should be written just as carefully and fully as if immediate publication was expected, and the manuscript filed in a safe depository and report of its existence made available. This, he stated, was the very least that should be expected of every fieldworker. Dr. Moorehead commended Dr. Wissler for referring in his address of the morning to the type of paper prepared and published by Harlan I. Smith for the American Museum many years ago, in which much ethnological data was reconstructed from the archaeological evidence. Dr. Moorehead suggested this type of report for the careful consideration of the younger men in the field.

The wisdom of publication of state archaeological surveys which contained maps showing accurate locations of pre-historic sites was discussed at some length. The opinion was expressed by Dr. Strong and supported by others that in states where there was still the possibility of destruction by "pot hunters," it was not wise to publish surveys showing exact locations of sites, as such information might be the means of destruction of valuable sites by calling attention to them. It seemed to be the general opinion that it was important to publish state surveys in abstract form, giving conclusions as to regional boundaries rather than exact location of sites.

The question was asked if any societies were regularly publishing on fieldwork in the South, other than the Alabama Anthropological Society. Mr. Brannon made a brief statement of the method used by this society and indicated some of the difficulties overcome.

[96] [Dinner and evening session, December 20, Dr. Guthe presiding.]

Dr. Guthe:
(Following informal remarks) We recognize today the importance of the pre-history of the South and the significance which many of the local problems have in relation to similar problems in other regions, such as the northern Mississippi Valley and the Southwest or Texas area. During the past few years, our knowledge of the detailed materials, of their relationships to the cultures of the past, and of the forces which they represent has increased tremendously, due in large measure to the accumulated experience of fieldworkers in securing evidence. Yet we must not forget that the foundations for the present research are to be found in the work of the pioneer archaeologists in the Mississippi Valley. It was their discoveries and their interpretations which were the incentives for further archaeological study.

Today we are the scientific descendants of such men as Cyrus Thomas, whom you all know as the man who made the survey of the mound area in the latter part of the nineteenth century, and of W. H. Holmes, whose monumental work on the pottery of the Mississippi Valley, contained in the 20th *Annual Report of the Bureau of American Ethnology,* laid the foundations for the interpretation of the ceramic industries of this area. I know that we all wish we might have met these men and their colleagues, and have had them here with us at this conference. Since, however, that is impossible, I have asked two of our number who have worked with these older men to tell us something about them.

Dr. Guthe then called on Dr. Wissler, who spoke in appreciation of Cyrus Thomas and his work, and on Mr. Judd, who told of the work of W. H. Holmes.

At the end of the evening, Dr. Guthe called on Mr. Dellinger, who had asked to be allowed to speak. After expressions of appreciation, he stated that, in view of the success of this meeting, a number of the members had indicated a hope that similar conferences of southern workers might be held in the future. He pointed out the obvious advantages that such conferences would have and mentioned the impetus that had been given to the study of Plains archaeology by the Great Plains conferences of the last two years. He called for comments and suggestions, and Mr. W. B. Colburn moved that an informal committee be authorized by the group to make the necessary arrangements for another conference in the fall of 1933. He suggested S. C. Dellinger as chairman, with H. E. Wheeler and Fred B. Kniffen as the other two members of the committee. The motion was seconded and passed by the group.

[97] The conference closed with the adoption of the following resolutions, presented by Dr. Knight Dunlap, a member of the National Research Council:

Resolved: That we the members of the Conference on Southeastern Pre-History, gathered in Birmingham, Alabama, December 18 to 20, request the chairman of the Division of Anthropology and Psychology to transmit to the National Research Council and the Committee on State Archaeological Surveys our appreciation of the impetus given to scientific archaeology in this region by this conference.

Resolved: That we congratulate Dr. Poffenberger, Dr. Guthe, and Mrs. Britten on the successful organization of this conference; we congratulate the local anthropologists on the excellent provisions which were made for our comfort, and we are profoundly appreciative of their warm hospitality; and we congratulate ourselves on having participated in a conference which has been highly pleasant and greatly profitable.

Adjournment, at 9:30 P.M.

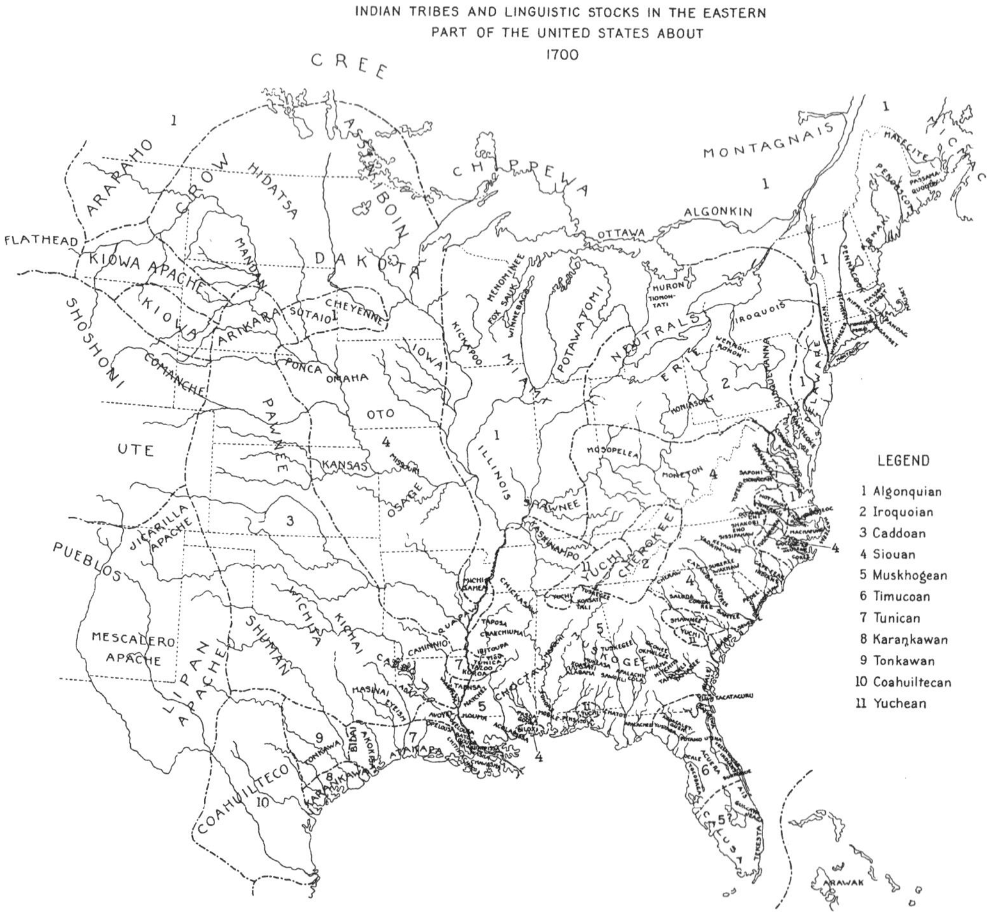
INDIAN TRIBES AND LINGUISTIC STOCKS IN THE EASTERN
PART OF THE UNITED STATES ABOUT
1700
LEGEND
1 Algonquian
2 Iroquoian
3 Caddoan
4 Siouan
5 Muskhogean
6 Timucoan
7 Tunican
8 Karankawan
9 Tonkawan
10 Coahuiltecan
11 Yuchean
CREE
ASSINIBOIN
CHIPPEWA
MONTAGNAIS
ALGONKIN
OTTAWA
ARAPAHO
CROW
HIDATSA
FLATHEAD
KIOWA APACHE
MANDAN
DAKOTA
KIOWA
ARIKARA
CHEYENNE
SHOSHONI
COMANCHE
PONCA
OMAHA
IOWA
OTO
PAWNEE
UTE
KANSAS
MISSOURI
OSAGE
ILLINOIS
MIAMI
POTAWATOMI
NEUTRALS
HURON
IROQUOIS
ERIE
MOSOPELEA
SHAWNEE
YUCHI
CHEROKEE
CHICKASAW
JICARILLA APACHE
PUEBLOS
MESCALERO APACHE
LIPAN APACHE
SHUMAN
WICHITA
KICHAI
CADDO
CHOCTAW
MUSKOGEE
TONKAWA
ATAKAPA
KARANKAWA
COAHUILTECO
CALUSA
TEKESTA
ARAWAK

[Figure 1]

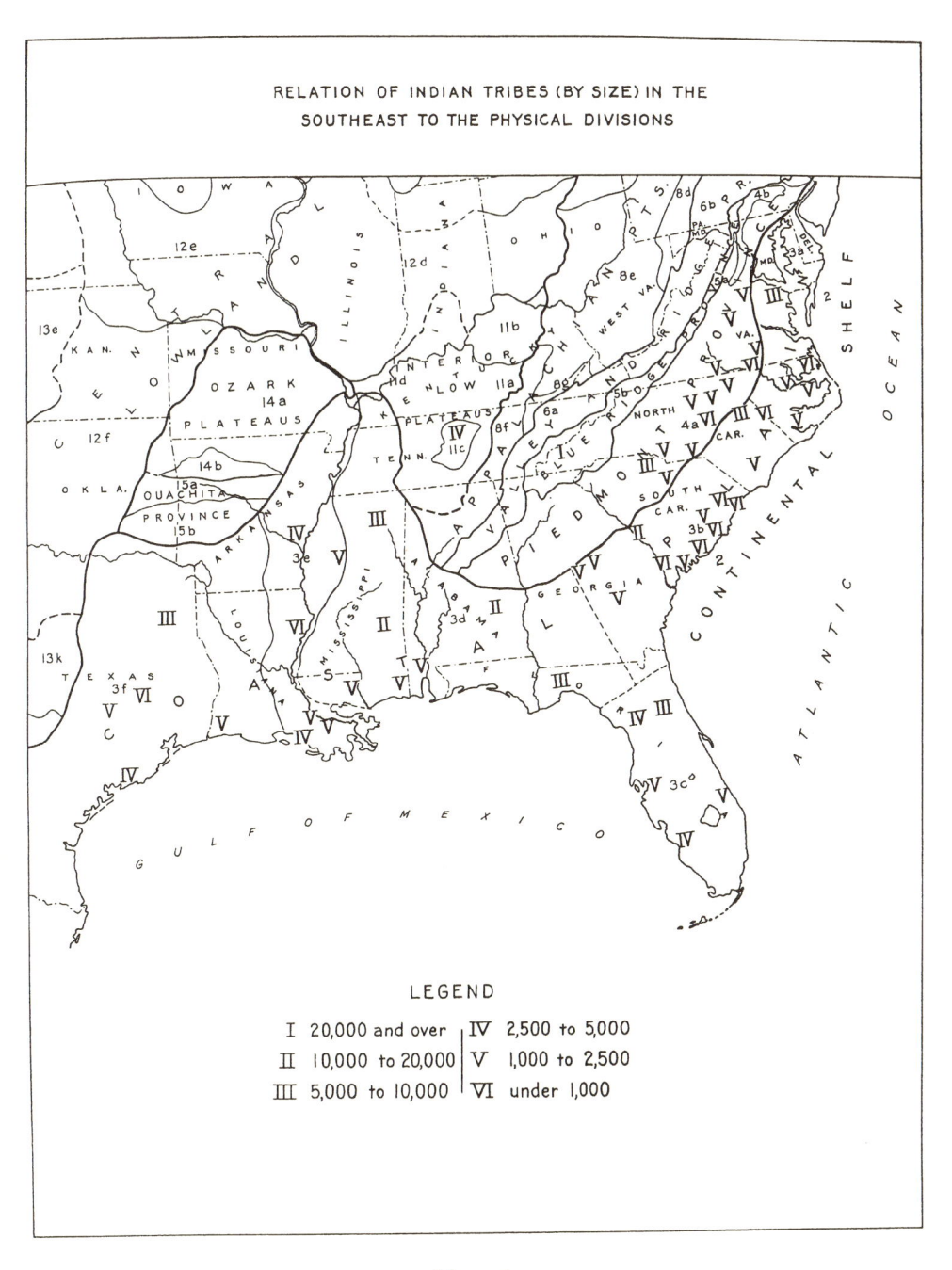

[Figure 2]

Atlantic Plain
- 3. Coastal Plain
  - a. Embayed section — 3a. Submaturely dissected and partly submerged terraced coastal plain
  - b. Sea Island section — 3b. Young to mature terraced coastal plain with submerged border
  - c. Floridian section — 3c. Young marine plain, with sand hills, swamps, sinks, and lakes
  - d. East Gulf Coastal Plain — 3d. Young to mature belted coastal plain
  - e. Mississippi Alluvial Plain — 3e. Floodplain and delta
  - f. West Gulf Coastal Plain — 3f. Young grading inland to mature coastal plain

Appalachian Highlands
- 4. Piedmont Province
  - a. Piedmont Upland — 4a. Submaturely dissected peneplain on disordered resistant rocks; moderate relief*
  - b. Piedmont Lowlands — 4b. Less uplifted peneplain on weak strata; residual ridges on strong rocks
- 5. Blue Ridge Province
  - a. Northern section — 5a. Maturely dissected mountains of crystalline rocks; accordant altitudes
  - b. Southern section — 5b. Subdued mountains of disordered crystalline rocks
- 6. Valley and Ridge Province
  - a. Tennessee section — 6a. Second-cycle mountains of folded strong and weak strata; valley belts predominate over even-crested ridges
  - b. Middle section — 6b. The same, but even-crested ridges predominate over valleys except on east side
- 8. Appalachian Plateaus
  - d. Allegheny Mountain — 8d. Mature plateau of strong relief; some mountains section due to erosion of open folds
  - e. Kanawha section — 8e. Mature plateau of fine texture; moderate to strong relief
  - f. Cumberland Plateau — 8f. Submaturely dissected plateau of moderate to strong relief section
  - g. Cumberland Mountain — 8g. Higher mature plateau and mountain ridges on eroded open folds section

Interior Plains
- 11. Interior Low Plateaus
  - a. Highland Rim section — 11a. Young to mature plateau of moderate relief
  - b. Lexington Plain — 11b. Mature to old plain on weak rocks; trenched by main rivers
  - c. Nashville Basin — 11c. Mature to old plain on weak rocks; slightly uplifted and moderately dissected
  - d. Possible western section — 11d. Low, maturely dissected plateau with silt-filled valleys

Interior Highlands
- 14. Ozark Plateaus
  - a. Springfield-Salem Plateaus — 14a. Submature to mature plateaus
  - b. Boston "Mountains" — 14b. Submature to mature plateau of strong relief
- 15. Ouachita Province
  - a. Arkansas Valley — 15a. Gently folded strong and weak strata; peneplain with residual ridges
  - b. Ouachita Mountains — 15b. Second-cycle mountains of folded strong and weak strata

*Degrees of relief are herein spoken of as low, moderate, strong, and high. As used here, *high* relief is measured in thousands of feet; moderate relief in hundreds of feet. *Strong* relief may be anything approaching 1,000 feet with a wide latitude on both sides.

[Ed. note: See previous page for locations.]

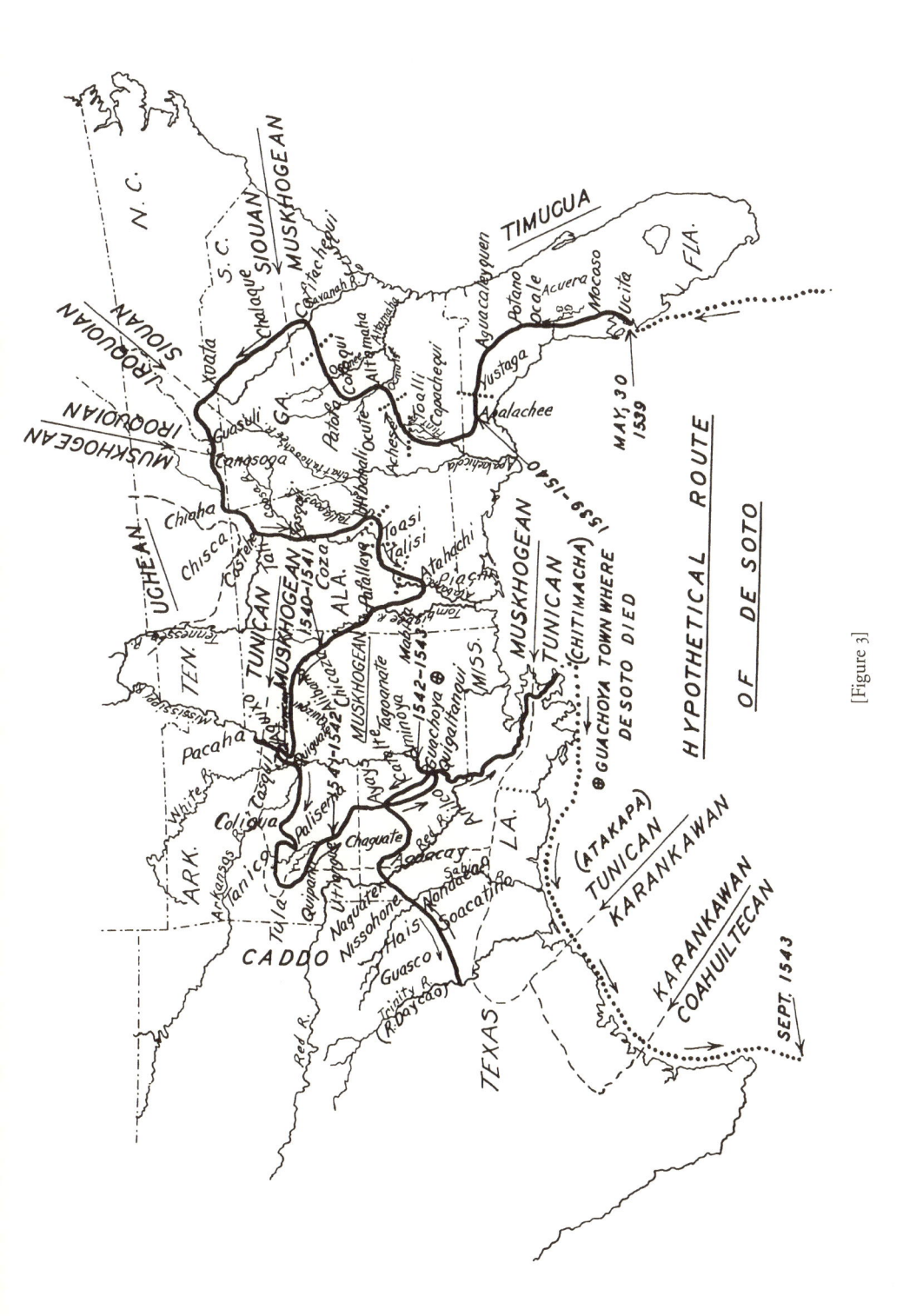

[Figure 3]

TRAITS IN THE SOUTHEAST

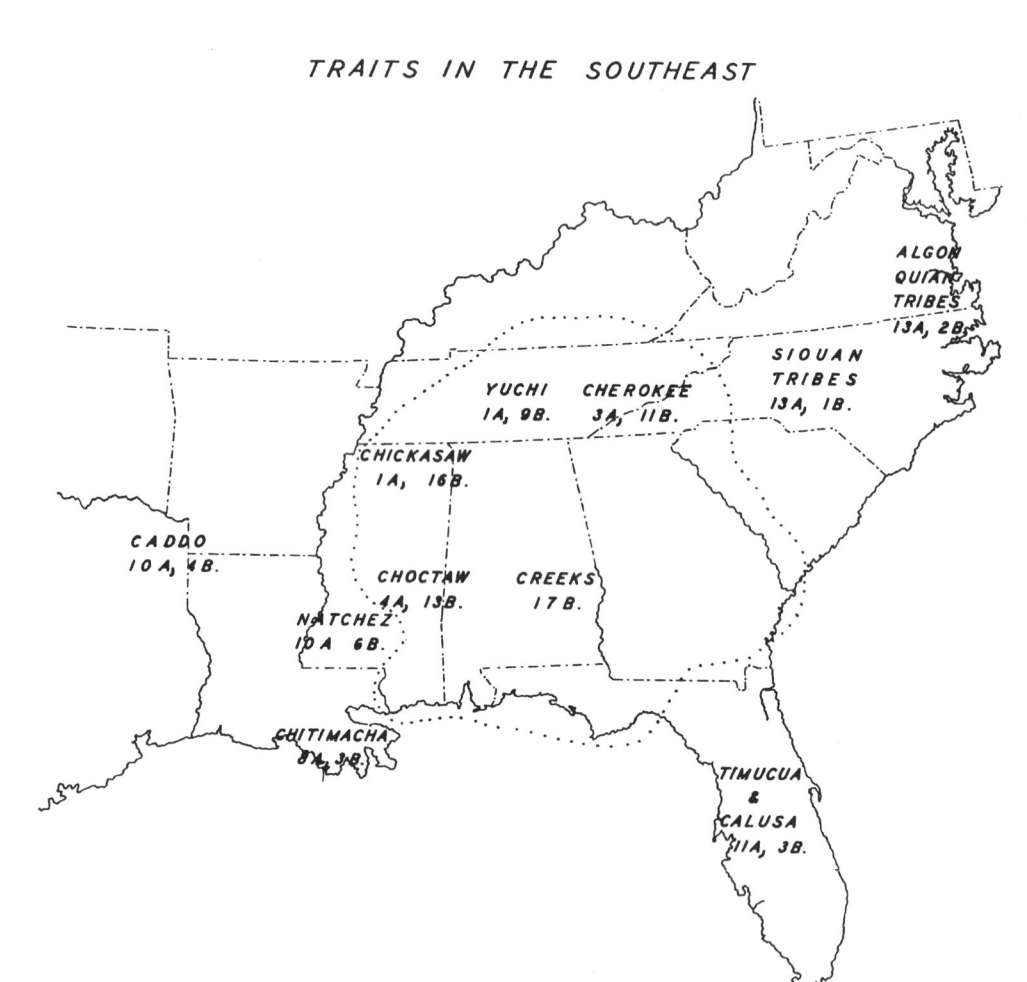

[Figure 4]

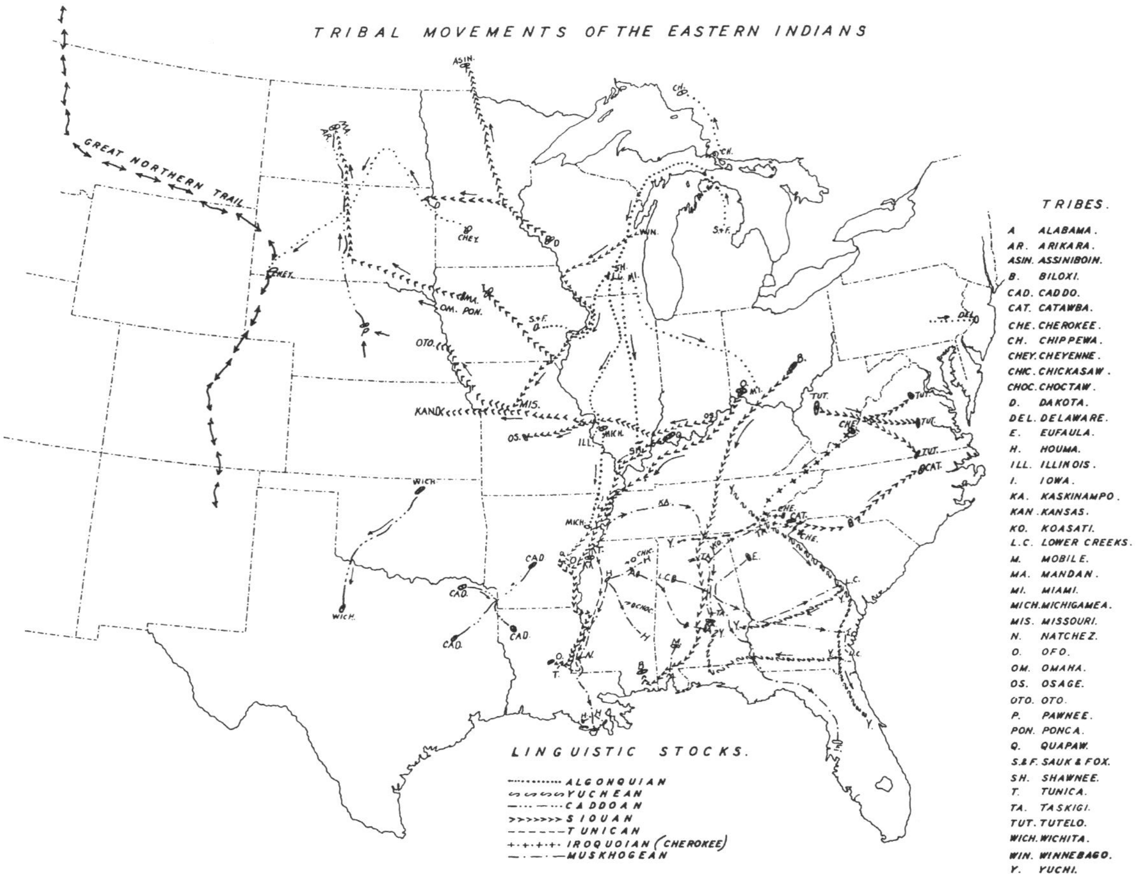
TRIBAL MOVEMENTS OF THE EASTERN INDIANS
GREAT NORTHERN TRAIL
LINGUISTIC STOCKS.
ALGONQUIAN
YUCHEAN
CADDOAN
SIOUAN
TUNICAN
IROQUOIAN (CHEROKEE)
MUSKHOGEAN
TRIBES.
A. ALABAMA.
AR. ARIKARA.
ASIN. ASSINIBOIN.
B. BILOXI.
CAD. CADDO.
CAT. CATAWBA.
CHE. CHEROKEE.
CH. CHIPPEWA.
CHEY. CHEYENNE.
CHIC. CHICKASAW.
CHOC. CHOCTAW.
D. DAKOTA.
DEL. DELAWARE.
E. EUFAULA.
H. HOUMA.
ILL. ILLINOIS.
I. IOWA.
KA. KASKINAMPO.
KAN. KANSAS.
KO. KOASATI.
L.C. LOWER CREEKS.
M. MOBILE.
MA. MANDAN.
MI. MIAMI.
MICH. MICHIGAMEA.
MIS. MISSOURI.
N. NATCHEZ.
O. OFO.
OM. OMAHA.
OS. OSAGE.
OTO. OTO.
P. PAWNEE.
PON. PONCA.
Q. QUAPAW.
S.&F. SAUK & FOX.
SH. SHAWNEE.
T. TUNICA.
TA. TASKIGI.
TUT. TUTELO.
WICH. WICHITA.
WIN. WINNEBAGO.
Y. YUCHI.

[Figure 5]

[Figure 6]

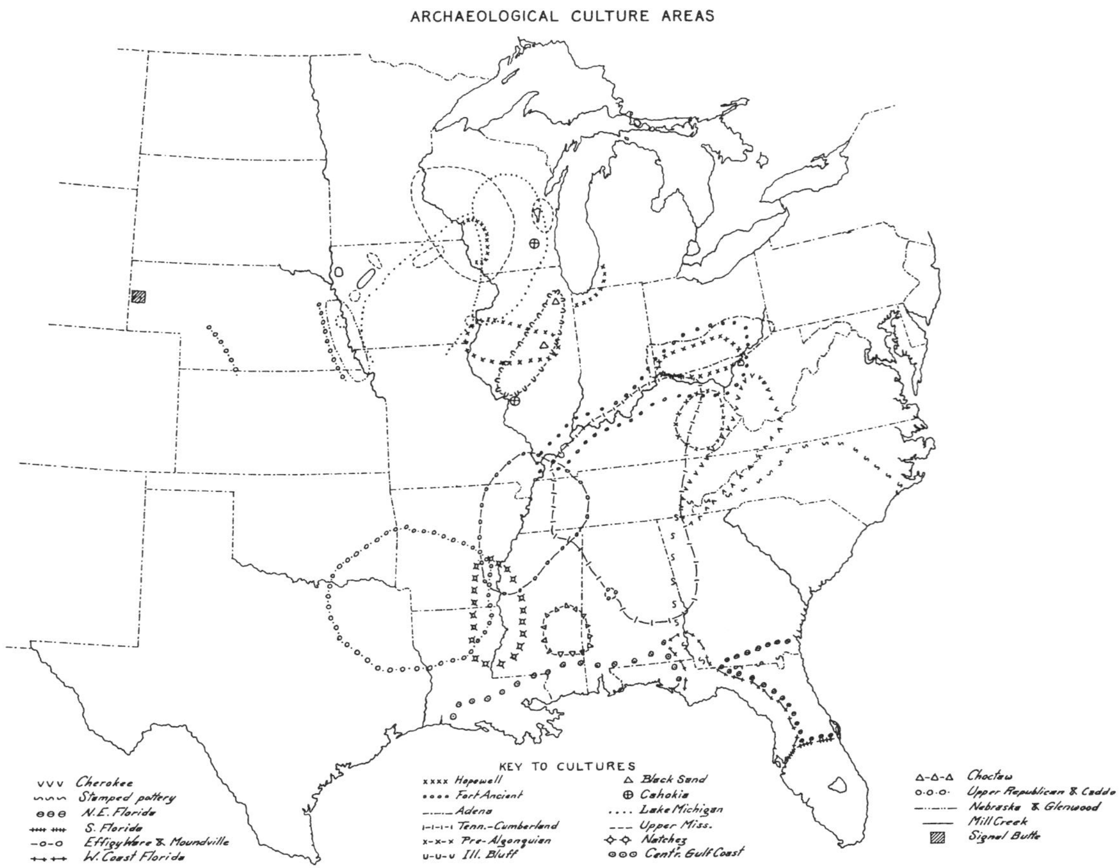
ARCHAEOLOGICAL CULTURE AREAS
KEY TO CULTURES
Cherokee
Stamped pottery
N.E. Florida
S. Florida
Effigy Ware & Moundville
W. Coast Florida
Hopewell
Fort Ancient
Adena
Tenn.-Cumberland
Pre-Algonquian
Ill. Bluff
Black Sand
Cahokia
Lake Michigan
Upper Miss.
Natchez
Centr. Gulf Coast
Choctaw
Upper Republican & Caddo
Nebraska & Glenwood
Mill Creek
Signal Butte

[Figure 7]

The Indianapolis Archaeological
Conference

A Symposium upon the Archaeological
Problems of
the North Central United States Area

Held under the Auspices of
the Division of Anthropology and
Psychology Committee on State
Archaeological Surveys
National Research Council
Hotel Marrott, Indianapolis, Indiana
December 6, 7, and 8, 1935

Issued by the
Committee on State Archaeological Surveys
Division of Anthropology and Psychology
National Research Council

Table of Contents

List of Delegates [iv]   337

Preface [v]   339

Friday Morning Session
Wisconsin—W. C. McKern [1]   341

Iowa—Charles R. Keyes [3]   343

Minnesota—Lloyd A. Wilford [8]   350

Illinois—Thorne Deuel [9]   351

Friday Afternoon Session
Indiana—Glenn A. Black [12]   354

Indiana (continued)—E. Y. Guernsey [14]   356

Ohio—E. F. Greenman [16]   358

Michigan—Carl E. Guthe [17]   359

Kentucky—Wm. S. Webb [18]   360

New York—William A. Ritchie [19]   361

"Siouan Tribes of the Ohio Valley"—John R. Swanton [24]   366

Saturday Morning Session
Frank H. H. Roberts, Jr. [35]   379

Fay-Cooper Cole (letter) [39]   384

Frank M. Setzler [42]   387

General Discussion [43]   389

Saturday Afternoon Session
James B. Griffin [48]   393

General Discussion [50]   395

Saturday Evening Session
Formulation of tentative classification for
the area under discussion [58]   403

Sunday Morning Session [63]   407

Tentative Classification Chart [69a]   414

Appendix
    "Certain Culture Classification Problems in Middle
    Western Archaeology," by W. C. McKern (reprint of
    Circular Series #17 of the Committee on State
    Archaeological Surveys) [70]   415

List of Delegates and Guests

Glenn A. Black, Indiana Historical Society

Amos W. Butler, Indianapolis

Thorne Deuel, University of Chicago

Emerson F. Greenman, University of Michigan

James B. Griffin, University of Michigan

E. Y. Guernsey, Indiana Historical Society

Carl E. Guthe, University of Michigan

Charles R. Keyes, State Historical Society of Iowa

Eli Lilly, Indiana Historical Society

W. C. McKern, Milwaukee Public Museum

William A. Ritchie, Rochester Museum of Arts and Sciences

Frank H. H. Roberts, Jr., Bureau of American Ethnology

Frank M. Setzler, U.S. National Museum

John R. Swanton, Bureau of American Ethnology

Carl F. Voegelin, Yale University

Erminie Voegelin, Yale University

Wm. S. Webb, University of Kentucky

Paul Weer, Indiana Historical Society

Lloyd A. Wilford, University of Minnesota

Preface

During recent years the amount of data relating to the archaeological cultures of the northern Mississippi Valley and the Great Lakes region has grown considerably, due in large measure to the increased field and laboratory research within this area. The attempts to define the several cultures and to determine their relationships demonstrated the need for a conference of the students actively concerned with the archaeological problems of this area to establish, if possible, a uniform methodology and a greater correlation of the investigations. In recognition of this need, and in response to a request from the Committee on State Archaeological Surveys, the National Research Council granted the funds which made possible this conference at Indianapolis.

The conference was called for the specific purpose of discussing the technical problems relating to the comparative study of the archaeological cultures in the upper Mississippi Valley and Great Lakes region. Detailed descriptions of the results of the investigation of individual sites were not pertinent to the meeting. The group of delegates was purposely kept small in order to ensure the freedom of informal discussion, and was confined to research students who were interested either in the archaeological problems of a restricted part of the area or in the comparative significance of these problems with relation to similar ones in other areas.

The first day of the meeting was devoted to a series of informal statements from the several delegates outlining briefly the cultural problems arising from a study of the archaeological materials occurring in their own state. These reports were arranged in a rough geographical sequence from west to east, as will be noted in the table of contents. The manuscript record made by Mrs. Dorothy Schulte of the Committee on State Archaeological Surveys was submitted to the speakers subsequent to the conference. In some instances extensive alterations and additions were made, which have served to increase the value of the statements as they appear in this report. It is inevitable that these revisions may contain interpretations and opinions reached by the speakers as a result of the conference.

With this general review of the current archaeological situation in the several states before them, the delegates devoted the second and third days of the conference to an illuminating and instructive discussion of the comparative significance of the problems presented, led by those delegates who were especially interested in this phase of the work. These statements received the same opportunity for revision as those made on the first day. [vi] The results of these deliberations were embodied, during the evening of the second day, in the formulation of a table of culture relationships within the area, which crystallized and recorded the combined judgment of all the delegates, based on individual experience and free discussion.

This table of cultural relationships is the outstanding contribution of the conference. It is based on a method of cultural classification first suggested by W. C. McKern many months prior to this conference and which had been generally accepted by the archaeologists of this region as a serviceable working tool. No attempt was made to define accurately the cultures listed in the table. Their relative position and classification were established after discussion, by the judgment of the delegates. It is expected that detailed technical reports will, from time to time, furnish the data by which these cultures may be defined accurately and their position in the cultural scheme established with reasonable precision. It must be emphasized that this table of cultural relationships has the status of a working hypothesis, which it is hoped may prove useful to archaeologists working in the area. It is subject to constant revision. In fact, during the preparation of this report, certain changes in the names and groupings of some of the aspects and foci were made in order that the table might represent more adequately the cultural classification as it is used at the present time.

There has been added as an appendix to this report a copy of a paper by W. C. McKern titled "Certain Culture Classification Problems in Middle Western Archaeology," which was read as the presidential address before the annual meeting of the Central Section of the American Anthropological Association in Indianapolis in May 1934. This paper constitutes the first concise statement of the principles on which this classification is based and the detailed methods by which it may be applied. It is included here because the discussions at the conference assumed that the delegates had a knowledge of its contents. No editorial revisions have been made in the original paper, although in the opinion of the author and others, time has made necessary certain changes. The most noteworthy of these is the substitution of the word "pattern" by the conference for the word "basic culture," which appears both in McKern's paper and in the discussions of the conference itself.

The Indianapolis Conference holds a significant place in the history of the development of Middle Western archaeology. It [vii] stimulated an increased coordination of research, through having made possible extended informal discussions among the leaders in the work; it recorded, through this report, the status of the problems of the region in the winter of 1935; and it made possible the formulation of the first comprehensive table of archaeological cultural relationships within the area.

In closing this preface, I wish to record, in the name of all the delegates, our appreciative thanks to the National Research Council and its Committee on State Archaeological Surveys for having made the conference possible; our gratitude to our Indianapolis colleagues for their hospitality; and our appreciation of the courteous treatment accorded the conference by the management of the Marrott Hotel. Finally, as an individual I wish to record my apology to my colleagues for the delay in the issuance of this report.

Carl E. Guthe, Chairman
Committee on State Archaeological
Surveys
April 1937

[1] **Friday Morning Session, December 6**

WISCONSIN (W. C. McKern)
I can't discuss local Wisconsin problems without touching on general problems. These center around an inadequacy of analytical and systematic methods and terminology. Our major problem is determining how to cooperate to mutual advantage with students of cultures similar to those in Wisconsin. We have great difficulty understanding each other because we do not do things in the same way and lack a systematized terminology. My specific problems relate to cultural manifestations and their place in the classification. I am going to start with Woodland because that is the primary interest in my area.

We speak of Woodland basic culture in Wisconsin, realizing that we don't know exactly what it is. There must be certain basic determinants which characterize this culture in the state, but these have not yet been identified. We cannot even define the determinants of the Lake Michigan phase in Wisconsin. A serious difficulty is that Woodland sites produce so little cultural material in proportion to the effort expended. We have worked for years on effigy mounds, which are classified as a Woodland aspect. Yet we know less of this complex than we do of any other manifestation in which we have worked. Another Wisconsin aspect which at present is placed in the Woodland basic culture is called Central Basin. In considering the more general underlying manifestation (probably a phase which includes Ohio Hopewell), we first recognized the Trempealeau sites as a local Wisconsin aspect, and now we have determined another, Cedar River, to which may tentatively be added a third local aspect situated in the southeastern part of the state. Since we find so many Central Basin traits which, except in pottery, do not conform with our present conception of the Woodland pattern, I would like to see the problem discussed thoroughly.

We are also concerned with the distribution of the divisions of the Mississippi basic culture. Both Middle and Upper Mississippi phases are found in Wisconsin, and we realize that their distribution in our area is not fully known. Components of these phases are widely distributed in the eastern United States as well as in Wisconsin, a condition which presents a serious problem. Not only do Upper Mississippi components occur much farther north than we had thought, but scattered materials definitely related to the Middle Mississippi phase are to be found within 60 miles of the northern border of Wisconsin. These consist of only three or four traits but serve to establish a definite complex.

[2] In my opinion, a great problem is bound up with the classification of specific types of materials such as pottery, stone implements, bone, shell, etc. Some of these, such as pottery, are sufficiently complex to require a classification before they may be adequately described. Comparative research will be greatly facilitated by the use of a standard classification, which should be worked out as soon as possible. That briefly sums up the problems that we have encountered in Wisconsin.

Discussion

Keyes: How many aspects of the Upper Mississippi phase do you now recognize?

McKern: Only one—the Wisconsin aspect. We have three foci: Lake Winnebago, Grand River, and Oneota.

Setzler: What objection would there be to eliminating the Woodland basic culture entirely and using a single basic culture in the classification, such as the Mississippi? Let us try to define two basic cultures that are distinct enough to be readily differentiated. Woodland in its essential details doesn't seem to be sufficiently different from the Mississippi basic culture. By accepting a single Mississippi basic culture, all the more general varieties could be classed as phases. A second culture might be established to include nomadic groups, such as the Folsom complex.

McKern: As originally proposed, the taxonomic scheme had five classes of cultures rather than four. The most general one was omitted from the final suggestion because we thought we would have difficulty in establishing determinants for four. The two basic cultures should be retained because, in my opinion, the differences between Woodland and Mississippi are marked. There is not so much as 10 percent similarity between the Woodland and Mississippi manifestations in Wisconsin.

Setzler: The term "basic culture" seems to be too fundamental in its implications.

Ritchie: Are there not two general classes of pottery which merge into one another? The cultural environments in which each developed were different. One type of pottery was made by people with a hunting-fishing complex and the other by a southern agricultural group.

Setzler: Woodland pottery may occur throughout the entire Mississippi Valley. I doubt whether the general differences referred to may be attributed to such variation in cultural environments. Adopt a single basic culture, like Mississippi, and have more phases and aspects.

[3] Griffin: One of the difficulties in this matter is that McKern recognizes a number of different aspects in Wisconsin which are distinct in his own mind. Some of us who have been in Wisconsin understand these distinctions. Except for Upper Mississippi and perhaps the Trempealeau aspect, these have not been defined so that they can be compared with materials in other states. Also, the characterization of cultural complexes found in other states lacks sufficient definition to establish comparable distinctions which are apparent to all of us.

IOWA (Charles R. Keyes)
In Iowa the Woodland culture is of widest distribution, being found on more than 200 inhabited sites in all parts of the state except for a half dozen counties in the extreme northwest. Even there, however, typical Woodland artifacts are found on the surface, so that probably at least a few inhabited sites existed in this area.

"Woodland" is a good descriptive name, for the sites are all closely associated with the timbered areas along the streams or lake margins; in fact, they are almost without exception within these areas. The kitchen refuse shows, moreover, a dependence on the plants and animals of the forest. House remains are lacking, as are, for the most part, such implements as pestles, mortars, spoons, and shallow dishes. The inference is that all of the former and most of the latter were made of wood or similar perishable material.

The work in stone of the Woodland people is of an astonishing variety and frequently passes beyond the utilitarian and enters the field of fine art: stemless, shouldered, side-notched, and barbed arrowheads, spearheads, and knives of almost infinite range of types and proportions; grooved axes of almost equally wide range of forms, sizes, and materials (these much more numerous than the rather wide and flattened celts); and many forms of ornamental and ceremonial objects such as gorgets, banner stones, boat stones, plummets, and the like.

Woodland pottery is generally distinguished from other pottery complexes without great difficulty. While there is considerable range in form, the grit tempering, rather soft paste, crumbly texture, dull red or rich brown color, absence of handles, and surface either plain or decorated with cord impressions are likely to prove definitive.

On the pottery basis, two phases (unless they prove to be aspects) of the Woodland seem rather clear. First, we have wide-mouthed vessels nearly always higher than wide and generally with conoidal bases. As a rule, they show cord-impressed or rocker-technique ornamentation over most of the body, while stamped or cord-impressed designs and single or double rows of rounded bosses are the usual decoration of the outer, and often the inner, rim. For this phase the term "Central Basin" has recently been suggested. In Iowa, it follows the Mississippi River along almost the entire eastern border, but typical potsherds occur on sites [4] in nearly all parts of the state, including the region of the Mississippi loess hills. The best examples of this pottery are from the mounds near the Mississippi, found associated with secondary burials and generally in subfloor pits.

The question rises as to where the Iowa Hopewell comes in. Several mound groups from Davenport to Toolesboro, at the mouth of the Iowa River, have produced some Hopewell artifacts with typical Hopewell burials: curved-base, plain or effigy-bowl pipes; copper axes and ornaments; and pearl necklaces, sheets of mica, and large marine shells. The pottery does not differ greatly from the Central Basin except that a few pots have flattened bases and the body decorations are sometimes laid out in definite areas within trailed lines.

Several facts seem to me to suggest more or less close connection between the Iowa Hopewell and the Iowa Central Basin. As stated above, the two pottery complexes have much in common. Further, a few artifacts that by themselves might well be called Hopewell have been taken from mound groups that apparently call for Central Basin classification: a curved base, plain-bowl pipe along with bear canines and large chalcedony knives from a mound on the middle course of the Turkey River in northeastern Iowa, the mounds of the group otherwise showing secondary burials without artifacts; a bowl with large decorative elements and two perforated copper "buttons," or ornaments, from a mound of the New Galena group in the Upper Iowa Valley; fragments of a Central Basin pot and three perforated bear canines from a second mound of the New Galena group (while all five of the excavated mounds of the group contained secondary burials only, three of these without artifacts); a pot and tubular copper beads from the Harvey mound group on an island in the Mississippi above Guttenberg, both of which would cause no surprise in a Hopewell mound but are associated only with secondary burials; a deposit of a girdle of rounded, massive copper beads and of seven large, thin, finely chipped blades from a mound of the Pleasant Creek group on a Mississippi terrace south of Bellevue, all the known burials in two of the mounds of this group being secondary, the one with the copper beads having been removed by amateurs and its nature uncertain. Further, the widespread occurrence on Central Basin sites of a highly specialized type of flint blade (wide, barbed, and with an expanding stem with convex base) also suggests a trailing off of the Hopewell into the more general Woodland, seven blades of this identical type having been found in the Franz-Green mound near Valparaiso, Indiana, associated with artifacts and burials strongly suggesting Hopewell. Other intimations of the Hopewell, but with associations that appear closer to Central Basin, could be cited. In Iowa the Hopewell might rather easily be looked on as a climax of the Central Basin. Making a decision here meets the usual obstacle—the material on which to base a judgment does not exist in sufficient quantity.

The second phase of the Woodland is found in mounds and rockshelters and on village sites along the interior streams of the [5] state. So far as I know, nothing clearly diagnostic appears except in the pottery. This differs markedly from the Central Basin in form, is thinner, and apparently the vessels are of smaller than average size. Almost no use is made of embossing or the rocker technique. The vessels are generally as wide as high, the base rounded or bluntly conoidal, and the shoulders flare quite sharply from the rim. Four lugs in two opposite pairs are a common feature, and the drawing out of these frequently gives the opening a squarish outline. The bowls are plain or, more often, they show irregular cord impressions. The rims are plain or, more frequently, decorated with horizontal or diagonal lines of cord impressions. Vertical or diagonal notches on the outer lip are a common feature.

A number of mounds on the upper Cedar River in northeastern Iowa have produced this ware; it is found in a majority of the rockshelters of east central Iowa, and several village sites in the Des Moines River valley are characterized by it. Similar to it, but generally distinguishable in texture and decorative designs, is a ware from the ravines between the Missouri hills in southwestern Iowa. This may possibly have to be recognized as a third Woodland phase, especially as it is associated with some characteristic types of artifacts. The material available does not justify a judgment.

In several published papers I have been incautious enough to refer to the Iowa Woodland as Algonquian, believing that, with its entire complex of Woodland type, it could hardly be anything else. It must be admitted, however, that as yet the Iowa Woodland has nowhere been connected definitely with history and that there is no proof of Algonquian origin except on archaeological evidence. For this reason, and also because of the existence of counteropinion, it might have been safer and better, for the present at least, to stick to the noncommittal term "Woodland." McKern's inclusion of Menomini material culture within the Lake Michigan phase argues, of course, for the Algonquian. On the other hand, Jenks and Wilford are convinced that the Woodland-type pottery of the old Kathio site on Lake Mille Lacs, Minnesota, is Sioux in origin.

Setting itself off sharply from the Woodland is the Oneota, to which I first applied this name in 1921 after the type locality, the valley of the Upper Iowa, or Oneota, River, in the northeastern corner of Iowa. Sites of this culture are almost continuous on the high terraces, or benches, of the Oneota across Allamakee County, though some 14 other sites are scattered widely over the state, with concentrations only on the Little Sioux below Correctionville and the Mississippi on both sides of Burlington. These sites are uniformly large as compared with the Woodland, from 10 to over 100 acres each, and occupy the high, open terraces or broad, rounded hilltops bordering the rivers. They appear always to have stood on prairie areas, hence always out in the open. The house type is unknown so far as archaeological evidence is concerned. [6] For the Oneota, a rather rich supply of entire pottery vessels and usable potsherds is available. The pots are generally wider than high, either round or elliptical in horizontal cross section, and the tempering is uniformly of crushed shell. More often than not there are one or two pairs of loop or strap handles. Decoration is of incised or trailed lines and occasionally of simple punctate designs, these confined mostly to the shoulder or upper body. The lip is usually notched or with finger-impressed crenulations. Small mortuary vessels have from about a half-pint to 2 quarts capacity, while the large storage vessels from the village sites frequently contain 4 gallons or more and sometimes have a diameter in excess of 20 inches.

Some characteristic artifacts are the triangular unnotched arrowheads, bun-shaped hand mullers, rather small shallow mortars, the thick, narrow celt, the grooved hammer or maul, and the disc-bowl pipe. Bone and antler implements are common, including perforated deer or elk ribs (or arrow shaft straighteners), scapula hoes, and socketed projectile points. The end scraper is the artifact of commonest occurrence in the village sites, and it is symmetrical and more finely finished than those from Woodland sites.

Burials are usually in cemeteries or (in Allamakee County) intrusive in mounds of Woodland origin. They are generally fully extended on the back with arms straight at the sides. Rather more than half of the burials show from one or two up to a fairly large number of mortuary offerings. The mounds of the large Blood Run site on the Big Sioux in the northwestern corner of the state appear to have been built by the Oneota, though the present amount of evidence is perhaps inconclusive.

As matters stand at present, a part of the Oneota sites must be put down as pre-historic, others are protohistoric, and several, including the Blood Run site, the Gillet's Grove site south of the Okoboji lakes, and the O'Regan site in the Upper Iowa Valley, are probably historic, the ultimate appraisal awaiting further study and comparison of the De Lisle map of 1718, the seventeenth-century accounts of Perrot, Andre, and perhaps others. The amount of trade material is not large. Most of this is found with burials and a part of it in direct association with the native pottery; a few necklaces and ear ornaments of blue Venetian glass beads; a few knives and bracelets of iron; and a number of tubular beads, coiled-spring ear ornaments, and bracelets of brass.

Last June I made a special trip to the Blue Earth Valley, Minnesota, to see whether evidence could be found there of the presence of the Ioways, where, according to some early accounts, they were resident in the second half of the seventeenth century. Numerous Oneota sites were found along the middle course of this river, a large amount of collected material examined agreeing trait for trait with that familiar to me on Iowa sites, and the sites themselves similarly located.

[7] At different times it has been my privilege to examine Oneota materials from South Dakota (near Vermillion), from Nebraska (near Rulo in the southeastern corner), from Kansas (the Fanning site in the northeast and the White Rock in north central), from Missouri (a site on the Missouri River in north central), and from Illinois (a site about 20 miles east of Cahokia). I am not very familiar with the Upper Mississippi of Wisconsin, but McKern tells me that his Mississippi Uplands is identical with my Oneota, that the Lake Winnebago focus is not far removed, and that the Grand River seems also to be connected. Between the materials from all these sites there appear to be no more than focal differences. As far as I know them from actual handling, I believe one could not, with any assurance at all, separate the materials from these widely distributed sites if they once got mixed without identification marks. It is rather difficult not to think of Chiwere Siouan when one studies the Oneota collections.

Discussion

McKern: The Lake Winnebago does seem to be a focus of the Oneota aspect, and it is rather definitely tied up with the historical Winnebago. In Iowa the tendency is to tie up one focus with the Ioway Indians, and in Wisconsin, the Lake Winnebago focus with the Winnebago. Strong evaluated the Nebraska culture as perhaps Chiwere Siouan also, and Wedel is inclined to agree with this.

Keyes: The distribution of the Oneota sites corresponds pretty well with what we know concerning the ranges of the Ioway, the Oto, and Missouri, as well as that of the parent Winnebago. Perhaps even the Dhegiha Siouans may have occupied some of the known sites, the distribution of these seeming to be a bit wide for the Chiwere alone.

McKern: What I formerly called the Mississippi Uplands is Oneota, identical with the Upper Mississippi as found in the Upper Iowa Valley. There is some evidence that it overlies the Woodland.

Keyes: In and near the Upper Iowa Valley we found, last year and this, the Oneota overlying the Woodland on five different sites. Thus the time sequence of the Woodland and Upper Mississippi is clear for a considerable area at least.

McKern: It is important to find these stratified sites or, lacking these, to connect any sites revealing a primitive culture with history. In Wisconsin, we so frequently find large Upper Mississippi sites in sections where Woodland people did not live, giving thus a pure culture without any opportunity for stratigraphy. The Menomini were pure Woodland (Lake Michigan or Effigy Mound) and have occupied the shore of the Green Bay since the time of first white contact. The only conclusion is that the surface pottery represents Menomini pottery, which thus, with these people, brings Woodland into the historic period.

[8] Keyes: Brief mention only need be made of two other cultures known to Iowa archaeology. What I have called the Glenwood is identical, so far as I know at this time, with the Nebraska culture of Gilder, recently treated in detail by Strong. In Iowa, it is known only along the front range of hills facing the Missouri River floodplain from a little north of the Missouri state line to the southern part of Monona County, a total distance of about 100 miles. The culture is known in Iowa mostly from the presence, and a few partial excavations, of house sites and from the artifacts found on cultivated fields or in the deep ditches eroded in the ravines. No excavations using modern techniques have been made as yet.

The Mill Creek culture, on 16 sites occurring in 2 foci, one on the Big Sioux River and its tributary, Broken Kettle Creek, and the other on the Little Sioux River and its tributaries, Mill Creek and Waterman Creek, increases in interest as the facts concerning it come to light. During the fall of last year and a short period in September of this year, some five weeks in all, Mr. Ellison Orr with two helpers carried out test excavations on seven of the sites, most extensively on the Broken Kettle site of the Big Sioux focus. The richness of the bone, antler, and pottery complexes is especially noteworthy, as is the fact that certain features, such as the occasional presence of small effigies on the rims of pottery bowls, calls to mind the Middle Mississippi. Nothing has been found to connect the sites in any way with history and, while this trait or that might seem to have Mandan, Pawnee, or Woodland affiliation, as others suggest the Middle

Mississippi, the total complex remains quite without any assured relationship. As the Mill Creek is thus far known only to northwestern Iowa, this is not the place to enlarge on it in any detail.

MINNESOTA (Lloyd A. Wilford)

We have not been working long in Minnesota archaeology. In southeastern Minnesota we have excavated a group of mounds at Rushford and a large rockshelter at Peterson, both in the Root River Valley. In the latter we found definite stratigraphy, with pottery of the Oneota aspect of Upper Mississippi overlying pottery of the grit-tempered Woodland type. The mounds were burial mounds on the tops of bluffs overlooking the river valley. The burials are extended primary burials in shallow excavated pits. A pile of stones was placed over the body, and a layer of stones was placed over the mound. Pottery buried with the bodies was definitely Oneota. The extent of the Upper Mississippi in Minnesota has not definitely been worked out, but is found in much of the state south of the Minnesota River.

In the southwestern part of the state are found sherds comparable to those of the Mill Creek culture in Iowa. Most of the material in the state belongs to the Woodland basic culture. The great settlement of the Dakota Sioux at Lake Mille Lacs, visited by Hennepin in 1680, has pottery that is definitely Woodland. It has [9] conoidal bases, and the bodies are decorated with the cord-wrapped paddle. About two-thirds of the vessels have additional decoration on their upper portions made by cord markings, roulette markings, and indentations. This site belongs to the Kathio focus. One component of this focus near Forest Lake, Minnesota, makes use of the trailed line in pottery decoration in 22 percent of the rim sherds, but in a manner that resembles the Elemental Hopewellian type rather than the Oneota. Kathio burials are secondary bundle burials in mounds, at or above the ground level.

Around the large lakes at the headwaters of the Mississippi River is the Blackduck focus with distinctive Woodland pottery. It is round-bottomed with pseudo-cord impressions and indentations, and is more ornate than the Kathio ware. The only burials thus far associated with this ware were primary.

At the northern boundary of the state is the Laurel Mound on Rainy River. This mound had over 100 burials, mostly in bundle form, but the long bones had been tapped to remove the marrow. The pottery has undecorated bodies, but the upper portion is decorated chiefly with roulette markings. Blackduck-type sherds are found on the surface near this mound.

The triangular type of projectile point is the dominant type of the Upper Mississippi culture, and is an important but minor element at Kathio. It is absent from, or negligible in, the sites farther north.

ILLINOIS (Thorne Deuel)
In Illinois we have under the Woodland, the Central Basin phase and another phase which has tentatively been called Tampico. This has some traits in common with the Lake Michigan phase in Wisconsin. Under Mississippi, we find the Upper and Middle phases represented. In addition to these phases we are inclined to include provisionally another called the Red Ochre. This manifestation seems to have more traits in common with Adena than any other group—leaf-shaped projectile points, copper objects, shell gorgets, Marginella shell beads, etc., in caches alone or associated with burials. Caches and burials, often covered with red ocher, frequently lie in shallow pits below mounds. Fully flexed burials constitute the burial position in Illinois. Skeletal material is poorly preserved: in some instances only the enamel shells of the teeth are present. A few calcined human bones and a single crematory basin comprise the only evidence of cremation.

[10] We place this manifestation in the Woodland because projectile points (in small numbers), drills, and copper forms are found and which occur on other Illinois sites that we designate Woodland. A pottery ware or type found in three Fulton County sites yielding this complex is characterized by very thick sherds, a light buff-colored paste and surfaces, and very coarse grit tempering.

It has been placed in a separate phase because of a specialized type of truncated leaf-shaped or lanceolate blade (in large numbers), shell gorgets, the use of red ocher with burials, and the burial complex, all of which seem to distinguish it from other Woodland phases in Illinois. Sites exhibiting these traits or some of the more characteristic ones have also been noted in other states besides Ohio and Indiana.

The Tampico phase is characterized by round bottomed, often flattened globular pots with vertical necks and "squared" rims with vertical lugs (without perforations) at the "corners." Beside the usual Woodland type, projectile points made from thin, curved flakes seem to be imitative of

Mississippi forms but are easily distinguishable from them. Burials are in mounds in the flexed or semiflexed position, associated rather rarely with strings of Anculosa shell beads, typical Woodland associated with characteristic small-flake projectile points, and pots containing mussel-shell spoons. Tampico sites occur in northwestern Illinois, in Peoria and Fulton counties. Some of the typical artifacts have been noted in Rock Island County collections.

The best known aspect of Central Basin in Illinois was mentioned many years ago in a Smithsonian report by Henderson—the complex from the mounds near Naples. Sites of a similar content explored by the University of Chicago include two in Fulton and one each in Tazewell and Mason counties. Two others in Fulton are known from the notes of amateur diggers. These approach closest of any Illinois complex to the Hopewell of Ohio, but with tremendous differences, judging the complex of the latter by the existing reports, including Willoughby's excellent account of the Turner group. Extended and secondary bundle burials usually occur beneath mounds, on the floor or in subfloor pits, unlined or lined with logs (or stones) and accompanied by a limited amount of "typical Hopewell" artifacts, including the platform pipe of pipestone, shell beads, and copper axes. Prepared sand floors are generally present beneath the mound, crematory basins rarely. A very small number of sherds of excellent pottery have finely crosshatched collarlike rims outlined below by punchmarks, polished necks, and bodies ornamented with alternately decorated and plain areas. This pottery type has been called "typical Hopewell pottery" by certain writers. On sites of Illinois "Hopewell" three or four sherds only (out of several hundred) are of this type. On some of the sites, none occurs.

[11] There are two foci in Illinois that we class as Woodland and that appear related to the Central Basin and perhaps may represent early forms of that phase. Apparently they lack all the traits usually listed as "typical Hopewellian," such as pipes, axes of copper, ear spools, stamped pottery, etc. One focus, the Liverpool, is represented by a single component, the Black Sand. The complex occurs beneath a mound of Illinois "Hopewell." The pottery is grit tempered. The characteristic decoration commences near the lip and consists of one or more series of broad lines incised over cord roughing, intersecting each other. Fingernail impressions and gouged-out depressions and embossing occur occasionally. Trailing, punching, and stamping are lacking. The decorated area includes the neck and a portion

of the body. Grooved axes are frequent; Woodland-type projectile points are numerous. Some varieties approach specialized or commonly occurring forms of Central Basin points; others are rare in other known Illinois sites. Anvils, drills, and flint scrapers are present.

The other focus mentioned is represented by three sites, one of which is stratified. The Middle Mississippi complex occurs in the uppermost levels of the site. The manifestation is characterized chiefly by the pottery and chipped flint. The pottery is in some respects transitional between that of the Black Sand component and the Illinois "Hopewell." Incising-over-cord-roughing appears but is easily distinguishable from the Black Sand type; punching is common, but the dentate stamp is rare. Embossing is even rarer than in the Black Sand site.

Upper Mississippi sites include those of Fisher and Blue Island. The Fisher site appears to be closely related to one of the components of the Madisonville (Ohio) focus. The Blue Island site does not connect closely with any of the sites in Ohio as reported in the literature. It may be related to the Wisconsin aspect. Kelly's Plum Island site has an Upper Mississippi component which may be closely related to Fisher. Upper Mississippi influence occurs in certain of the Middle Mississippi sites, which appear to be late. Pottery decoration and handles seem to me to show Upper Mississippi contacts.

A Middle Mississippi manifestation occurs at the site near Metropolis and is characterized by the richness of the remains, variation in pottery form, pottery pipes, and the use of cane in houses. I was inclined to place this with the Lower Mississippi group at first. The closest affiliations are the Gordon-Fewkes (Tennessee), Tolu, and Wickliffe (Kentucky) sites. These are more or less marginal to the southeastern center. F°14 in Fulton County (Illinois), the Dickson site, and the Cahokia group are also considered Middle Mississippi.

[12] Foci and aspects are difficult to define until a number of sites with a fair representation of traits have been explored. Of the Woodland sites excavated which are thought to be of the same focus, the material is scanty and the overlapping apparently slight. Even in the four Middle Mississippi sites investigated, only two seem sufficiently related to include in the same focus. To date the sites known of each manifestation are insufficient in number, and some are known only through incomplete records of amateurs. For this reason I am not prepared to make any further division into smaller units.

Discussion

McKern: Our incised (Central Basin) ware was hard and tempered differently from utility ware found in the same burial. The pottery was characterized by a black surface and trailed designs (painted in one case).

Setzler: You get a definite type of stamp (elliptical and notched) on Ohio Hopewell pottery. At the same time you do get pottery that is comparable to some of the pottery from Marksville. That seems to be one of the outstanding Hopewell traits of decoration—the different types of small stamps.

Deuel: You find punching and embossing without stamping in some Illinois components. In others, the three are in association, with stamping and embossing predominant.

A general discussion of types of pottery manufacture followed. It was felt that there should be a meeting devoted to an attempt to straighten out Hopewell and Central Basin. The meeting adjourned for luncheon.

Friday Afternoon Session, December 6

INDIANA (Glenn A. Black)
I think it might be well to sketch rather briefly what we have, or think we have, here in Indiana. Woodland-type material is found throughout the entire state, with a suggestion of Iroquois in the northern portion of the St. Joseph River valley.

We have historic Algonquian sites at Ouiatenon and Fort Wayne, which most certainly would bear investigation in the near future.

We have a site in north-central Indiana which appears to be early Iroquois, fitting in somewhat with Greenman's work in [13] northwestern Ohio but somewhat distinct from Parker's Iroquois or Ritchie's Early Algonquian.

Central Basin is well defined in northern, south-central, and southern Indiana. Contrary to the conditions outlined for Central Basin by Keyes and McKern, we do have geometric earthworks at Anderson and New Castle, both in central Indiana. It is impossible, due to almost total absence of data, to relate these two sites with similar Ohio groups.

Setzler's work in the Whitewater Valley indicates a preponderance of Adena or Red Ochre. The Nowlin mound, just completely excavated, proved to be Adena in affinity and carries the culture south to the Ohio River in the southeast corner of the state.

Along the Ohio River in Dearborn and Ohio counties, we have a great manifestation of Fort Ancient. It extends north into the Whitewater and Miami valleys and possibly northwest as far as Indianapolis.

In the southwest portion of the state we have Woodland, Central Basin, and a site which somewhat resembles Etowah in outline and superficial material remains.

At the mouth of the Wabash we have a site, the large collection of pottery from which resembles Meyer's Gordon-Fewkes and Webb's Tolu sites.

In southeastern Indiana are the stone mounds in which I am personally much interested and about which so little is known. They are found along the Ohio River and its tributaries in Switzerland, Ohio, Dearborn, and Franklin counties. They are found bordering the river in Kentucky and continue up the Ohio on both sides as far as Washington, Pennsylvania.

Discussion

Swanton: There is an ancient Shawnee village site at the mouth of the Wabash, but there is no reference for the Quapaw site at the mouth of the Wabash. The Ohio River as conceived of at that time was also called the Quapaw River.

Black: With reference to Indiana, I would like to concur with McKern and Keyes as to the distinction between Woodland and Mississippi. Although we lack plains, where Mississippi material occurs the sites are found on broad terraces, whereas the Woodland mounds and habitation sites tend to be confined to the terraces on the tributaries and in the wooded hills.

[14] Setzler: Is there Woodland comparable to the Adena material in Indiana?

Black: Possible Adena pottery was found in the mound we finished this year. The west mound was built first and then overlapped by the east mound. There were two sherds in the dirt from the west mound and more in the east mound. This suggested that the area might well have been strewn with village debris and that the debris composing the earth in the west mound was due to residence on that site while building the mound. Forty or 50 sherds were grit-tempered, plain, and cord-marked. This is not a village site. It is 360 feet above water. There is no copper in the village site, which might have been associated with the mound. The points and pottery are the same. No excavation has been done on the village site.

In Woodland Indiana, points are basically triangular but not isosceles nor T type, and beautifully chipped, unnotched with curving sides. It is extremely difficult to separate Red Ochre or Adena from Central Basin or Hopewell in this area. The work which McAllister did in Porter County produced what he considered typical subfloor Hopewell burials. In Ohio burials are on the floor or above. Adena has well prepared, subfloor tombs. There must be some place in Indiana where fusion took place. In the mound nearly completed, the burials were all on the floor or above and contained 34 stemmed projectile points. Ohio Adena points are also stemmed, as are Illinois Hopewell. Marginella beads, which occur in Central Basin material, were found in the Nowlin Mound. One segment of a gorget, not associated with a burial, seemed to be typically Adena. No copper objects were associated with burials, although one copper bracelet was in the mound. A rectangular sandstone tablet and three bone awls were other traits.

It seems to me that in pre-historic times, Indiana must have been a buffer section in which many of these basic groups were represented. The problems as they exist are extremely complex.

INDIANA—continued (E. Y. Guernsey)

Clark County is one of the most interesting and important archaeological regions in southern Indiana. It lies south of the Dearborn County region recently studied by Mr. Black, with three counties intervening between these culturally divergent locations. Both counties are contiguous to the Ohio River, which forms the entire eastern and southern boundary of Clark. In the northern sector there are a number of stone mounds, some quite large. None has been excavated or studied. South of this region there are a number of extensive village sites, with occasional low and flat mounds, apparently indicating a rather long occupancy. There are, as well, a number of related cemeteries.

[15] The Prather site, partially excavated last year, involves three large mounds and one of the largest village sites in this region. A test pit sunk in the largest mound indicated that it was of "temple mound" character. In the next largest we found postholes indicating a rectangular house structure which had been burned. Many of the framing timbers were well preserved. Both water and food jars accompanied the burials. With the most important burial there were shell beads, a large circular shell gorget, implements of bone, the complete skeleton of a large fish, and a very unusual bird effigy carved of wood, the claws in relief, the whole overlaid with copper. Considerable pottery, all shell tempered, was recovered from the site. It suggested a considerably modified Middle Mississippi origin.

Of major importance in the county are the Clarksville sites, of which there are four of evident relationship. They are opposite the city of Louisville, occupying the site of the extinct town of Clarksville. Three of these sites, which we have called the Clark's Point, Newcomb, and Elrod sites, have been carefully explored during our survey. They are so large, however, that several years might be devoted to excavation. Other lesser, but obviously related village sites extend for some 15 miles along the Indiana shore in this area. Both vertical and horizontal stratigraphy are represented at Clarksville. The lowest stratum appears barren of artifacts or human skeletal material. Above it are burials probably Woodland in placement. The principal culture represented is doubtless identical with that of C. B. Moore's "Indian Knoll" site. The burials are usually flexed and accompanied by red ocher. Banner stones of marble, plasma, etc., are sometimes ceremonially broken—as are usually the long roller pestles. Artifacts of bone, stone, and shell are abundant and well executed. Bone netting needles, awls, needles, projectile points, and fishhooks are common burial accompaniments. There is no pottery and no pipes. The upper stratum, also containing numerous burials, probably belongs to the Upper Mississippi. In it the burials are rather consistently on the right side, with legs sharply flexed. Funereal pottery is usual. Pipes are often found.

The western part of the county involves a highly elevated "knob" or karst topography, and is heavily wooded. The scattered village sites of this area probably belong to a later period than those just noted. The flint material is commonly the blue-gray hornstone taken from aboriginal quarries in Harrison and Crawford counties. Projectiles are exceptionally well chipped. The "knob" area abounds in deposits of impure hematite and limonite, usually occurring as lenticular concretions. These materials have been extensively used in making axes, celts, pestles, and other articles of utility or ornament. The New Albany shale has been utilized for various digging tools, and at Clarksville is sometimes found worked into an oval pillowlike headrest accompanying burials. One of the most spectacular walled fortifications in the Ohio River area is at the mouth of Fourteen Mile Creek. Not far from this site is a notable circular enclosure.

[16] South of Clark County are the well-known flint deposits just described, occupying an area some 15 miles square. The material is a blue-gray hornstone of exceptional chipping quality. Projectile points of this material occur at certain Hopewell sites in Ohio and in the vicinity of Long Island, New York. The several hundred quarries, large and small, were probably worked over a very long period. Our survey suggests, how-

ever, that the earliest inhabitants of the Clarksville sites were unfamiliar with it. In fact, it appears rather certain that the presence or use of this so-called Wyandotte flint may provide a medium for time fixation in at least a limited way.

OHIO (E. F. Greenman)

Ohio archaeology seems to be well known to this group, and rather important. I don't know if there is much new to say. We have three main cultures—Fort Ancient, Hopewell, and Adena—the Fort Ancient tying in with the Upper Mississippi; the Hopewell groups with the Central Basin; and the Adena with the Red Ochre that Deuel is getting in Illinois. I am very much interested in the Hopewell problem and in determining just what it is and how it ties in with any historically known people. We must define it more clearly first. What are the significant features between the major mound groups in Ohio now called Hopewell? The Seip, Turner, and Hopewell sites are not identical. The material differs markedly. Some sort of a statistical study must be made of these several well-known groups before we are able to say what Hopewell really is. Then, of course, some of you know that I have been interested in the possible relationship between Fort Ancient and Ohio Hopewell. The latter seems more clearly tied in with the south than it is with the Woodland, or with what has been found in Illinois, Wisconsin, and Iowa.

Of course, there has been a great amount of digging in Ohio, but mostly in the southern part, in the Scioto and Miami River valleys. Nothing is known about the eastern part of the state. I believe you will find something more like the Wisconsin and Illinois Hopewell than what you get in southern Ohio, illustrating some of the transition between Adena and Hopewell, if there should be such a thing.

There are sites in northern Ohio which are similar to the Iroquois, and I do believe that one or two things tie them quite closely to Fort Ancient—for instance, curvilinear lines resembling the guilloche on pottery. From east to west there seems to be a gradual emergence from Iroquois to Woodland. We have dug 12 or 15 sites between the Ohio-Pennsylvania line and at Sandusky, all of which are Iroquoian with predominance of flexed burials, fortified promontories, narrow-based triangular points, and much bone material. These are not historically known and are classified [17] on the basis of material found in New York. Mussel shells are found, but no stone or pottery discs. West of Cleveland there is an acceleration of the merging from Iroquois to Woodland. This goes up into Michigan. You also get the Hopewell in northern Ohio, notably near Huron, where the contents of two mounds excavated in 1930 suggest a relationship to a village site a mile away that is difficult to class but is either Woodland or Upper Mississippi.

It is important to realize that in Ohio you know very little about anything outside of Ross County, Hamilton County, and the border of Lake Erie. Between Ross County and the northern tier of counties, there is a big gap. There are very few mounds. Gravel-kame burials predominate.

Discussion

Ritchie: How many of the so-called geometric enclosures have been tested thoroughly?

Greenman: The Newark enclosures have been dug into rather thoroughly, also that of the Turner Group in Hamilton County.

McKern: Outside of distribution, you have no criteria to associate these earthworks with Hopewell cultures.

Greenman: The Seip Mound is part of a very ornate system of geometric earthworks. So also was the Hopewell group of mounds, and there are geometric earthworks in connection with the burial mounds of the Turner Group. No, I would not hesitate to say that the geometric enclosures are Hopewell.

MICHIGAN (Carl E. Guthe)

In a sense, Greenman ought to make this report, but it just occurred to me that Michigan ought to be mentioned. It is only this last summer that we secured funds to enable us to do actual excavation in our state. Very briefly, I think the picture is like this. On the western side of the Lower Peninsula we have evidence of Central Basin, running up into the Newaygo region and possibly two-thirds of the way into the Peninsula. From the pottery and the stone material in general, the major part of the state appears to be Woodland in character. On the eastern side, the variation in the pottery indicates some Iroquois influence, perhaps a secondary one. The Neutrals were actually in that region. We have a very good collection of surface material from the Saginaw district. Other than that, we know practically nothing about Michigan.

This summer, Greenman, digging in the Thumb, uncovered a site of great interest to us. We don't know what it means yet. The [18] pottery is related to the Woodland, but there are several characteristics that are unusual. There are two long, narrow enclosures indicated by postholes, scattered through which are fire pits; two extended burials; and rearticulated burials in which the bones were cleaned and put back into what the Indians thought was anatomical order. The most interesting discovery was postmortem perforated skulls. In some skulls a circular portion of the occiput had been removed, making discs 2 or 3 inches in diameter, which were buried with the long bones in the secondary burials. They are apparently related to the Woodland culture. There are some perforations in the long bones. Postmortem perforation also occurs in Ontario, Canada, just east of Michigan. Sometimes the skulls are buried alone, but usually with the skeleton.

KENTUCKY (Wm. S. Webb)

The question was raised this morning about the possible effect of environment on cultural patterns. Kentucky seems to show definitely such a relation. About half of the earth mounds of the state are concentrated in the Blue Grass region. These mounds fall into two classes. Some are associated with earthworks of large size and some with village sites which we call Fort Ancient. The Fox Farm, explored by Smith, is typical of this latter group and is most closely related to the Baum site in Ohio.

Within the area of great mound density there are many earth mounds not associated with village sites. Of this type only a few have been thoroughly excavated. One was found to correspond closely to the Adena in Ohio. From one there has been taken a tablet with Hopewell-like design and other carved tablets suggesting the Etowah eagle. Burials in these mounds have yielded tubular pipes, copper bracelets, and much mica.

In western Kentucky in the Cumberland-Tennessee area, large village sites are numerous, usually associated with stone-grave cemeteries. Pottery from these sites showed a great variety of form and type of manufacture. The mounds on the village sites are domiciliary. Burials in most cases are in cemeteries adjacent to the mound. About one-half of the burials are in stone graves. The availability of stone had nothing to do with the form of burials. Domiciliary structures were indicated by post molds. On such sites the quantity of pottery sherds is excessive. The salt pans are mostly textile marked. The textiles were squeezed on the pottery while the clay was wet and the vessel was turned upside down. Real textile-marked pottery on salt pans occurs in a limited area, and the patterns seem to indicate cultural divisions. These sites are characterized by rectangular post-mold patterns, multiple occupancy of sites, stone graves, and much pottery. Burials in stone graves [19] are usually extended on the back, one to a grave. Bundle burials were added in some cases. In cemeteries burials would sometimes amount to 300 or more per acre.

In the central part of the state there is a great cave region containing many rockshelters which show habitation. There is a second area in the mountains of Kentucky also characterized by evidence of occupancy under the cliffs. Hominy holes are numerous. In these, food material was ground with bell-shaped pestles. Under these rockshelters there are great ash middens in which we find the material apparently as well preserved as if it were put there yesterday—vegetal material, textiles, sandals, and wood (a cradle board for an infant and arrow points with shaft attached). These people had no pipes and very little pottery. (Note: There is no evidence to connect directly the pottery found in the caves and rockshelters with the burials, vegetal, and textile materials. This is also true of the Norris Basin. JBG 1937.) The sherds are gravel tempered, very rough, and fairly heavy. While the quantity found is small, it is enough to tie in with sites in east Tennessee and the Norris Basin. In listing as many as 60 traits from the cavern sites of Norris Basin, there is an 85 percent correlation with the top layer of the Stallings Mound in Georgia and the rockshelters of eastern Kentucky. This pottery was malleated with grass-wrapped paddles and does not fit in with other Kentucky culture complexes. The wood and charcoal are too fragmentary to get any dendrochronological information from them. There were no pipes or evidence of smoking in 40 shelters in Kentucky.

In Ohio County, on the Green River, there are many shell mounds from 4 to 10 feet deep. These seem to be quite different from other prehistoric remains in the state. They have been but little studied. C. B. Moore dug one of these mounds at Indian Knoll. It was about one-half shell and one-half black dirt and showed a unique culture complex.

NEW YORK (William A. Ritchie)
Our problems are many and diverse, but essentially similar to those already outlined here. Dr. Parker and I are trying to define and extend the relationships of New York cultures, and if possible to correlate some of them with ethnic groups.

We have vestiges of several well-differentiated occupations preceding the Iroquois. What is probably the most ancient we call the Archaic (Archaic pattern, Lamoka focus). It differs in every particular from Willoughby's Pre-Algonkin as described in his new book. A nomadic hunting-and-fishing culture, it is widely distributed in lower Ontario, whence it spread through the Ontario peninsula over [20] most of New York, except the northern and eastern sections, and southward to central Pennsylvania.

We have excavated three village and some half dozen campsites of this occupation. The diagnostic traits have been found to be the beveled adze; a rude celtlike "chopper"; narrow-bladed straight or notched-stemmed projectile points; and a singular serrated antler-tine object, sometimes striped with hematite. These are combined with such traits as the roller pestle, pitted and unpitted hammerstone, net sinker, large shallow mortar, rude rectangular celt, plano-convex adze, and a variety of bone and antler artifacts, embracing awls of many types, fishhooks, gouges, scrapers, whistles, punches, etc. The harpoon is absent. Not a sherd of pottery or fragment of a steatite vessel has come to light, nor have any pipes, copper or shell articles, grooved axes, gouges, hammer stones, gorgets, or other problematical forms. Moreover, no trace of agricultural products exists, but many carbonized acorns and nuts have been found. A small terrierlike dog was known and sometimes carefully buried.

Burials are simple flexed inhumations without grave goods. The skull type is long and narrow, with high narrow face and narrow nose. It is rather small, not very robust, and closely resembles the Basketmaker type of Hooton, which he regards as one of the oldest elements in the aboriginal population of America.

Apparently following the Archaic and pressing southward into eastern New York from New England came a culture focus, the Orient, characterized by steatite vessels (but no clay pottery); narrow-bladed, somewhat lozenge-shaped and fishtailed projectile points; the oval two-holed gorget; rude celts; and plano-convex adzes. Liberal burial offerings accompany the cremated skeletons. (It is still too early to outline the content or establish the provenience and range of this manifestation, which is being described by Mr. Roy Latham of Orient, Long Island. It may be found to be an independent aspect rather than a focus of the Vine Valley aspect, where it has been provisionally assigned.)

A second and stratigraphically superior focus of the Vine Valley aspect is the Coastal, a diffusion apparently from the Chesapeake Bay region into New England and inland through the great river valleys—the Susquehanna, Delaware, Hudson, and Connecticut—into Pennsylvania, New York, and New England, becoming progressively attenuated in its westward migration. Its most advanced development in New York occurred on Long Island. This horizon introduced such new increments as agricultural beginnings; a coarse pottery with conoidal bottom, straight sides, and roughly punctate, cord- or fabric-impressed embellishments; ocean-shell ornaments; the bone harpoon; crude obtuse-angle pipes of clay; stone pipes; the grooved axe; banner stone; relatively broad-bladed points of flint, argillite, rhyolite, quartzite, and jasper; and probably the gouge. (The gouge [21] may have been a northern element which was absorbed by this culture.)

The burials are predominantly flexed, but bundle and ossuary interments are reported by Charles F. Goddard on Long Island. Not infrequently, mortuary offerings occur. Judging from the scanty material available, the skull type is dolichocephalic, or narrow, with a vault of medium height, a rather low, broad face, and a nasal aperture of medium breadth.

Probably contemporaneous in northern, central, and western New York with the Coastal focus farther south were the Middlesex and Point Peninsula foci of the same aspect (Vine Valley). These foci are interrelated to an undetermined extent, and traces of both appear in a Coastal form in certain parts of the state. The new influence was from the west, pressing both north and south of Lake Ontario, eastward to the sea. Chief among the determinants are the platform pipe, native-copper implements, the bird stone, and a distinctive type of carving or engraving tool consisting of a beaver incisor tooth hafted in antler. The gravel-knoll burials and the Intrusive Mound culture of Mills in Ohio produce practically the same complex.

New York graves of this period are usually rich in artifacts. Bundle and cremated burials are known, but the majority are flexed. The crania are large and robust, with pronounced supraorbital ridges. Mesocephaly or brachycephaly definitely prevail, coupled with a vault of medium height (orthocephalic), a broad, low face (chamaeproscopic), orbits of medium size (mesoseme), and a broad nasal aperture (platyrrhine).

Not many centuries prior to the Iroquois invasion, the Owasco aspect appeared in New York, coming northward principally by way of the Susquehanna Valley and spreading east to the Hudson, west to the Genesee, and north to the Mohawk and Lake Ontario. The culture congeries are well known from our excavation of half a score of sites, and there are reasons to believe that we may succeed in linking this aspect with an historic Algonkin group.

Two foci are distinguishable, the Canandaigua and the Castle Creek, the latter exhibiting clear Iroquois infiltrations. A ceramic progression, based on the styles of form and decoration seen in the Coastal focus, is discernible. The pots become gradually more globular with constricted neck, everted rim, and collar, and elaborate decorations in pointille emerge, often employing the herringbone design. The obtuse-angle pipe, first seen in the Coastal focus, reaches its mature development here. Other differentiating traits are the broad, usually equilateral triangular arrow point and a great deal of fine bone work, some of it engraved. Polished slates, the grooved axe, gouge, shell ornaments, and copper never appear. House types, both circular and rectangular, have been [22] traced on the large village sites, some of which at least were fortified with ditch and palisade. The burials are of the usual flexed type and are very rarely accompanied by a pipe, pot, or other object.

Little difference is observable between the crania of the Owasco focus and the Iroquois. The form is predominantly dolichocephalic, hypsicephalic, or high vaulted, with a moderately narrow face, orbits of medium size, and a generally broad nasal aperture. In size and ruggedness it stands midway between the Archaic and the Vine Valley series.

Two other less prominent culture phases should be mentioned, the Hopewellian and the Ground Slate. The former is known from mounds in the southwest section and from one fine tumulus on the Genesee River south of Rochester. This contained, within a stone cist, two skeletons accompanied by a curved-base monitor pipe of Ohio fireclay, native copper earplugs of the "double-cymbal" variety, pearl and shell beads, and Flint Ridge chalcedony blades. Nearby a flat copper axe and a second platform pipe were dug up.

The Ground Slate phase is most prominent in the St. Lawrence Basin. Its source lies to the northeast, and in it there are rubbed slate ulas and points much to suggest the Eskimo. It is present in Ontario, the Maritime Provinces, Labrador, and over the greater part of New England. For all we know, it may be more ancient than the Archaic.

Dr. Guthe then introduced Dr. John R. Swanton as one who would discuss the ethnological approach to the archaeology of the Ohio Valley.

Dr. Swanton: Speaking as an ethnologist, and using linguistics in the main as a background, I would first make certain generalizations. Each linguistic stock gives us a particular impression, although, of course, that is produced by its condition during one short segment of time immediately after the whites came in contact with it. After that, many rapid changes took place, most of them produced by the white invasion. There are some stocks that seem to be modifiers of culture, while others are themselves centers of culture. Thus the Algonquians impress us as a people who were modified rather than a people who modified, and the northern tribes are regarded as most representative. In the Southeast the Delaware and related tribes had a social organization resembling in some respects that of the Iroquois, while the western Algonquians, such as the Illinois, Miami, and Sauk and Fox, inclined toward the organization of the southern Siouan tribes. It is a question of how much was [23] Algonquian and how much Siouan. In the Chesapeake region we seem to get a hybrid culture due to southern contacts.

When we come to the Iroquoians, we find something entirely different. The Cherokee seem to have been modified by foreign groups, but the others—including the Iroquois proper, the Huron, Erie, Susquehanna, Neutrals, and so on—seem to have had a social organization of a relatively uniform type. Speck has commented on differences between Iroquoians and Algonquians noticeable down to the present day. We should therefore look for something very distinctive in connection with Iroquoian peoples.

It is a surprising fact that the Muskhogean and Siouan stocks, and even some of the smaller ones, do not exhibit the same internal uniformity as the Iroquoians. In the Muskhogean stock, for instance, the Creeks and Choctaw, while about half of their vocabularies are the same, had widely divergent social and ceremonial organizations, that of the Choctaw being relatively loose and simple and that of the Creeks highly complex. The Natchez and allied Taensa, though related to the above, had still another type of organization and a diverse ceremonial pattern. Other differences are exhibited among the Timucua of Florida, who appear to have been of the same connection, though they have usually been classed by themselves.

The Natchez Indians seem to have shared the culture that has been called "Lower Mississippi," or perhaps we should say of that part of the lower Mississippi found south of the Arkansas River, and it appears to me that there may have been a connection between certain other Muskhogean tribes, particularly the Creeks, and the culture found in the St. Francis region.

Along the eastern margin of the Plains were tribes of the Caddoan stock, which seem to have been anciently connected as to habitat with the woodlands, in spite of the common superstition that they entered the Plains from the southwest. It has been suggested that they were anciently connected with the Iroquois, and this presents a problem for future study.

When I was a graduate student at Harvard in 1897, I was given the job of investigating a village site just outside of Cincinnati, Ohio, the famous

Madisonville site. Although our work was done in a rather crude pioneer manner, it is certain that we found blue glass trade beads in three or four "ash pits" and with one of the skeletons, and also one or two pieces of iron. During earlier excavations on this site, Professor Putnam found European objects in close association with pottery of the prevailing Madisonville type. From the nature of these occurrences, Willoughby concluded that the site must have been occupied by one tribe down to the first period of contact with European traders, but not until the country was [24] settled permanently by whites. As the Shawnee and Miami entered that section in relatively late times and were from then on in continuous contact with white traders, it would seem that had the Madisonville village been one of theirs, it would show European trade objects throughout. The setup is what we should expect had there been a tribe settled in the section for a considerable period before white contact, but driven out very shortly afterward. These specifications are perfectly met by a tribe known as Mosopelea.

(Dr. Swanton then showed slides illustrating his reasons for believing the Mosopelea were the people who once occupied this site.)

Siouan Tribes of the Ohio Valley

When European explorers and missionaries encountered the various branches of the Siouan linguistic family, it was divided into two main sections: an eastern one in the Piedmont country of Virginia and the Carolinas but extending to the Atlantic Ocean between Cape Fear and Santee rivers; and a western on the western woodlands and eastern plains between Lake Winnipeg and the mouth of the Arkansas River, its eastern boundary coterminous for the most part with the Mississippi River. The Winnebago formed a detached group at Green Bay, Wisconsin, and the Biloxi were on the lower course of the Pascagoula River, Mississippi. Before 1700 a small Siouan tribe called Ofo, or Ofogoula, of which we shall have much to say presently, had settled on the Yazoo River in the latter state.

The relationship of these tribes proves, of course, that the several bodies had formerly been in contact. In the case of the two major divisions we must conclude that the tribes formerly occupying the territory between had died out; that the eastern tribes had migrated farther east or the western farther west; or that both easterly and westerly movements had taken place. There is evidence that the last supposition is correct, but also that the two southern tribes, the Biloxi and Ofo, filled the gap in part.

Specific traditions of a movement of the Virginia and Carolina Siouans from the west are preserved by Lederer and Lawson, and there is one regarding a similar movement of the Catawba Indians preserved by Schoolcraft, though this last is filled with exaggerations and reflects many of the ideas of a later period. When Mooney gathered into one bulletin the results of his own researches and those of Hale and Gatschet, and attempted to fix their western boundary, he failed, in part because he did not note a distinction between the Nahyssan and the Tutelo proper, the latter a tribe in the western parts of Virginia which escaped notice until the journey of Batts and Fallam, and partly because he did not have access to the [25] narrative of Needham and Arthur, which shows that in 1673–1674 the Moneton, erroneously called Mohetan by Mooney, were on a westward-flowing stream, probably the Kanawha. There is also some evidence derivable from place names indicating that the Catawba once extended farther up into the southern Appalachians. On several early maps, indeed, Kentucky River is called by their name. By some it has been supposed that this was derived from the Cherokee Kituwha, a word which they applied to themselves, or at least to a part of their nation. Mooney has shown, however, that this was originally the appellation of an extensive town site, and it is possible that the site itself received its name because Catawba once lived there, the phonetic alteration being due to the fact that there are no *p* and *b* sounds in Cherokee, although it is true that the later Cherokee name for the Catawba was Ani'ta'gwa. We feel sure also that Siouan tribes covered all of the northwestern parts of South Carolina until after 1567.

On the other hand, most of the western Siouan tribes retained until very lately traditions of an eastern origin, which James Owen Dorsey assembled in a rather too systematic form. Iowa, Oto, and Missouri legends point back to a former residence with the Winnebago, and this is borne out by their languages. The story of Mandan migration from the east was recorded by several early travelers. The data which particularly concern us, however, is that relating to the Quapaw and Osage. Migration of the Quapaw from the Ohio, meaning that part of the Ohio above the mouth of the Wabash, was so fresh in the minds of the Indians in that country when the French entered it that several refer to the fact, and on the De l'Isle map of 1702 "Acansea-sipi où Rivière d'Acansea" is the name given to a stream which appears to have been the Cumberland. Possibly the tribe stopped for a while on the Cumberland River during the journey south. Of particular interest is the following quotation from Father Douay, a companion of LaSalle: "The Akansas (Quapaw) were formerly stationed on the upper part of one of these rivers, but the Iroquois drove them out by cruel wars some years ago, so that they, with some Osage villages, were obliged to drop down and settle on the river which now bears their name, and of which I have spoken."

Here it is evident that the river referred to was an easterly branch of the Mississippi, though immediately before he had been discussing western branches. It would have been absurd, however, to speak of the Iroquois dislodging these tribes from the headwaters of the Missouri. The Osage here mentioned may be the Osage of history, who accompanied their kindred only as far as the great river, or they may have been the Ofo.

Enough has now been said to indicate that shortly before the appearance of white men, the eastern Siouan tribes had extended [26] farther west and the western Siouans farther east. I will now turn to the two southern tribes, the Ofo and Biloxi.

Ofo was the name given me by Rosa Pierrette, the last individual of the tribe who had any knowledge of the language. This is, of course, a part of the name Ofogoula or Offagoula of French writers, whose history I have traced in Bulletin 47 of the Bureau of American Ethnology. Either, as I have assumed, the Choctaw *okla*, "people," has been suffixed to the native name of the tribe, or Ofo is an abbreviation of Ofogoula. Whatever may be the truth, Ofogoula was supposed by the surrounding Indians and the French to be from Mobilian or Choctaw *ofi*, "dog," plus *okla*, and the tribe was called "Dog People," as by Du Pratz in 1758.

The apparently Choctaw or Mobilian origin of their name and the fact that Du Pratz states that, unlike their neighbors the Yazoo and Koroa, they did not have an *r*-sound in the language led me to believe at first that they were merely a small Choctaw or Chickasaw band like those on the upper Yazoo. In 1907 this seemed to be confirmed by Volsine Chiki, chief of the Tunica Indians, who said that he was able to recall one word in the language, *feskatcakĭ*, "opossum." As *f* is a characteristic Muskhogean sound and no Siouan dialect then known contained it, the Muskhogean relationship of the language seemed assured. In 1908, however, when I met Rosa Pierrette, I discovered that she knew in some form or other, a considerable number of words, and it was immediately clear that the language was a new Siouan dialect.

Our first notice of this tribe under the name Ofogoula is by Iberville in 1699, native informants giving it as one of the tribes on Yazoo River. It remained on Yazoo River until the Natchez uprising when, unlike the neighboring Yazoo and Koroa, it refused to join the hostiles and later sought refuge near the rebuilt French Ft. Rosalie. Subsequently it united with the Tunica and followed them up Red River to Marksville Prairie, where I located my informant.

In a short description of the country along the lower Mississippi, Tonti mentions this tribe under the name Chongue, the word for "dog" in Ofo (atc-huñki) or, more likely, from Quapaw (Cuñki). It is of some importance to note that this term is repeated again by Daniel Coxe in his text and on his map. He says: "Ten leagues higher (than the 'Matchicebe' river, on which lived the Mitchigamea), on the east side (of the Mississippi), is the river and nation of Chongue, with some others to the east of them." Remember that this is above the mouths of both the Arkansas and the St. Francis, and far above that of the Yazoo.

The Tunica, and probably the Yazoo and Koroa as well, called this tribe Ûshpī, as I was informed by the Tunica chief himself. This word in a corrupted form (Onspik) was used by Gravier, who visited the Tunica mission of Father Davion in 1700 and evidently derived his information through Tunica channels. He also mentions the Yazoo but not the Ofogoula. Three other writers, however, speak of the two as if they were distinct. [27] Iberville in 1699 called them Ouispe and Opocoulas, the latter name evidently a misprint; Pénicault in 1700 calls them Oussipés and Offogoulas, and La Harpe (or rather Beaurain) in 1722 has the forms Onspée and Offogoula. In two of the three cases the names are given in conjunction, and there is no reasonable doubt that they are synonyms for the same people.

Coxe, whose reference to the Chongue has been alluded to, also supplies information regarding another tribe of which we know nothing more, under that particular designation. He says: "Ten leagues (above the Arkansas) is a small river named Cappa (probably the St. Francis), and on it a people of the same name, and another called Ouesperies, who fled to avoid the persecution of the Irocois, from a river which still bears their name, to be mentioned hereafter."

An account of this other river appears a little further on: "South of the Hohio is another river, which about 30 leagues above the lake—a mythical lake into which the Wabash, Ohio, and the southern branches of the Ohio were supposed to flow—is divided into two branches; the northerly is called Ouespere, and the southerly the Black River; there are very few people on either, they having been driven away by the aforementioned Irocois."

There is no earlier and no subsequent mention of a tribe under this name. Under a somewhat different designation, however, we do hear of a tribe formerly resident on the upper Ohio which sought refuge in the south about the same period. In 1682 when La Salle and his party were ascending the Mississippi after having traced it to the sea, and while they were stopping at the landing place from which the Taensa villages were reached, Tonti says: "The next day (May 1) a chief of the Mosopelleas, who, after the defeat of his village, had asked permission of the chief of the Tahensa to live with him, and was living there with five cabins, went to see M. de La Salle and, when he said he was a Mosopellea, M. de La Salle restored to him a slave belonging to his nation and gave him a pistol."

The Taensa villages were then on Lake St. Joseph in northeastern Louisiana, not many miles below the mouth of the Yazoo.

The enemy to whom these people owed their discomfiture is not named, nor is the location of their former home, but this last is supplied for us by the Franquelin map of 1684. Here, on the north side of a river, evidently intended for the Ohio, and above its junction with the Wabash, we find the name Mosopelea and under it the words "8 vil. détruits." The inference is, therefore, a perfectly fair one that the Mosopellea encountered by La Salle were part of the tribe which had been driven from the upper Ohio, and the probability that their enemies were the Iroquois, or at least Iroquoians, is proportionately increased. Indeed, in a letter of La Salle's dated in 1681 or 1682 the "Mosopelea" [28] are listed among those tribes overthrown by the Iroquois.

Marquette gives us a glimpse of these same people in 1673 at a point on the east bank very much higher up the Mississippi, in fact somewhere between the mouths of the Ohio and Arkansas. He thus describes the encounter he and his companions had with them after passing the mouth of the Ohio on their way south:

> We were compelled to erect a sort of cabin on the water with our sails as a protection against the mosquitoes and the rays of the sun. While drifting down with the current, in this condition, we perceived on land some savages, *armed with guns,* who awaited us. I at once offered them my

plumed calumet, while our Frenchmen prepared for defence, but delayed firing, that the savages might be the first to discharge their guns. I spoke to them in Huron, but they answered me by a word which seemed to me a declaration of war against us. However, they were as frightened as we were; and what we took as a signal for battle was an invitation that they gave us to draw near, that they might give us food. We therefore landed and entered their cabins, where they offered us meat from wild cattle and bear's grease with white plums, which are very good. They have guns, hatchets, hoes, knives, beads, and flasks of double glass, in which they put their powder. They wear their hair long, and tattoo their bodies, after the Hiroquois fashion. The women wear head-dresses and garments like those of the Huron women. They assured us that they were no more than 10 days journey from the sea; that they bought cloth and all other goods from the Europeans who lived to the East; that those Europeans had rosaries and pictures; that they played on instruments; that some of them looked like me, and had been received by these savages kindly. Nevertheless, I saw none who seemed to have received any instruction in the faith; I gave them as much as I could, with some medals.

These Indians are not named, but on his map Marquette calls them "Monsoupelea."

Hanna, from whom the above is quoted, assumed that the source of the European objects these people possessed was the Spaniards of Florida, and he is probably right. In the latter part of the seventeenth century the Shawnee and other Indians of the Cumberland and Tennessee rivers were in the habit of trading with the Spaniards of St. Augustine, and this trade probably began at a much earlier period. In theory, Spaniards were not supposed to sell firearms to Indians, but theory and practice in such matters are not always concordant and, moreover, these guns may have been obtained otherwise than in trade. In September 1673, a party of Tomahitans (Yuchi) came to the house of Abraham Wood on the Appomattox River in Virginia, and Wood says: "Ye Tomahitans have about sixty gunnes, not such locks as oures bee, the steeles are long and channelld where ye flints strike." I do not know enough about the firearms [29] of the period to tell from this description whether the guns were Spanish or not. At this period, articles of European origin were also being introduced by an established trading path from the Dutch settlements in New York and New Jersey across the Susquehanna River, as the Jesuit fathers discovered when they reached the Huron Indians and the Neutral Nation. Wood says, however, that when Arthur returned to him in June 1674, he was accompanied by "a Spanish Indian boy."

This Gabriel Arthur, Wood's indentured servant, visited the Moneton tribe, the one mentioned above, in company with the Tomahitans, and on their return they went three days out of their way to attack a numerous tribe on a river which I take to have been the Ohio. These were probably the Shawnee but, if my supposition is correct, they occupied the territory of the very Indians met by Marquette the year before in their town on the Mississippi. As the Siouan Moneton were on terms of friendship with the Yuchi, it is probable that the Mosopelea were friends also, in which case they could have been supplied with guns from Florida through the Yuchi, although the Indians who took their place were without firearms when Arthur met them.

Before passing on, note should be made of Marquette's apparent assumption that the Mosopelea were related to the Huron, but the evidence is at most inconclusive. The Mosopelea had lived close to Iroquoian tribes from which they could readily have adopted the method of wearing the hair similar to that of the Huron. It is to be noted, also, that although he addressed them in Huron, he misunderstood their answer. Unfortunately he leaves us in doubt whether the information which they afterward gave him was obtained through the medium of Huron or Illinois, but indeed there is little that they might not have communicated by signs. Huron and Seneca, the extant Iroquois dialect spoken nearest to the ancient home of the Mosopelea, are not mutually intelligible, but we do not know whether this was true of Erie, Susquehanna, and Black Minqua or not. In any case, mistakes in identifying the language of a tribe are not uncommon. The Yuchi were supposed to have spoken Shawnee by two different explorers. In the present instance, any argument for an Iroquoian connection is more than counterbalanced by arguments pointing toward affiliation with the Siouan stock.

The following facts have now been developed:

1. At the end of the seventeenth century there was a small tribe on the lower course of the Yazoo River which spoke a Siouan dialect, was known as Ofo (and by the Choctaw or Mobilian term of Ofi Okla, meaning "Dog People," sometimes translated into a Siouan language as Shonki) and was called by the Tunica Indians Ûshpī.

2. At the same period Daniel Coxe speaks of a tribe called Ouesperie

or Oùespere which lived at one time on the Cumberland River, had been expelled from its ancient home by the Iroquois, and had settled near one of [30] the Quapaw towns on a river probably identifiable as the St. Francis.

3. A tribe known as Monsopelea, Mosopelea, or Monsouperea, which had once occupied eight villages on the north side of the Ohio above the mouth of the Wabash, was expelled from this territory by the Iroquois before 1673, and probably before 1669, when La Salle received a description of the Ohio country and its former inhabitants from some Seneca, and in the latter year was living on the east side of the Mississippi between the mouth of the Ohio and the mouth of the Arkansas. By 1682 at least part of them had sought refuge among the Taensa, a tribe friendly to the Quapaw, living some miles south of Yazoo River.

While we have no historical evidence that the Ofo or Ofagoula ever lived farther north than the Yazoo River—except for the fact that Coxe locates his Chongue on a river flowing into the Mississippi east of the St. Francis—the testimony of their language points in that direction. All of the other Siouan tribes except the Biloxi are far to the north, and the Ofo language itself is closer to the languages of the Tutelo and Dakota than to that of the adjacent Quapaw, as is shown in the following comparative vocabulary:

Comparison of Words in Five Siouan Dialects

English	Catawba	Ofo	Tutelo	Dakota	Quapaw
ax	pas^c	a^{n}fhepi	nisēp	o^{n}spe	i^{n}spe
bear	nume	u^{n}thi	mūnti	mato	ma^{n}tu
bone	sap	aho	wahōi	hu	wahi
day	yap	no^{n}pi	nahambe	a^{n}pe	hă^{n}ba
dog	ta^{n}si	atc-huñki	tcoñki	cuñka	cuñkĕ
ear	duksa	nas-husi	naxōx	noghe	na^{n}tă
father	nane	at-hi	eāti	ate	edćate (his)
fish	yī	ho	wihōi	hogha^n	hu
foot	yipa	*ifhi*	ici	siha	si
four	paprere	topa	tōpa	topa	tuwa
ghost	yī^{n}we	nă^{n}tci	*wanuntcī*	wanaghī	wanaxe
grandfather	tatewa	etiko^{n}so	eku^{n}i	tuñka^{n}cida^n	etikă^n

grandmother	istcū	ĭkoni	higūⁿ	kuⁿsitku	ĕkăⁿ
great	patki	*ithoⁿ*	*itáⁿi*	tañka	tañka
house	sūk	*àthi*	atī	tipi	ti
iron	dorob	*amõⁿfi*	*maⁿs*	maza	mâzĕ
mother	istci	oⁿni	ina	ina	ehaⁿ
mouth	sumu	*ihi*	ihī	ī	ā
one	nĕpĕ	nûfha	noⁿc	waⁿji	miⁿxti
six	diprare	*akăpĕ*	*akāsp*	cakpe	capĕ
tobacco	umpa	itcani	yēhni	tcaⁿdi	tani
tooth	yȧp	*ifha*	*ihī*	hi	hi
tree	yap	itcaⁿ	oni, wiēⁿ	tcaⁿ	jăⁿ, jûⁿ
water	yēhiĕ	ani	manī	mini, mani	ni

[31] The Catawba forms are taken from Gatschet's unpublished vocabulary as revised by Speck but without Speck's last revision; the Ofo is from my own material; the Tutelo is from Hale's vocabulary; the Dakota from Riggs; and the Quapaw from Dorsey.

The great divergence of Catawba from all the other dialects is at once apparent. The closer connection between Ofo and Tutelo than between either of them and Dakota or Quapaw is also indicated. It is particularly evident in the cases italicized, in some of which the tendency of both languages to retain a vowel before the stem consonant is evident. The position of Dakota is not so clear, and it may be altered by later investigations, but the present material allies it somewhat more closely to Ofo and Tutelo than to Quapaw, though the Quapaw must once have been near neighbors of the former tribe.

Aside from the Ofo, there is no tribe on the lower Mississippi which may be suspected of having descended from the north during the early historic period, excepting of course the Quapaw. There is reason to think that the Tunica once inhabited villages in northwestern Mississippi, but no indication carrying them beyond that point, and their language and culture belong to the region in which the French found them.

Not only Marquette but several later students have looked for remnants of such Iroquoian tribes as the Huron, Erie, Neutrals, and Susquehanna in the Gulf region, but in all my reading of historical documents I have come upon only one reference that might indicate such an event had perhaps taken place. An enumeration of Indian tribes within the French sphere of influence dating from an early period in the eighteenth century does speak of Andaste, i.e., Susquehanna, living among the Cherokee, but this remains without corroboration of any sort.

Evidence by exclusion therefore points strongly to the identity of the Ofo or Ofagoula, the Ouesperie, and the Mosopelea, and if we assume that to be true, we can fit all references to them into one pattern. Driven from their original seats on the upper Ohio before 1673, we may assume them to have stopped on the Cumberland River long enough to have given it their name on the lips of certain of the hunters and explorers relied on by Coxe in making up his narrative. By July 1673, they were on the east bank of the Mississippi below the mouth of the Ohio. This corresponds rather closely to the place where Coxe locates his Chongue. The fact that Coxe inserts "Monsouperea," "Chongue," and "Ouesperie" independently need cause us no concern since he drew his information from all sources without attempting any checkup. There is every reason to think that he copied French writers in using the name Monsouperea, and he was so careless as to put it above the mouth of the Ohio instead of below it.

After 1673 we may suppose that the tribe in question moved to the west side of the great river and settled near the uppermost Quapaw town. [32] In 1682 they had pushed on to the Taensa, who were on friendly terms with the Quapaw, but when Tonti prepared his short account of the lower Mississippi, they had shifted once more to the Yazoo. If we do not accept such a reasonable and consistent story, we must leave the later fate of the Mosopelea and Ouesperie, and indeed the identity of the tribes themselves, a profound mystery. At the same time, we are forced to assume a northern origin for the Ofo without attempting to identify their ancient home or the steps by which they reached their later one. Moreover, I think that we may indicate with high probability that Ûshpī is merely an attenuation of the name Mosopelea. My argument for this is shown on the following chart:

Authority										
Marquette	Mo	n	s	ou	p	e	l	e	a
La Salle	Mo		s	o	p	e	l	e	a
Tonti	Mo		s	o	p	e	ll	e	a
Douay	Ma	n	s	o	p	e	l		a
Franquelin	Mo		s	a	p	e	l	e	a

Marquette	Mo	n	s	ou	p	e	r	e	a
Thevenot	Mo	n	s	ou	p	e	r	i	a
Coxe	Oue		s		p	e	r	ie	
Coxe	Oue		s		p	e	r	e	
Gravier	Ou	n	s		p	ik (ie?)			
La Harpe	O	n	s		p	ée			
Pénicaut	Ou		ss	i	p	é			
Iberville	Oui		s		p	e			
Swanton	Û		sh		p	ī			

A word or two should now be said regarding that other isolated southern Siouan tribe, the Biloxi. Although the Biloxi language is quite distinct from that of the Ofo, it clearly belongs among the northern dialects, including Tutelo and Dakota, rather than with the Dhegiha group and certainly not with the Catawba and its allies. Therefore an immigration from the north is to be expected in the case of this tribe as well as in that of the Ofo. We seem to have a suggestion regarding the route this may have taken in an entry on the De Crenay map made in 1733. Here, at a point on the Alabama River readily identifiable as the mouth of Bear Creek in Wilcox County, in the state of Alabama, we find a town laid down called "Bilouchy." While this may have been a summer camp of the Biloxi Indians, it is too far away from their own territories on the Pascagoula River to render that probable. I rather regard it as indicating a stage in the southward movement of the people. It would suggest that, instead of descending the Ohio and Mississippi like the Mosopelea, they had moved overland as part of the Shawnee did in later times, perhaps crossing the Cumberland at Nashville, reaching the Big Bend of the Tennessee, and picking up the trail from that point across to the Coosa, which they could then descend to the Alabama. As the Mosopelea movement was in relatively late times, and we find at least traditional remembrance of one Biloxi location on the journey, this change of position may have been relatively modern, and I will here make a suggestion which must remain for the present [33] very much in the air but may yet be confirmed in unexpected ways. Some of our early French writers on Louisiana speak of a Biloxi town, or perhaps a tribe associated with the Biloxi, as Capinans. I used to think that this might be merely the name of a French concession near which some Biloxi village stood, but such seems not to have been the case. Is it within the range of possibility that it could be a form of the name "Capitanesses," which on very early maps of the Susquehanna River we find applied to a tribe living west of the Susquehanna and apparently reached by way of the Juniata. The nearness of this tribe to the Susquehanna as represented on the map need not disturb us, as the distances are very much foreshortened, and apparently part of the Erie of northern Ohio are also included under another appellation. I might add as a curious fact that in the Catawba migration legend, it is stated that part of their nation during the removal south had gone off with the Choctaw and Chickasaw. But the Catawba language is so utterly different from Biloxi that I regard this tradition as of minimal value. It may at least be cited as a curious coincidence.

Let us now turn to the possible archaeological bearing of the facts developed above. Twenty-six years before my discovery of the Ofo language and identification of that tribe with the Mosopelea, I happened to have been set to work on an archaeological site which there are circumstantial reasons for believing had been occupied by these people before they had been driven down the Ohio River. This is the well-known Madisonville village and burial site just outside of Cincinnati. Explorations were carried on here by the Peabody Museum at Cambridge from 1882 to 1911, by Dr. Charles L. Metz and Professor F. W. Putnam, the former having had general supervision during most of that period, though assisted successively by Harlan I. Smith, Roland B. Dixon, the writer, R. E. Merwin, and B. W. Merwin. The report of this work was finally prepared by Dr. Ernest A. Hooton, assisted by C. C. Willoughby, and was published by the museum in 1920 (Vol. 8, No. 1, of the Papers on Archaeology and Ethnology). Willoughby, who prepared the concluding section of this paper, says regarding the period of occupation of the Madisonville site (and it is to be remembered that this was printed three years before my own investigations on the Mosopelea were given out):

> It is evident from the foregoing pages that the occupation of this site covered an interval immediately preceding the first intercourse of the Indians of the region with Europeans, and extended into the protohistoric period, at which time the inhabitants were able to secure a small amount of European iron, brass, and copper, together with a few glass beads, either directly from the early missionaries or traders, or indirectly through their Indian neighbors.

That these later inhabitants were of the same group as the earlier dwellers on the site is evident from the finding of a cross and other trinkets of brass in a grave containing also a pottery vessel of a type common throughout the cemetery. The site, however, appears to have been [34] abandoned long before the arrival of European settlers in the Ohio Valley. (p. 135)

These European objects consisted of brass ornaments found by Professor Putnam in 1882, iron objects found by myself, and blue glass beads found by myself and Mr. B. W. Merwin.

I submit that the Mosopelea tribe is the only one whose history fits in at all satisfactorily with the conditions set down by Mr. Willoughby. We know that they lived at least in this general region and that they were driven out by other Indians, probably the Iroquois, before their country had been visited by white men but after they had been sufficiently in contact with whites to provide themselves with guns and "flasks of double glass." It is probable that Marquette wishes us to understand that the "hatchets, hoes, knives, and beads" of which he speaks were also of European origin and that the Indians told him they had traded with the Europeans sufficiently to buy cloth.

From the Arthur narrative it is probable that by 1674 the old Mosopelea territory had been occupied by the Shawnee or perhaps the Miami. Yet even supposing the actual site to have been preempted by one of these tribes at that date, not sufficient time elapsed between then and the period when we first have a clear view of the country to produce a cemetery as extensive and ash pits as numerous as those on the Madisonville site. It is also evident that had the settlement been started after white contact we should find objects of European manufacture throughout the site instead of with a few skeletons and ash pits.

As Willoughby notes in the quotation given above, Indian objects found associated with those of European manufacture are of the same cultural provenance as the Indian objects not so associated, and on the basis of these objects Madisonville has been classed as a Fort Ancient site. That being the case, I conclude that the Mosopelea Indians up to the time they left the Ohio possessed a culture of the Madisonville type, which means that at least one Siouan tribe possessed the Fort Ancient cultural patterns. From this fact we might confidently look for a similar association elsewhere, but this does not prove that all Siouan tribes were possessed of Fort Ancient culture, nor that Fort Ancient culture was confined to

Siouan tribes. The extent of the association is one of the things to be demonstrated.

The points of my argument may be summarized as follows:

1. When the French came in contact with the tribes on the lower course of the Yazoo River toward the end of the seventeenth century, they found among them one speaking a language different from the rest, which I identified in 1908 as a Siouan dialect. Comparison of this dialect with others of the same stock shows that it is closest to the Biloxi of the [35] Gulf Coast and to the Tutelo dialect formerly spoken in Virginia.

2. Sometime before 1673, a tribe called Mosopelea is known to have been driven from this region and to have settled on the lower Mississippi. There are strong circumstantial reasons for identifying it with the Ofo, partly due to the resemblance between the name Mosopelea and a name applied to the Ofo in later times and partly because what we know of the language of the Ofo calls for precisely the sort of movement which the Mosopelea underwent. If the two tribes are not identical, we must suppose the Mosopelea to have disappeared without leaving a trace.

3. Although the topography on Franquelin's map is very much distorted, the location of the Mosopelea corresponds as nearly as can be made out with southern Ohio or southern Indiana. The remains found at Madisonville must be explained by supposing that a tribe had lived here for a long period before white contact—had lived until the time when objects of European manufacture first made their way into the country—but had left before regular white settlement began. The other tribes known to have been occupants in this region, the Shawnee and Miami, were late intruders, and if Madisonville had been inhabited by them, we should find European objects scattered everywhere.

The meeting adjourned until Saturday morning, December 7.

Saturday Morning Session, December 7

GENERAL (Frank H. H. Roberts, Jr.)
May I say that I am glad of the opportunity of hearing these things discussed because one of the things I have stored up to do is to work up the material I found at Shiloh, Tennessee. I have been pretty much in a quan-

dary as to just what it might mean. I went through it very carefully, and according to the way things checked out, I had what looked to me like your greatly disputed Woodland basic culture. When you started to talk about it, it sounded to me like the Shiloh material, but as each man gave his views, I noted first one thing, then another, which differed until I am now more uncertain of it than ever.

From the standpoint of one who knows nothing about the Mississippi Valley problem, perhaps I should mention simply the things that stand out in my mind as the result of thinking over what you have said. The most striking, it seems to me, is the question of the Woodland basic culture. One speaker discusses it as a typical hunting phase. The next gives the idea that it is an agriculture–pottery-making complex. It seems you have two features working at cross purposes. Possibly your basic hunting-fishing group is something else—as Basketmaker II is a non–pottery-making group, while Basketmaker III is a pottery-making group. There may be a similar condition here. From the discussions, it has seemed to me there were two separate and conflicting things linked together that may be the cause of the confusion.

Another feature is the question of division. The implication was that the Upper, Middle, and Lower Mississippi were quite distinct. Also, we are handicapped because no one is presenting the side of the Lower. All of these divisions seem to be part of one basic thing. As I mentioned in the discussions, I personally feel that you must consider and make some provision for a general, widespread basic factor out of which all of the others develop. We find here that the general division is the Mississippi, and in that you have the Upper, Middle, and Lower Mississippi phases. As I see it from the broad point of view, you have only one basic culture, which includes the others as various phases. Not understanding at all the implications of your Woodland, I couldn't attempt to say whether it is truly basic or not. Is it sufficiently different to be basic? Archaic Algonkin I would consider basic, in that you have something different, as it has neither pottery nor agriculture. The next stage, as I gathered from the discussions, has entirely different features. Now which is the basic Woodland? It may be clear in your minds, but the discussions did not indicate where you would make the distinction. In some ways, this is just the broad general impression I have gained. It struck me that here you have the reverse of the attitude that we had in the Southwest at the time of the first Pecos Conference. Then our whole outlook was based primarily on knowledge from the San Juan area, but the general assumption was that that covered the whole field. As work progressed, we found a different feature in the south; in fact, a distinct subarea now called the Hohokam. That in no way vitiated the Pecos classification as far as the Pueblo area proper is concerned, but it made necessary a recognition of two patterns. Yet the two are definitely tied together by the basic house-building, sedentary, pottery-making features. Here I get the impression that you are considering the Upper, Middle, and Lower as distinct units in themselves, when they actually all belong together.

The question of using ethnological or linguistic names with reference to particular groups seems to be causing difficulty. One illustration of what may happen is demonstrated by Gladwin's interpretation of the Hohokam. He reached the conclusion that the Hohokam came into the Southwest with a fully developed pattern. He postulated their coming from the east. Looking to the east, the only thing he could see was the Caddo. Between his area and the Caddo was a group called the Jumano. They were in a very good position to have transmitted features from the Caddo area to the Southwest. In general, the Jumano for a long time [37] were regarded as the most western Caddoan group. Consequently Gladwin believed the Hohokam traits must have come from them. But from the ethnological angle, there was disagreement. The men who have made the most detailed studies of that area have reached the conclusion that the Jumano were a Pueblo group which moved east and took over a Plains culture. This was so late that nothing from the Caddo could have filtered over into the Pueblo area in time to affect its development. The actual chronological sequence and definite dates are against this theory.

Then, we have two or three illustrations of identifying specific sites with a definite group. I worked several times on the Zuñi reservation. We know they have been there since 1540. Archaeologically, you can go back from 1540 in a direct line to about 1300, which extends through Pueblo V and IV without a break. On this reservation are a number of definite Pueblo III ruins. Thus far, a transition from Pueblo III to Pueblo IV can't be demonstrated, so you can't actually say that these were Zuñi. The same thing holds true along the Rio Grande. The Keresan and Tanoan peoples definitely claim certain sites as their old villages. Archaeologically, they cannot be separated. There again is another example of what you may have happen. At Walpi is the village of Hano, which is not Hopi at all but Tewa, the people coming from the northern Rio Grande in historic times.

Yet all the material culture is so similar that the archaeological remains from the Hano and the Walpi villages would be such that you would call it all one site.

Another thing that struck me, which may explain some of the present difficulty, is the fact that in some of your phases or aspects you have a definite series of traits that come from one particular district. Wherever you get them, they are typical and the name of that phase comes from that group. In other cases you have a phase named from a certain group, but the phase name is not from the group containing the most characteristics. Is it right to designate a phase in that way? That may simply be my lack of understanding of just what is included in the group. It would seem to me that in using the term for a phase, there should be a pretty close connection between the sections containing the particular characteristics and their name; the name should be where the type series is found.

Insofar as the classification is concerned, I think you can't compare too closely the situation in the Mississippi Valley to that of the Southwest, because we have a feature there which makes things considerably more simple, and that is the chronological sequence. The Little Colorado and the San Juan sites, despite local differences, take their place in the broad picture by reference to the chronological chart. Pueblo II is a sequence stage in the development of the cultural pattern, not a time classification. The sequence was developed on stratigraphic evidence; the time element was a later addition. There has been some [38] confusion, and many think the cultural sequence names are synonymous with dates. This is not necessarily so.

Discussion

McKern: There was a parallel classification developed in the Southwest.

Roberts: Yes, in one sense. Yet they are different in another. You can get a full change in the culture in a single generation. This was overlooked in the Southwest until the time chronology was determined. There is a definite time lag of the various Pueblo horizons in the peripheral regions. The Hohokam classification profited from the faults of the Pecos Conference. They completely avoided the time-sequence fixing of I, II, III, and IV, etc. They took the names Pioneer, Colonial, Classic, Historic, and Recent. This avoids a definite tying-down to an implied chronological sequence.

Geographical grouping was for a period considered an important question—that was the main method of classification for a long time. In the drawing up of the sequence for the Pueblo area at Pecos, they left out all geographical groupings, actually without any definite decision to do so, because it wasn't essential. It has no significance outside of the location of the general features. You can say that in *a* certain general district, you get certain things, but it has no sequence importance. I can't see why this classification you have here won't work in the Southwest as well as in the Mississippi Valley. The only thing I would object to is the seeming omission of the broad basic feature. If you used the word "base" and left off the use of culture, that might suffice.

McKern: You mean, use base in the case of a group of an agricultural or a nomadic-hunting people, and then use culture for Mississippi, Southwest, etc.?

Roberts: Yes.

McKern: The criticism has been made of our reports. We haven't given Roberts the facts. There is a great deal of variation in the smaller divisions. Trempealeau is a term applicable to a focus, whereas Hopewell should be an aspect. The point Roberts has brought out in regard to basic culture is a good criticism, and the rest of the criticism is the result of our type of presenting the reports. Woodland as a term does not refer to geographical distribution or to an ethnological group. It may include groups which were not Algonkin. There is no special connection between Woodland archaeologically and the Woodland ethnological groups.

I think that conforms with what you have said about basic cultures. You have a base which includes all of the peoples living in the Mississippi Valley who made pottery and had agriculture. Under that, you [39] certainly couldn't object to dividing various groups living in the Southeast from each other on a finer cultural basis. There seems to have been a rather sharp division between a relatively early standardized northern culture extending from the Atlantic Coast to the Plains and the more complex influences which have come up later. The remarkable thing is that there is not more amalgamation than there is. Some of us have even questioned ourselves as to whether some of our Woodland and Southeast pottery had the same origin.

Guthe: In speaking of Pueblo II, we are not speaking of identical traits in every division, nor are we talking about the same time period in each division. In the Southwest, they talk about a culture complex, and even such a feature as an architectural complex need not be the same in different places.

Roberts: There is not always a correlation between Hohokam and

Pueblo. We don't know whether Pioneer Hohokam was earlier than Pueblo I or not. I think they are roughly parallel. The Pioneer Hohokam and Pueblo I, and Colonial I and Pueblo II were probably parallel. They are distinct patterns, yet they have the same basic features.

McKern: It has always been confusing to me to have a term which proposed being cultural, such as Basketmaker and Pueblo, which certainly implies a time rather than a cultural classification.

Roberts: As I said before, the Hohokam classification avoided this fault. Ultimately, the solution in the Southwest will be to say that in 1066 in Flagstaff, we had so-and-so, in Chaco Canyon, such-and-such, etc. The tree-ring studies may put it completely on a chronological basis and enable us to discard confusing terminology.

A group discussion of some length followed this. On its conclusion, Dr. Guthe read a letter from Dr. Cole, quoted below:

> It now seems certain that the doctor will not allow me to attend the Indianapolis meeting. I am mighty sorry, as I had expected to gain a great deal from the discussion. Since I cannot be there I am going to afflict you with some observations; they probably will not fit into the discussion, so do what you please with them.
>
> When I first began to take an interest in anthropology, the situation in southwestern archaeology was much like that in the Mississippi Valley five or six years ago. A considerable amount of excavation had been carried on—some good, much very bad. Collections had been made and certain similarities between sites and regions had been noted. Workers had begun to talk of San Juan and St. Johns types of pottery, Pajarito ware and so on. Cliff Dweller was an established term and the Wetherills had unearthed bodies and objects to which they gave the name Basketmakers. Theories and speculations were as numerous as the [40] excavators, but most materials were classed regionally without any satisfactory evidence as to age or relationship.
>
> Then came the work of Nelson, Kidder, Spier, and many others. Intensive excavation in restricted regions showed actual stratification, or gradual change through long periods of time. Pottery sequences, architectural periods, and a study of the total cultural objects of the sites began to indicate definite chronology. Meanwhile a wide survey showed the relationship of materials from one region to those of another.
>
> Out of this came the Pecos conference and a system of classification.

This was a unilineal series of "cultures" each with certain type differences, but grading one into the other. While certain traits appeared in several "cultures" some rose to importance in only one. Thus it was possible to cite the distinguishing characteristics of each, while recognizing the relationship of all. Description and classification made possible the entry into more intricate problems.

That the tentative classification set up at Pecos was one of the most important advances in southwestern archaeology is generally accepted. That it was not final is shown by the changes and modifications which have taken place. Now the surveys and excavations in southern Arizona indicate that further modifications must be made. It is not *necessary* that *a scheme be perfect to prove useful.*

With the application of the Douglass tree-ring method southwestern archaeology has entered into a new epoch. The tentative succession of cultural objects have, for the most part, been given actual dates. Contemporaneous occupation of nearby sites, as in the Chaco, has shown the great similarity but not actual identity of many sites, and, finally, cases of cultural lag have been established.

Here, it seems to me, is an important point. If we study existing pueblos, we find great similarities, but not absolute identity. This I believe is a worldwide condition in ethnology. What is true for ethnology is doubtless true for archaeology. We can expect to find such close resemblances that we can class groups and sites together, on the basis of material culture, but we cannot hope to find a people, or region, or group of settlements in which cultural manifestations will be actually the same.

I have cited the Southwest because I believe it will help us to visualize our own problems.

It is unnecessary to dwell at length with our own situation, say in 1930. Certain states had done sufficient excavation and comparative work so that they could define several "cultures." Ohio could boast of Hopewell, Adena, Fort Ancient, and other states reported their "cultures." [41] But when Wisconsin, Iowa, and other states began to find "cultures" with many resemblances to Hopewell the question was raised as to just what we meant by Hopewell. Langford's materials from the Fisher site looked, in many respects, like Fort Ancient. Yet it seemed clear that he was dealing with more than one occupation and that the different levels were not identical.

Statewide surveys indicated several different "cultures," while intensive work yielded several cases of cultural stratification. A pictorial survey of several states clearly suggested a regional distribution for many cultural determinants.

The situation was chaotic, but enough data were available to suggest a method of attack. At this time McKern proposed a classificatory scheme which led up to the one now in use.

Certain things seemed obvious. Some of these so-called cultures were related although differing in many respects; some of these related "cultures" were apparently separated by a considerable time interval; some "cultures" were succeeded by quite unrelated "cultures." Clearly we had a situation quite different from the Southwest. Here was no story of development, from one simple beginning through various steps. Rather it was a story of movements of peoples—often unrelated—and the frequent superposition of unrelated cultures.

To solve our problem, a purely objective scheme was devised. Briefly put, it was thought possible to show cultural relationships (when they existed) by using four groupings, which varied in the number of traits they had in common:

foci—cultural manifestations practically identical (a situation probably never actually encountered)
aspects—a predominating majority of cultural manifestations in common
phases—a near majority
basic culture—fundamental traits in common

To fill in such a scheme we should, *ideally*, excavate and describe such a number of sites in each area that we could clearly see all the relationships; then we should set up a *focus* or *foci*. These in turn would be placed in the larger groupings. But *practically*, if we wait for such data we will remain in the chaotic condition of 1930 for another generation.

Today we have broad surveys; we are in a position to set up a tentative scheme of *basic cultures, phases, aspects,* and *foci*. We are justified in such action *if* we keep in mind that this is a tentative scheme. It is a scheme to be tested by the evidence; it is to be changed, modified, or discarded at any time. Obviously we shall be adding new foci, and [42] regrouping the *foci* into more or different *aspects*. The recent tests of the scheme by Griffin and Deuel indicate that classification is possible. Doubtless many changes will be brought about in conferences such as this, but it is my opinion that Mississippi Valley archaeology has made a great advance in the acceptance of this classificatory system as a working base.

GENERAL (Frank M. Setzler).

The impression I have made here has led me to compare myself to an alumnus who has left college about five years ago. The president of the school asks for a donation for a new anthropological building. My donation has been extremely small. On the selected day I return for the opening. I had hoped to see the central hall portraying the Hopewell culture, but to my amazement I find it stuck away in a corner under Woodland. This upsets me, so I say tear the whole building down and start over.

First of all, let me say that I am interested in this classification. Every objection I make will go by the board if it is going to create the feeling that I am opposed to such a classification. Naturally my own fieldwork has caused me to be interested primarily in what has been termed, until now, the Woodland basic culture. Nevertheless, I am also interested in the Mississippi. I realize the differences between the so-called Mississippi and the Woodland—anyone can see the contrast between the Fort Ancient and Hopewell. The thought occurred to me, when the suggestion first came out in regard to the term "Woodland," that it might imply certain eastern Woodland cultures. This means Algonkin to me. I realized that certain of us felt that the Hopewell might be related to Algonkin. I could never agree that the material from the Hopewell and Seip mounds had any relationship with the true Algonkin.

I have developed a feeling against using the term "Woodland" in connection with the Hopewell complex. If it has no Algonkin implications, all well and good. Nevertheless, I realize that certain Woodland traits are related to the Algonkin cultures. I also have a feeling that during the time of Mills, Hopewell could not exist outside of southern Ohio.

We must use a certain number of comparable traits to affirm a relationship. I feel that if you have a sufficient number of artifacts that only compare with one group of cultures, why isn't it right to compare it with another group? Why is it that certain Algonkin traits can be segregated only into the Woodland? In Ohio I felt certain that the only evidence of a true Algonkin complex would be the surface elements actually left by the historically known Algonkin groups.

Another thing occurred to me. There were certain groups of us approaching this classification problem from two extremes. One has attempted to start with the components and work out every single trait known at a single site and build up by careful comparison to the phase and basic elements. [43] The other group starts with the basic cultures and works down to the components. I can see that in order to teach the archaeology of the Mississippi Valley to students, we must have some particular classification that will make things clear cut. You have to be arbitrary in teaching undergraduate students and the public because there isn't enough time to go into details. But aren't we apt to make a mistake in doing this? I prefer to analyze the individual sites, leaving the questionable sites in midair, not trying to explain them for the present. It seems to me that we will make less grievous errors and have less to correct in the future than by arbitrarily setting up the basic cultures. I suggest for want of a better name to cover the cultures in the Mississippi Valley, a term "agripo" (agriculture, pottery). We should have some term to take in this agriculture, maize, pottery complex, and another for the pure nomadic, hunting groups such as the Folsom people. If we are going to strike out the term basic culture, why couldn't we have all of the Mississippi Valley people (practically everything east of the Rocky Mountains) listed under such a term as Mississippi culture for our base? Under Mississippi culture, let us get such terms as those which will call to the minds of students a definite division of cultural determinants which will stand out from everything else.

For this reason, I would suggest the term "Hopewell phase." That will call to our minds a specific group of determinants. Under that we could say Mississippi culture; Hopewell phase; Louisiana aspect; Red River focus; Marksville component. How would that work for Ohio? Mississippi culture; Hopewell phase; Scioto aspect; Seip focus; Seip component. For Illinois—Mississippi culture, Hopewell phase, Illinois aspect, etc. For Wisconsin—Mississippi culture, Hopewell phase, Wisconsin aspect, Trempealeau focus, Red Cedar component. In this way we would immediately see the division made under phase for the general Mississippi cultures, and the other divisions give us the local geographical distribution without any linguistic or ethnological connotations.

I have talked to a number of people who are familiar with the ethnological features of the Algonkin, and none of them seems to feel that the Algonkin Indians could ever be considered true mound builders. In his book, Willoughby characterized the generalized Algonkin as having the grooved axe, celts, mortars, tubular pestles, steatite dishes, sculptured-stone heads, grooved and perforated stones, long-stemmed elbow pipes, and pottery vessels. Perhaps the exotic Hopewell culture represents only a ceremonial phase of the Algonkin. I doubt it! We have never found Hopewell village sites in Ohio. But in Louisiana the pottery and other artifacts from the village site are exactly the same as that in the mounds. (McKern stated this was also true in Wisconsin.)

Discussion

Ritchie: This Old Algonkin pattern as Willoughby presents it is very incomplete. If he included most of the material, he would have a [44] typical example of what we call Second Period Algonkin, and which we are not certain is Algonkin. His Generalized Algonkin has the crude punctate pottery. Hopewell seems to be a reintegration.

Setzler: This whole classification problem is to simplify, as far as we can, the archaeological work we are doing. I am wondering if we are not simplifying it too far, and in doing so losing sight of the essential determinants. Certain reports come out with statements which are not true and which will need to be corrected later on.

McKern: I don't know of the history of any scientific development which has not involved a constant changing in the terms and classification. There has never been any serious human effort which has not resulted in outright statements and classifications which will have to be corrected. The trouble is in holding back information until one has something he considers perfect. The exercise of controlled imagination should also come in. A man has a right to make tentative adjustments.

Webb: The difference between what a man saw, and what he thought, should be made clear.

Setzler: In the Southeast, you have certain characteristics which may be compared to the Hopewell phase of the Woodland. We also find variations in the Southeast which I can't help but feel are definitely part of the Marksville complex and belong just as much to the Marksville as to the Hopewell. From those variations, not found in the Ohio complex, we do begin to see a relationship between some of the other pottery groups in the Southeast. Ford has carried on work at Coles Creek, and Collins at the Deasonville site. The evidence is that at the bottom we have a pure type of Marksville components. Fifty percent of those traits are found in association with something that seems to have developed out of the Marksville group. Then we find more of the variations from the Marksville, and those characteristics at the top seem to tie in with the protohistoric groups in the Southeast—the Tunica, Natchez, and Caddo. In other words, it looks as if the Hopewell in the Southeast is something basic. I can't help but feel we have enough data from Marksville to be compared with your northern Hopewell. Five years ago, every Hopewell pot found in the Southeast was considered entirely due to trade. Now we are finding a definite complex. Whether the South affected the North, or the North affected the South, I can't say. We can pick out certain characteristics from the Marksville site which show definite amalgamation. Perhaps the Marksville people moved up the Mississippi and settled in Ohio, where they met a group from the North, before any Algonkin influence reached that region. This southern group met a people familiar with obsidian, copper, carving on stone; the southern people brought mound building, pottery, and platform pipes, and the resulting amalgamation may account for this exotic culture found in Ohio.

A discussion followed, during which Mr. Setzler commented in some detail on the Marksville pottery and other things related to the Ohio [45] complex—pottery traits, two fragments of copper, earthworks, mound burials. The burial customs were entirely different from those in Ohio. McKern mentioned the use of bark, rectangular pit, and covering of the pit. In connection with Marksville, Setzler mentioned platform pipes of clay, the use of bark, smoothing stones, and decorated areas on vessels roughened by rockers instead of by roulette. He stated, "You must go through the Hopewell traits as has been done for the Fort Ancient before we can definitely compare the features. . . . Now is this Hopewell related to the Algonkin?"

Ritchie: All we are saying is that Hopewell has certain northern as well as southern features. We are not uniting Hopewell with Algonkin.

McKern: We are trying to divorce our classification from any linguistic association. If it turns out later that certain Woodland manifestations can be linked with Algonkin, alright.

Setzler: We must make it as clear as possible. Cord-marked pottery, conoidal bases, and footed fragments seem to belong to a different group from the pottery which I have been calling typical Hopewell. We have never found that at Marksville.

McKern: We must leave all doubtful material out of consideration for the present.

Deuel: I am inclined to agree that there is a distinction between Woodland and Mississippi that should be indicated in the classification in some way. It seems reasonable to group these under one *base* on the strength of the work done by Setzler and Ford in the southern states. Woodland cultures are found over the area of the northern United States from northeastern states and southern Canada across Ohio to Wyoming. They extend as far south as the Mason-Dixon line, with a few sites noted farther south. The Mississippi culture appears in the region from Arkansas, northeastern Texas, and possibly eastern Oklahoma to the Atlantic, with the exception possibly of most of Florida and a narrow strip along the Gulf Coast through Louisiana and Texas. A site 5 miles west of Pascagoula, Mississippi, which includes numerous shell heaps as well as earth mounds, may indicate a third, or Southern, base. An amateur living near there has collected the material that washes from the shell deposits and falls on the beach where it is found at the water's edge. He has recovered straight-stemmed Woodlandlike projectile points (but not the small triangulars), grooved plummets, boat stones, two-holed stone gorgets, and potsherds. The pottery is not like that of the Woodland culture, but resembles Marksville—beakers, shallow bowls, and seed-bowl ollas all with flat bases. The decoration consists of broad incised lines. A burial uncovered in a mound by an amateur was in the flexed position. Ford thinks the complex extends along the Gulf Coast. Culturally it seems to stand between the Woodland and Mississippi and may have been [46] coexistent with northern Woodland manifestations. Both Woodland and Southern may have developed out of a common widespread culture, possibly the bluff dweller of Arkansas.

There is much to be said for the manner of classifying Hopewellian and Marksvillian manifestations advanced by Setzler. Another explanation should not be overlooked. Carved-bone objects, stone pipes, obsidian, pearls, the cruder Woodland pottery, and art typical of the Ohio Hopewell are missing from the Marksville focus. Apparently in Hopewell sites the finer pottery types are in the minority. On the Turner site, Willoughby reports that 75 percent of the poorly finished Woodland ware (scarcely represented at the Marksville site) occurred. Doubtless we cannot account for the close resemblance between the Marksville and Hopewell aspects or foci as the result of either pure trade or independent invention. One explanation might be that a Southern group migrated into the Ohio Valley, where they combined with a simple-cultured group already occupying the region. A highly developed culture resulted, and an exchange of traits north and south affected groups in both regions.

There seems to be a very good case for the development of the Mississippi base out of this so-called Southern one, of which Marksville is probably a representative. The pottery forms of the latter—the beaker, shallow bowl, and seed-bowl—are also found to some degree in the Mississippi, with the addition of the plate, the olla, and effigy forms. A new projectile point and the discoidal also appear in this complex. We must take into consideration that these three groups—Woodland, Southern, and Mississippi—could all have existed contemporaneously and may have continued to do so until the appearance of the Europeans.

Guthe: Where do you put the grit-tempered, paddle-marked ware of Georgia?

Deuel: In the Middle Mississippi phase. It also occurs on certain sites that appear to have been occupied by Cherokee settlements.

Setzler: We had better wait before we put the Southern groups in a basic category.

McKern: This discussion illustrates the error in our method. You find one site with a very limited complex of traits, and you start out with a basic culture on this basis. We should be satisfied at this time to determine the focus, or component, then we may later determine the basic culture.

Ritchie discussed the arbitrary indoctrination of students in archaeology, etc. Deuel thought students should see the connection, where it appears to exist, between sites, and not be confronted with a long series of components apparently without cultural relationships of any sort.

[47] McKern: I don't want to be too critical. My objection is to what seems to me classification by areas. We don't know the distribution of these cultures, which is apt to create a false impression that we are trying to establish culture areas in archaeology, which we should not do.

Guthe: Deuel has a clear picture in his own mind as to what these basic cultures are—specific traits, geographical distribution, and tentative time relationship. His psychological attitude is different.

Ritchie: What shall we include in a basic culture?

Setzler: I think all of us must agree that there are definite divisions in basic culture—the agricultural and the hunting-fishing groups. We can't work out aspects and phases without determining foci and components.

Griffin: Anyone working in the Northeast can't help but realize that he is actually classifying material that belongs in much larger divisions. The only difference is that some people see over a larger area and are presenting the picture they can see. People working from specific details to more general cultural divisions are not ready to characterize the basic cultures as Cole and Deuel have done.

McKern: I have perhaps overstressed the importance of starting in at the detailed end rather than the general. I didn't mean that we should entirely disregard the more general divisions. I simply meant that instead of starting there and working down, for the present we have a great deal less information about the higher divisions and should take our classifications tentatively. You can be much more absolute in your statement when you come to the detailed divisions. You should not place as much emphasis on the determinants of a culture we don't know, as on the details of a focus which we do know.

Deuel: I think the best thing would probably be a definite upper limit of the classification. But when we get into the field of agriculture in a region, we then have an agriculture-pottery complex arising, with its attendant complications.

[48]   **Saturday Afternoon Session, December 7**

GENERAL (James B. Griffin)

I am going to remind you of some of the things we know and suggest work for the future. There are at least two valid approaches to a clearer understanding of the pre-history of the area under discussion. The first is a classification of cultures on the basis of their determinant traits. I have worked with the Fort Ancient sites and have made an attempt to show my approach to the McKern classification in a brief statement published in Notes from the Ceramic Repository for the Eastern United States, University of Michigan, 1935, No. 1. What I want to do in the future is to limit the area of the Fort Ancient aspect, include sites not completely described in the literature, and show how they can be incorporated into what I believe is a true Fort Ancient aspect. The second approach is to determine the ethnological relationship of the archaeological cultures.

I think it has been demonstrated that Fort Ancient is probably an archaeological aspect. Shetrone pointed out that there were divisions within the Fort Ancient culture, and I think these can be correlated to some degree with the river valleys. The older workers in the field recognized that Fort Ancient was related to the Iroquois and perhaps to the southern Cherokee culture. Three years ago McKern and I worked out to our own satisfaction, excluding the pottery, that the Fort Ancient culture has approximately 60 percent of traits in common with his Upper Mississippi material and that the Fisher and Blue Island groups could also be included as Upper Mississippi. These relationships are not too accurate and need to be carefully reworked. Until that time the relation of Fort Ancient to Oneota can be suggested but not demonstrated. The same holds true for the relationship to Iroquois. While this latter aspect has been characterized fairly well in New York, it is not from the point of view with which we are now trying to work. The Iroquoian aspect will need to be reanalyzed in order to conform to the type of studies now being advocated in studying the related Middle Western aspects.

Upper Mississippi, then, extends from Iroquois in the East to Fort Ancient in the Southeast, and to the Oneota and other divisions in Illinois, Wisconsin, Minnesota, Iowa, Missouri, Kansas, and Nebraska. There is little doubt that there are certain sites over this whole area that can be correlated. The exact degree of relationship will have to be decided in the future.

The Iowa Mill Creek is a well-defined unit which can be separated into the Big and Little Sioux foci. This material includes certain pottery and other traits establishing a connection with Middle Mississippi. This is theory. That Mill Creek has a closer [49] relationship to the Upper Republican aspect of Strong and Wedel is again theory, but will probably be demonstrated in the future. As Dr. Keyes has said, his Glenwood and the Nebraska aspect are certainly very close. These seem to have definite relationships to the Mississippi area. Whether they will be shown to be Upper or Middle, or a different phase of the Mississippi basic culture, is a question that has not been settled.

I don't know much about the Middle Mississippi phase, and I would like to have someone define it. The Woodland basic culture is a problem with which I have not been extensively concerned. The first phase I can clearly see in the central states, however, is the Central Basin (eastern Minnesota, western Wisconsin, eastern Iowa, Illinois down to the confluence of the Mississippi and Ohio rivers, northern Indiana, western Michigan, Ohio, northern Kentucky, and up into New York state). There are certain cultural elements in this whole area which cause us to speak of the Central Basin phase. Its relationship to Hopewell in Ohio has not been clearly demonstrated, with the exception of one state, and since the existing picture of Ohio Hopewell is misleading, the whole thing will have to be readjusted.

What I have read about some of the stone mounds explored by Fowke in Missouri, Black and Setzler in Indiana, and other stone mounds and cave sites makes me believe that eventually a complex will be established which occurs from Kansas and Nebraska to Virginia. There is some evidence for the suggestion that this material is older than the Hopewell.

The other phase I can clearly see in Woodland is what McKern defined as Lake Michigan. It needs to be studied to point out the similarities and differences in the various localities where it occurs. We need, then, to prove our archaeological relationships and not assume them. We need a detailed site-by-site analysis to establish our foci and from them, our aspects and phases. It will be slow, tedious work, but the framework will be sound.

Another problem is to correlate these archaeological cultures with historic Indian groups. In the Middle West, this has been almost entirely ignored. I don't know if it is possible, but we should be able to find reasonably authentic historic sites and by digging there and finding recurring complexes correlate this material with ethnological data. It seems to me quite a number of groups in the Middle West are still rather vague as to the proper ends of archaeological work.

Another point I should like to make is that I don't see how one can speak of an archaeological culture until he has a complete picture. When we speak of Ohio Hopewell, it does not represent, to [50] me at least, to any accurate degree the culture of the people who built those mounds. Now I don't see how we can tie in what appears to be a ceremonial expression of a culture with the kitchen-midden refuse of a group in another area. I don't think 15 or 20 traits from a burial mound are sufficient to determine basic cultural relationships.

Discussion

Webb: Does he mean that some traits are more valuable than others for determining a culture?

Griffin: I don't see how one can take a few mounds, such as the Red Ochre group in Illinois where you find only burials and a few traits, and tie it in as an archaeological culture with the Adena group in Ohio, which is a burial complex as it is now defined, and arrive at a sound reconstruction of cultural relationships.

Deuel: You must take a site as it stands. Griffin wants a complete archaeological culture. McKern does not believe we get a fairly complete archaeological culture.

Griffin: I meant that those 15 traits or so of this small group seem to tie it in with another group, but you are not comparing archaeological cultures, you are working with burial complexes.

McKern: Can't you say that you have tentative evidence of similarity which would cause us to throw the two together until more evidence appears? We cannot say just what a burial complex is.

Guthe: In the Adena material, are you justified in saying that it is an Adena aspect of the Central Basin phase? The Adena culture as defined by Greenman had only 10 percent of the traits in common.

McKern: I do not believe for a minute that these people made a lot of materials for burial purposes only, in Ohio Hopewell.

A general discussion on the differences between burial-mound and village-site material followed.

McKern: I would like to bring out two points regarding the absence of good material in refuse deposits. At that little focus up in Barren County, all our implements, including one or two characteristic large blades, one pipe, and the best pottery, came from the campsite and not from any mound.

Greenman: There is very little campsite material in Ohio.

[51] Roberts: Would this have any application in the Southeast? We have in the Southwest a distinct group of culinary wares and decorated wares which were also utilitarian in general function. Cook pots form a very small percentage of grave material. It is almost always the decorated ware. Once in a while you will find a burial with a cook pot in it. Usually these are missing from the grave complex.

McKern: In upper California you get a certain type of basket which is made exclusively for burial purposes, with a peculiar shape and by an unusual technique. The woman does her best job on it. Here is a case of an actual element of a burial complex. But to say that all the Hopewell material was made for that purpose sounds extraordinary.

Swanton: I don't recall anything of that sort in ethnology. Of course, you have the "killed" ware in Florida.

McKern: That could be utilitarian ware.

Swanton: I think there is no difference except in the holes.

Greenman: Isn't it generally true that Woodland material is comparatively scarce when compared with Fort Ancient, where you get a great deal of village-site material? You don't get very much debris around a Hopewell mound.

Griffin: Then the Hopewell people must have all come in from Wisconsin, Illinois, and Michigan, bringing their nice material with them to bury in Ohio. The people who built the mound should have had a village site someplace.

Greenman: How big is a village site in Wisconsin?

McKern: It is difficult to say. A campsite is based on the extent of location and material. A village site shows long and extensive occupation. I have seen campsites where you would search hours for material.

Greenman: Are the village sites in connection with the mound debris, or do you think they contain many artifacts?

Deuel: We have a village site 8 feet deep.

Greenman: I don't mind saying that I don't believe that in the state of Ohio the village sites in association with Hopewell mounds have ever been properly looked for.

[52] Black: It seems to me that this point involves the distinguishing characteristics between Mississippi and Woodland.

McKern: You find historic Algonkin groups in Wisconsin that never, from a historical or ethnological standpoint, are known to have lived in extensive villages, but who lived in an area in which there was seasonal migration.

Ritchie: Would this have a bearing on the matter? In my own area there are plenty of places where we find practically no surface evidence, but we do find village refuse in pits made for testing.

Deuel: Keyes told me he found his Woodland village sites at some distance from the burial mounds. Do you still hold to that?

Keyes: All of the Woodland mounds I can think of have village sites near them on the neighboring hill, and the material is abundant, averaging 4 to 5 feet deep.

McKern: Don't you get the situation in Iowa as we do in Wisconsin, where we find that the whole area of the shoreline of a lake or a river to be one scattered campsite?

Keyes: A few years ago we had a Woodland site at the foot of Crooked Creek which was completely excavated from 3 to 4 feet by the current. There were lying on the surface about 300 flint implements, grooved axes, hundreds of potsherds, and other village debris. We figured that most of the rest went on down the creek. I think the Woodland sites contain plenty of material.

Ritchie: Is this not scattered to a great degree?

Keyes: What we have called the Woodland sites are usually small and with the material fairly well concentrated.

Setzler: Would you consider this material the same as Hopewell? Is there any village material connected with eastern Iowa Hopewell?

Keyes: Yes.

Setzler: Are those sites very concentrated?

Griffin: In northeastern Iowa a nonprofessional is digging in a cave which has refuse deposits 4 feet deep, with pottery from top to bottom and with no observable stratigraphy between Lake Michigan and Central Basin.

Keyes: Of course, the rockshelters are one locality where things are concentrated.

[53] Deuel: Most of the northern sites strike me as being sparse, while in the southern sites you get so much material.

Greenman: I have never seen any accumulation of village refuse that could not be attributed to the activities concerned with building the mound.

McKern then reviewed Griffin's points. "It seems to me that the whole purpose of our archaeological classification is to permit us to handle these materials conveniently until we can tie them in with the historic groups, but we can't force the issue. In the first place, I insist that the historic approach is a secondary method, for the simple reason that in a vast majority of cases, the first thing with which you come in contact is cultural debris, which cannot be related. As soon as you know you have a culture complex, other things being favorable, you have a chance to associate it with materials of some historic group. Until then, you have no right to assign it to a historical group. As a general thesis, the recurring complex method comes first, and then the direct historical method."

Setzler: Some Natchez, Tunica, and Yazoo archaeological sites are definitely known. A criticism of the East is that the Middle West has not done enough with historical sites.

Guthe: In Michigan, Dr. Hinsdale has identified many sites which seem to be historic sites. These should be traced for archaeological significance. I imagine this is true in every state.

Swanton: I think the difficulty has been due largely to the difference in the two sections when the whites came into the country. There was a discontinuity between the people in occupancy and the antiquities of the mound builders. In the South, you come upon these documented sites, and in several places you can begin establishing historical continuity. The difference in the North is due to historic and pre-historic causes. I agree that both methods should be pursued. The method depends on the area in which you are working.

Guthe: Shall we record it as the sense of this meeting that it would be very much to the point for the archaeologists to pay more attention to the historically documented sites of this area?

McKern: I would like to have it understood that the documentary evidence should be presented and supported by archaeological evidence. In other words, one should not say that there is documentary evidence without giving the source. I think we can agree that Griffin is [54] right that degree of relationship cannot be settled without a great deal more technical and analytical study.

Guthe: Are we sure that the various Woodland groups are related? Is there any mechanism by which we can settle that question?

McKern: We can at least apply the method used in Griffin's analysis of the Fort Ancient sites to the problem of Woodland.

Webb: How is one to know what a trait is? What has been the basis of the analysis of Fort Ancient traits?

Griffin: One of the first criticisms that came to us on my attempt to classify Fort Ancient was that archaeology was becoming too statistical, but I don't think that is actually true. The attempt was made to show objectively why Shetrone, Mills, and others thought that Fort Ancient was an archaeological culture. I wanted to make my comparison as clear as possible. In selecting the traits, a good deal of the reliability of the traits an individual selects depends on his archaeological experience. From the literature, and after visiting the Ohio State Museum, I tried to separate things which I thought would be significant. For example, the use of bone awls is a trait which could be broken up into subdivisions such as turkey-metatarsal awls. Then I also included notched turkey-metatarsal awls because at all five sites a certain percentage of awls made from that bone were notched. These became determinants of equal value for the Fort Ancient culture, but in a larger cultural division we may find them relatively unimportant.

Webb: You can't have notches in an awl until you have the awl. Naturally in this case one trait depends on another.

Griffin: Another objection was that the relative amount of materials was not of as much significance as their occurrence at all. I tried to show this difference in the percentage of the material, which was difficult to obtain and to present.

Ritchie: There is a difference in trait determinants and link traits. Bone awls are found all over the world. So are arrowheads and celts. I wouldn't say they attain diagnostic value. On the other hand, take the double-cymbal ear plug, occurring in Hopewell. Where is a beveled adze found except with our Archaic Algonkin, where it becomes a trait determinant? I say the link trait is the bone awl and the beveled adze is a determinant trait.

McKern: This is the difference between general and specific traits. It applies to the culture classification as a whole. A bone awl becomes a diagnostic trait for a specific division. When [55] you come to an aspect of that phase, you have subvarieties of this same implement. The subvarieties are not determinants for the whole area, but for the smaller group. What you have to consider is the cultural significance of any trait—either as a link or a diagnostic trait. If its application is limited, it is a determinant for a smaller group.

Webb: If Griffin is going to talk in terms of statistics, can he say which trait is more diagnostic than another? One should be more significant than another. You must set up your own standard, and you should give one trait more emphasis than another.

McKern: As regards mathematics, I believe I am perhaps one of the first to use the term "percentage" in archaeology. I don't think you can say that 75 percent means one thing and 60 percent something else. We must take into consideration that we haven't a complete series. You must rule out the statistical method in the strict mathematical sense.

Webb: I feel that one group of 10 traits may be more diagnostic than some other group.

McKern: You have complex objects which comprise a group of traits, such as pottery, and you also have simple objects with perhaps only one trait.

Webb: What are you going to do with burials?

Griffin: I think that perhaps burials can be used as traits defining foci or aspects. I don't believe they can be used to define a phase or a basic culture.

McKern: A burial is a complex of traits.

Webb: Would burial customs in general tend to be fixed, or would they vary more than any other custom?

In the discussion that followed, Dr. Swanton mentioned finding five types of contemporaneous burial customs among the Haida. The Tlingit cremated their dead. Yet they were close neighbors geographically and belong to the same culture province. In the Southeast, the Choctaw separated the bones of the dead, put them in a house, and placed them in mounds, whereas the related Creeks buried their dead under the houses, at first flexed. Griffin mentioned the facts he encountered when preparing his master's thesis on burial traits. Swanton stated that over a certain area, a certain type of burial is likely to prevail and that it may persist for a considerable length of time. There is evidence that the Creek Indians came in from the northwest in relatively late times and the people they encountered buried like the Choctaw (bundle reburial). The Creeks had a [56] different custom. Guthe mentioned the whole question of burial customs and their significance in defining culture determinants.

Black: Shouldn't every worker include every trait he finds? These should be listed whether they are diagnostic or not.

Setzler: May I suggest that we make the descriptions purely objective without drawing conclusions?

McKern: Why not offer conclusions in a separate section?

Black: Traits can be listed purely tabularly.

Guthe: In a publication on a given site, every trait should be objectively described and listed. We can start from there. There are two ways of following the argument. How are we going to describe these traits? How to differentiate between the character of the traits? My suggestion is that we try to define a linked trait and a diagnostic trait. Can we define a diagnostic trait in general terms?

McKern: *A diagnostic trait is a trait which applies only to that class and is unique for that class. A link trait is one which ties it in with another group.*

Ritchie: The number of diagnostic traits are few.

Griffin: You have diagnostic and link traits which make up the culture complex.

McKern: We have what we are calling a diagnostic trait for a phase which is not found anywhere else. But within that phase, that diagnostic trait may become a link trait for a smaller group.

Ritchie: We have seven components in our area. These perhaps form an aspect—all linked together by the presence of ground-slate points.

Black: How about preponderance of type of points for a single site? Would you regard the percentage of points as diagnostic?

Ritchie: I would say that the broad triangular point is a diagnostic feature for a particular focus.

Deuel: You might have link traits between two comparable units in the same group, to units outside of the groups, and vertically and temporally. Are each of these link traits?

[57] McKern: In comparing Grand River with the Effigy Mound, the triangular arrow point is a diagnostic trait for Grand River. But in comparing Woodland in general with effigy-mound burials in the Upper Mississippi area, it is not a diagnostic trait.

Griffin: In the Upper Mississippi phase we will have a Fort Ancient aspect made up of a number of foci, and we will also have a focus in

Nebraska that belongs in the Upper Mississippi. You might find traits in the foci which would be linked traits between them but could not be determinant traits in the phase.

Guthe: This applies to any comparison you are making. A diagnostic trait is a trait that is unique for the particular class of culture that is being discussed.

Wilford: How would that work in a negative way?

Ritchie: Can the absence of a thing be diagnostic?

Guthe and Deuel discussed this point at some length.

Ritchie: Diagnostic traits of components form a focus and become link traits between the components.

McKern: There is a certain idea of historical sequence which enters into Ritchie's culture classification of the New York state groups. I think this should be divorced from the classification even if the groups follow each other historically. The term "Algonkin" should be eliminated.

Guthe: A diagnostic trait of any culture class is one which occurs only in it and not in the culture class of the same order with which it is being compared. In other words, a focus would be compared with a focus, and a phase with a phase.

McKern: We are going to encounter many hypothetical groups. We might want to compare a phase with a focus. I suggest omitting the phrase "of the same order."

Keyes suggested using "group" instead of "class."

Guthe: Can we define a link trait? A link trait of any culture is one which occurs both in it and in the culture classes with which it is being compared. There is one thing that stands out, and it is that everyone of us yesterday recognized differences in culture groups. We all agree there are two different major groupings. Can we give the diagnostics of those two groups? I would like to have every [58] person here send me the diagnostic traits in his own area for the differences between Woodland and Mississippi. Out of a tabulation of these traits we would get an idea of what constitutes Woodland and what Mississippi.

The committee office can very well take care of this tabulation. Is it possible in your own minds to objectively work out diagnostic differences between the two major divisions in our own region?

McKern: Each man will speak in terms of a limiting area.

Keyes: What does this group think of the matter of considering the proposition as applicable in this area? For example, in the Woodland we have the triangular, unnotched arrowhead in small numbers. In the Upper Mississippi, we have this type of arrowhead predominating by a large majority. We have the notched arrowhead in large numbers in the Woodland and in very small numbers in the Upper Mississippi. Is it properly diagnostic to say that the Upper Mississippi has the unnotched triangular arrowhead as predominant?

McKern and Black thought this tabulation would bring out the difficulties in terminology. Deuel suggested more publication of nomenclature in papers.

Roberts: In the Southwest the workers in the Hohokam area have prepared a classification of pottery which is useful for that cultural group only. In the bulk of your material you are all using the same terminology. When a man knows he is disagreeing with the general use of a term, couldn't he explain the synonymous term used?

Guthe: I think the only way to do that is to have the committee work up a tentative statement for criticism and frequent revision.

Roberts: Don't you think it would be very helpful if people occasionally referred to the dictionary definition of the word?

The meeting adjourned until evening.

Saturday Evening Session, December 7

Guthe: Is it advisable to set up a series of cultures? Is it advisable to take stock of this whole problem of phase and aspect, and the question of the so-called basic culture and the elimination of the term "basic culture" as it has been used? Most of us seem to agree that it is a good idea. We have been trying to simplify the terms of the concept because of the use of basic cultures. If [59] we think of culture as an agriculture-pottery base as opposed to a nomadic-hunting base, we can eliminate some of our difficulties.

McKern: The word "basic" as used in the terminology of our classification has led to a lot or misunderstanding. You can't get away from a classification with such fundamentals as pottery-agriculture-sedentary life or nomadic hunters. Those determine the base. As we have been using the term "basic culture," why shouldn't we eliminate the word "basic" from our present terminology? We are not dealing with anything basic at present. My idea was that we might use the word Dr. Cole suggested in the first place. Dr. Cole suggested culture, and we called it base, so we compromised on basic culture.

Guthe: I remember very distinctly that the objection to the term "culture" was that we use that term now in a general way. If we retain the term "culture" for one of our classes, you will get confusion.

Deuel: What about the term "pattern" that you had originally?

After further discussion, Guthe suggested base, pattern, phase, aspect, focus, and component for the culture terminology.

Setzler: We will be putting in another division which we have not used until now. Under your pattern you are going to divide the Woodland and Mississippi.

Ritchie: What are you going to call this hunting-fishing or agriculture-pottery group?

Discussion on this was postponed.

Webb: If there is evidence of a hunting group at a particular site, that should be mentioned.

Setzler: Can we decide on what divisions we are going to make in the pattern? Most of you seem to want Woodland and Mississippi.

Ritchie: Where would you include Great Lakes?

McKern: "Woodland" is an ethnological term which has a different meaning.

Setzler: Do we have to make a division into the two patterns?

McKern: As a matter of fact, I can't see much similarity from a phase standpoint between what we are calling Woodland and what I [60] would call Mississippi. They both use pottery, but that is too broad a distinction, and they were both agricultural to some extent. Most of those things they have in common are so fundamental that they would apply to a base.

Setzler: Don't you think Woodland and Mississippi were indigenous?

McKern: No, I don't. Whatever evidence we have indicates Woodland was very old. It shows Woodland to be as old as anything we have in the whole region.

Setzler: I want a single pattern called Mississippi, with all pottery-agriculture divisions listed under it. Can't you make your divisions under phases instead of the pattern? We should make the Mississippi pattern include everything else—even Woodland.

Deuel: It seems to me what is bothering Setzler is the fact that he sees a genetic relationship between the Gulf cultures and the Mississippi cultures, which should be if the two are classified on the basis of their inherent traits. There should be no difficulty in putting in a third division.

McKern: Ritchie's Archaic does not seem to have developed into his so-called Second and Third Period Algonkin.

Setzler: What would the group think if we abandon the term "Woodland"—use only Algonkin and put in the term "Hopewell" to describe what we consider Central Basin?

McKern: In this part of the country, the underlying phase is Woodland, with the other cultures coming in later.

Setzler: I feel that if I were to weigh all the traits of my Marksville site, the Lower Mississippi traits would overshadow the Hopewell traits.

Guthe: What is wrong with saying one base followed another historically?

Setzler: Do you feel we could discard the term "Woodland"?

McKern: Yes, if we can find another satisfactory term.

Guthe: All this discussion has been caused by the term "Hopewell." Why couldn't we use the term "Central Basin" in place of "Woodland" and make it a pattern for the time being?

[61] Swanton: Does Setzler think Hopewell is being divided and half assigned to Mississippi and half to Woodland?

Setzler: My original argument is that you are putting Hopewell in this Woodland phase and showing a definite relationship between Hopewell and other cultures which I don't think are related.

McKern: It seems to me that the majority of Hopewell traits are un-Woodland.

Deuel: Outside of Ohio, our Central Basin largely consists of Woodland characteristics, and there are a number of sites called Hopewell that have traits like Marksville and others which cannot be placed.

McKern: There is about a 20 percent similarity between any manifestation of Woodland in Wisconsin and Hopewell.

Roberts: Would you say that in your Central Basin, except for Ohio, you have about an equal division of Woodland and Mississippi traits? It seems to me that your separate pattern here is the Hopewell. You may find out that, that is the northern extension of your southern pattern. Why not make the pattern Hopewell?

Guthe: As a matter of convenience, what is there wrong in thinking in terms of aspects and phases and putting Central Basin for the present as a pattern? Include a Hopewell phase under the Central Basin pattern.

McKern: Why can't we say an unnamed pattern under which we get Hopewell?

Setzler: Why not use Hopewellian phase instead of Hopewell?

McKern: Hopewell is also a component in itself. Use the Scioto Valley as focus.

Swanton: (referring to a tentative classification under Central Basin) I thought it was understood that ethnological terms would be excluded. Why not exclude Iroquois as well as Algonkin?

Ritchie: We know definitely there is an archaeological culture known as Iroquois. If we can definitely prove that Algonkin sites exist, we can call them that. We work back from well-documented sites in which all the traits occur. If the pattern remains stable, aren't we justified in saying that this is Iroquois?

[62] McKern: When we can establish a connection with an ethnic group, then we can change the name to that group or tribal name. When an aspect is concerned, we would not have a tribal name to apply. Whenever it is possible to substitute a definite historic connection for this tentative terminology, it should be done.

Swanton: In that case, I feel we are justified in referring to archaeological sites known to be Iroquois by that ethnological term. I suggest we can also substitute Natchez.

Guthe: In the Plains, Strong has identified Pawnee with an archaeological culture.

Swanton: As long as you use this term in such a way, make it clear that you don't mean the people at a particular site used the Iroquois language. As I understand it, there are reasons why some of you think Fort Ancient sites may have been occupied by Iroquois peoples.

Griffin: Madisonville and Fort Ancient pots were found in an Iroquois site known to have been destroyed in 1680.

Swanton: Archaeological occupancy does not necessarily indicate Iroquois-speaking peoples.

Ritchie: Iroquois pottery was found on a typical Algonkin site, but the people did not speak the Iroquois language.

Griffin: In Ritchie's area, the material will break down into foci. Greenman found some sites in northern Ohio which are obviously related to the Iroquois foci, but they belong to a larger order, perhaps an aspect. We would hesitate to say they were occupied by Iroquois-speaking people.

Greenman: What are we going to do with material that we call transitional?

Guthe: We have two opinions on this matter of classification. One group is talking in terms of foci and determinants, and the other is trying to create a relatively simple set of relationships for students. Can the teachers perhaps stop talking in terms of patterns and begin thinking in terms of phases as a group of working hypotheses? Can the others start grouping their aspects into phases as working hypotheses? Then we would have a middle group into which the two could come together.

Deuel: It seems to me that you can't take the sites and just affiliate them on the basis of components alone. Fundamental traits [63] determine basic cultures. If that is true, suppose you did find in a component compared with another component, some of the basic traits which would put them under two different divisions. I think of Hopewell as being a hybrid.

McKern: We cannot be too arbitrary at present.

Guthe: We should all talk in terms of phases.

McKern: We must be rather careful not to have the students think this is a final classification.

Guthe: We have found that there are definite problems that we didn't see until we began our discussions and that each of us has certain detailed information which the others lack. Is there any way in which we can circularize the group—broadcast the detailed information of interest to the whole group? Is there any point in tentative circulars of information? Another thing worth talking about is attempting to see what the immediate problems are confronting the central part of the whole region—the lower part of Ohio, Illinois, Indiana, and Kentucky. Can we come to any conclusions as to the immediate problem in this central area which will help clarify the situation?

Setzler: I have a feeling that if we could get someone assigned to working out the details of the southern Hopewell, it would be a great help. We should have a detailed survey on the lines of the Fort Ancient aspect. We speak very glibly about the Hopewell spreading from the north to the south, or south to the north. Now there is an area of about 800 miles where we have no definite evidence of Hopewell features. I would like to cover that area on the east or west side of the Mississippi which would show this transition period. That happens to be my own particular problem. That is something I think would help to clarify greatly the things we are trying to determine.

Black: Let us find out what the problem is and have some concerted action.

The meeting adjourned until the morning of December 8th.

Sunday Morning Session, December 8

McKern: I think another of our big problems is the Tennessee-Cumberland complex and what it involves. Webb is working in the [64] midst of this region and should be able to help.

Webb: These rivers were the highways of the aborigines. For miles and miles along the banks are a continuous series of middens, mounds, and campsites. It is going to take many years to get at the facts of this complex. I thought four or five years ago that only a stone-grave culture was represented.

Griffin: If an archaeological survey was made down the Mississippi, it might help extend the geographical limits of the Hopewell, among other things. If the University of Chicago Department of Anthropology could do so, it might be a good idea to send a survey party from Cairo up the river to see if they could locate evidences of Hopewell.

Webb: Perhaps the Smithsonian Institution or the Committee on State Archaeological Surveys could formulate three, four, or five general problems, and decide the strategic points to be worked. Our principle is cooperation. I don't believe the Woodland groups stop at the Mason-Dixon line.

Guthe: The Hopewell phase seems to be the major problem of this area. Our tentative classification has many gaps in it, particularly regarding Middle Mississippi, Woodland, and Adena.

McKern: We don't know just what Middle Mississippi is.

Swanton: How are you going to get anywhere with Middle Mississippi until you investigate the Arkansas–west Tennessee district?

Webb: Lewis is working on Duck River in Tennessee and is making some advances. Dellinger is doing work in northeast Arkansas.

Black: That brings up the dissemination of unpublished data.

McKern suggested that *American Antiquity* would be glad to publish some of these articles.

Guthe: It is obvious from the discussions that Middle Mississippi is also a major problem. It is found in Illinois, Indiana, Kentucky, Tennessee, Wisconsin, and possibly in Nebraska and Iowa. The Upper Mississippi is probably our best known phase. Another thing that seemed impressive last night was our inability to handle the Woodland problem.

Keyes: Will the conference care to outline an aspect or two? The manifestations in Iowa could be either phases or aspects.

[65] Guthe: There is also a Woodland pattern in this area.

Ritchie: This whole pattern is perhaps more simple than Upper Mississippi, and there may not be as many variations.

McKern: We must work this out and settle the details.

Deuel: I would suggest a Lake Michigan phase and a Tampico phase. They may ultimately turn out to be the same thing.

McKern: We need more research in the Woodland pattern. The whole complex needs to be studied. I have had a student working on Woodland pottery designs, and a variety of designs and motifs has been found for local sites. The impressing or punctate treatment is in direct contrast to the Mississippi pattern. The startling thing to me is the differences between two types of pottery that are so close together.

Black: Can you see a connection between Lake Michigan and Tampico?

Guthe: Woodland seems a more important problem than the other two. It shows us our blind spots.

McKern: I think we have gone further with Hopewell than we have with any other pattern.

Setzler: We need more information on the Woodland pattern from Illinois, Kentucky, Indiana, and Wisconsin.

Webb: How are we to know what Woodland is? We should have a list of things we regard as having Woodland characteristics.

Guthe: The Committee on State Archaeological Surveys has a power mimeograph. If we can work out some scheme by which you fellows will send us lists, we will send out the material to the whole group.

McKern: There are two things we need. One is factual material. The other is discussion about this factual material, and it seems to me that is an ideal subject for the columns of the Correspondence Section of *American Antiquity*.

Deuel: Most of us would be glad to put on paper what we consider type sites and their diagnostic characters. I would be glad to list what I consider Tampico. It would help if McKern would pick out what he considers a type site for Lake Michigan.

[66] Roberts: Each of us should be willing to list what he considers at present to be the diagnostic traits of a phase or component or site, in a purely tentative way.

Guthe: Yesterday we thought it would be a good plan for each of us to send in what he considered the diagnostic traits of Mississippi and Woodland. Why couldn't each of you send in a list of all the traits you can think of that belong to the Woodland pattern in your own territory? If we could exchange that among ourselves, we could see where we agreed and disagreed. Then, instead of asking for the diagnostic traits of the two patterns, we would want the complete traits for the culture as it is in each area.

Wilford: I think in Minnesota we find the three patterns quite distinct on the basis of decoration of pottery. Some of the pottery of New York and Minnesota is so much alike that the two seem to be the same thing.

Guthe: We have then an agreement on the mechanics of the next step concerning Woodland. We must outline the complex which each considers to be Woodland in his own area. Yesterday the idea was to send in the diagnostic traits differentiating between Mississippi and Woodland. Can't you send in the diagnostic traits for Woodland and Mississippi? And now send in all the traits for Woodland.

McKern: Let us be clear just what a trait is. It is not just an object such as pottery. Make the statement clear.

Ritchie: Why not have a few drawings or pictures?

Guthe: The whole point is that we are not issuing it for publication to the general archaeological field. It is to be sent to the group here at the conference.

McKern: Why not include the archaeologists from Nebraska and Arkansas?

Guthe: In the first place, the group here knows what we are talking about. If we bring in others, their ideas will not be the same. They have never worked out this problem and do not understand this classification scheme. Let's understand among ourselves what we are trying to do and then ask the others for their opinions. I think we have settled the question of what to do with Woodland. Is it better to do anything with Hopewell and Woodland, or should we leave that for the time being?

[67] There are two other problems that come to my mind. One is the question of historically documented sites. Another is fieldwork as opposed to laboratory work. Should we discuss them? We all know there is a great deal of material in laboratories and museums on which reports have not been made. On the other hand, we realize there is a great deal of work that needs to be done on new sites in order to clarify our knowledge.

McKern: Local problems determine what one must do. Museum material needs studying. Unfortunately, most of us have many other duties. A museum's policy is controlled by the board of trustees. Their policy is that fieldwork should continue when funds are available. There are many duties in the museum other than laboratory work, for which we cannot get men or equipment. It seems to me we can't dictate a policy, although I think we all realize the importance of laboratory work over fieldwork at the present moment.

Ritchie: Sometimes we have to get material out of the ground before it is destroyed.

Guthe: Both sides must be considered and carried on. Many older reports are inadequate.

Deuel: A meeting like this might make evident what some of the major problems are.

Guthe: The most important single problem in the area is already evident. It is a specific field problem that confronts us right now—tracing the manifestations of the Hopewell phase along the Mississippi and Ohio rivers. Is there another specific field problem that we can select for investigation?

McKern: Why not trace out the possible routes of cultural diffusion or migration that account for Upper Mississippi? A mapping of sites by culture would be an important thing. If this distribution of what we call Upper Mississippi represents a migration, we should be constantly collecting data.

Swanton: Do you suppose it would be possible to get some sort of advance report regarding the region in Tennessee and Kentucky for distribution among archaeologists? I have a suspicion that many of my Muskhogean people will be traced back to that particular section.

Guthe: Couldn't we ask for the same thing regarding Middle Mississippi as we plan to do for Woodland?

[68] Webb: The Wheeler Basin material is going to be important; also Lewis's material and that from the Shiloh, Page, and Tolu sites, which should give us a picture of what is in that region.

Keyes: I have an idea that the problem would be further clarified if we could have a conference of men from both areas in a year or two.

Swanton: I believe this Cumberland-Tennessee culture is important. The Tennessee River has been a channel of movement of western tribes toward the east. Many of them apparently moved up the river and founded sites at the headwaters of the Coosa and then went farther south.

Guthe: I think we can all agree that if we can first get this Woodland idea clear in our minds, the next thing to do is to apply the same methods to the Middle Mississippi.

McKern: A more intensive search for, and working of, stratified sites should be carried on. These are rarely encountered but would help in associating sites with ethnic groups.

Deuel mentioned Kinietz's work for the University of Michigan, going through records in libraries and definitely identifying sites known as historic.

Setzler: Has anything been done in the North similar to Swanton's work in the Southeast, such as tracing La Salle's and Hennepin's routes and determining sites at which they stopped along the way?

Swanton: Identifying sites is not an easy job even when we have the documents. By archaeology and ethnology working together, we can make more progress.

Roberts: Everyone appreciates the value of work in historic sites, yet nobody seems anxious to do it. There is the same situation in the Southwest. Aren't there students in some of the graduate departments who could be interested in this particular problem?

Black: We have rather definite contact records in Indiana along the Wabash. We also have a later one dating an old Moravian mission on the White River. Isn't it true that aboriginal records might be mixed up at the time of the contact? Wouldn't it be better to work at a nearby site?

Webb asked about the Shawnee site in Clark County, Kentucky, and whether it would be worthwhile to explore it carefully. Swanton [69] thought it might be. In most of the southern sites, pottery is associated with trade material for a long period.

Guthe: I think the discussions bring out clearly that there is a great gap in our knowledge. We can profit from the experience with historical sites in the Southeast, in New York, and at Ouiatenon, Indiana. I can't help but feel that this field offers worthwhile possibilities in the Middle West. A project in Michigan is being carried out by W. V. Kinietz, which includes a study of documents of the Great Lakes region, primarily of French origin. He is going through the material as an ethnologist, transcribing information on traits. We may have some valuable information, including the specific location of sites all over the Great Lakes region. More of that can be done.

McKern: It seems to me that before we depart the point should be stressed that we have gotten a great deal out of this meeting. It seems advisable that we should have such meetings at least once a year. Personally, I know I have gotten a great deal more out of this than I have from many meetings.

Guthe: We are confronted with several problems regarding further meetings of this sort. The National Research Council is trying to withdraw from projects it has supported for a long time. According to present plans, the Committee on State Archaeological Surveys will go out of existence in June or July 1937, which means that that machinery will disappear. The other thing is that each one of us is going to have to convince his particular institution that it is a justifiable expense sending men to such a conference as this. The machinery will probably not exist so that we can get money from a central organization.

Keyes: I wonder if an explanation of this kind to the heads of the various institutions might not be worthwhile in preparation for such things in the future.

Roberts: At the southwestern conferences, all of the men paid their own way. Of course most of them were within several days drive of Pecos, but they didn't let that fact handicap them.

Setzler: I would like to submit a motion that we instruct the chairman to extend to the National Research Council our sincere vote of thanks for making this one of the most important conferences all of us have ever attended, and also that we as a body should extend to the group in Indiana (Mr. Lilly, Mr. Black, and Mr. Weer) our thanks for their very kind hospitality. (The motion was seconded by Webb and passed unanimously.)

Tentative Archaeological Culture Classification for Upper Mississippi and Great Lakes Areas

PATTERN—Mississippi
PHASE—Upper

   Aspect I—Fort Ancient

      Focus 1—(Baum) Gartner
         "   2—Madisonville
         "   3—Feurt
         "   4—Anderson

   Aspect II—Iroquois

      Foci: the various tribes

   Aspect III—Oneota

      Focus 1—Orr (Iowa, etc.)
         "   2—Blood Run (Iowa)
         "   3—Correctionville (Iowa)
         "   4—Grand River (Wis.)
         "   5—Lake Winnebago (Wis.)
         "   6—Burlington (Iowa)
         "   7—Blue Earth (Minn.)
         "   8—Rulo (Nebr.)
         "   9—Fanning (Kans.)

   Aspect IV—

      "Floating" Foci

         1—Blue Island
         2—Fisher
         3—Big Stone Lake

PHASE—Middle

   Aspect I—Monks Mound

      Focus 1—Rock River
         Component—Aztalan
      Focus 2—Spoon River
         "   3—Kingston

PATTERN—Woodland
PHASE—Lake Michigan

   Aspect I—Effigy Mound

      Focus 1—Buffalo Lake
         "   2—Sheboygan
         "   3—(Grant River)

   Aspect II—Wolf River

      Focus—Shawano

   Aspect—III

      "Floating" Component—
         Stearns Creek

PHASE—Northeastern

   Aspect I—Owasco

      Focus 1—Castle Creek
         "   2—Canandaigua

   Aspect II—Vine Valley

      Focus 1—Pt. Peninsula
         "   2—Middlesex
         "   3—Coastal

PATTERN—(Unknown)
PHASE—Hopewellian

   Aspect I—Ohio

      Focus 1—Scioto
         "   2—
         "   3—

   Aspect II—(Elemental)

      Focus 1—Trempealeau
         "   2—Cedar River
         "   3—Goodall
         "   4—Greene
         "   5—Ogden
         "   6—Utica
         "   7—
         "   8—Sandusky
         "   9—Henderson
         " 10—New York

   Aspect III—Southern

      Focus 1—Marksville
         "   2—St. Andrews
         "   3—Crystal River

PHASE—(Adena?)

   Aspect I—Adena

      Focus 1—Westenhaver
         "   2—Whitewater
         "   3—Kanawha
         "   4—Athens

PATTERN—(Unknown)
PHASE—Ground Slate (N.Y.)

PATTERN—Archaic

   "Floating" focus—Lamoka

[70]

Appendix
Certain Culture Classification Problems in Middle Western Archaeology
By W. C. McKern

During the past two years a group of students with fields of interest representing practically the entire Mississippi Valley has been considering (1) the need for culture classification in North American archaeology and (2) the advisability of adopting a certain culture-type classificatory or taxonomic method submitted to them through the channels of the Committee on State Archaeological Surveys of the National Research Council. At present, thanks to the critical interest and cooperation of workers in all sections of the area involved, a revised method has been tentatively accepted by a sufficient number of these students to ensure a thorough trial for it.

As a matter of fact, the method has been applied in classifying divisions previously apparent in several provinces, including Wisconsin, and it now seems appropriate to consider some of the problems encountered in these experiments. However, before attacking this phase of the subject, I wish to consider certain general aspects of the purposes and means of classification itself.

At the very beginning, something should be said of the need for cultural classification in archaeology. It may be advanced that we already possess an adequate taxonomic method. I have received such questions as this: Why call the cultural manifestation of the preliterate Iroquois, Upper Mississippi, or any name other than Iroquois? In some instances we may have sufficient data to verify identification with a known historic group, such as the Iroquois. However, in most instances, we cannot immediately bridge the gap between preliterate and historic or protohistoric cultural groups; and in many instances we cannot hope to ever be able to do so. Yet we perceive that there are archaeologically collected data that warrant cultural segregation. The only taxonomic basis for dealing with all cultural manifestations, regardless of occasional direct historical links, is that of culture type as illustrated by trait-indicative materials and features encountered at former habitation sites. If in the future it becomes possible to name the historic ethnic group of which the preliterate group is the progenitor, no more confusion should result from the statement that, for example, Upper Mississippi Oneota is Ioway Sioux; no more so than from the statement that *Elephas primigenius* is the mammoth.

[71] Aside from the inadequacy of the direct-historical method in supplying the archaeologist with a means of attachment to the ethnological classification, the latter, even if applicable, would not ideally answer the needs of the archaeologist. One ethnological classification divides the aborigines into linguistic stocks, which are first subdivided into more specific linguistic groups and, finally, into sociopolitical groups. The criteria for classification are social, primarily linguistic. The major portion of the data available to the archaeologist relates to material culture, and in no instance includes linguistic data. Consequently this ethnological classification does not satisfy archaeological requirements.

It may be said that we have the ethnologically conceived culture areas to supply a basis for archaeological classification. However, these so-called culture areas involve two factors which the archaeologist must disregard in devising his culture classification if he is to avoid hopeless confusion; these are the spatial and temporal factors. First, the culture area attempts to define, or at least limit, geographical distribution. Unfortunately, the American aborigines did not always succeed in confining their cultural divisions within a continuous area or in keeping culturally pure an area of any important size. Second, the archaeologist considers the American Indians from the standpoint of all time, and certainly there can be no cultural areas devised which can include an unlimited temporal factor.

In brief, the archaeologist requires a classification based on the cultural factor alone; temporal and distributional treatments will follow as accumulating data shall warrant. Moreover, the archaeological classification must necessarily be based on criteria available to the archaeologist.

Any statement that the archaeologist has no need for a culture-taxonomic method is in conflict with facts which all students of the subject must have encountered. One has only to consult the reports on research in almost any American province, particularly where identification with historically known groups has not been attempted, and note the indefinite use of the word "culture" to signify anything from the manifestation of a general pattern influential over an area a thousand or more miles in extent, to the highly specialized manifestation of a culture apparent at a cluster of closely localized sites. The confusion of unstandardized cultural terminology forces recognition of the need for simplifying the complexity of cultural data and concepts through the establishment of systematic order. In men's affairs, as should be particularly apparent at this time, chaos does not reduce itself to order without a plan. The accomplishments of science stand as a monument to planned orderliness.

[72] Unlike the student of ethnology, the American archaeologist has not been influenced appreciably by the initial complexity of his subject to specialize in some certain aspect of that subject. He is more inclined to embrace in his studies all apparent aspects of his subject within the area available to him for investigation. As his problems lead toward comparative studies over wider areas, his conceptions of cultural manifestations take on broader interpretations. Starting with cultural differentiation, he begins to observe evidence of cultural affinities, not only as regards specific complexes but involving distinctive types of complexes. He lacks a specific terminology that is standard with his fellow students, by means of which he can clearly express his maturing concepts. He stretches old meanings to apply to his new needs and finds himself justly criticized, primarily by students limited to ethnological experience, for his extraordinarily indefinite, inaccurate use of the term "culture," which, for want of a more specific term, is made to serve a multitude of purposes for which it never was intended. Incidentally, the ethnologist should exhibit some hesitation in offering severe criticism, since he has nothing adequate to offer of a constructive nature that might aid the archaeologist in surmounting his difficulties.

The point is that the student of archaeology is greatly in need of a standardized culture scheme such as can be realized only through the medium of a taxonomic method.

There are some who have hesitated to cooperate fully in this classificatory experiment on the grounds that the time for classification has not yet arrived. They feel that we lack adequate information to warrant wholesale classification. With due respect for the caution exhibited in this attitude, I cannot but feel that this caution is based on a false conception of the very nature and purpose of classification and a misunderstanding of the intentions of those endorsing the taxonomic method in question.

Classification is nothing more than the process of recognizing classes, each class identified by a complex of characteristics. At the present time we are all active in identifying cultural classes, no matter by what name we may call it, but we are not performing efficiently because we lack the necessary equipment. In the Encyclopedia Britannica, Dr. Abraham Wolf says, "Classification is one method, probably the simplest method, of discovering order in the world.... In the history of every science classification is the very first method to be employed." We have tried to get along without it too long. It is classification that makes it possible for one student to describe phenomena in terms readily comprehensible to another student versed in the taxonomic method. [73] It reduces a multiplicity of facts to simplicity and order and supplies a standardized terminology without which students encounter difficulty in conversing intelligently on a common subject.

The adoption and use of a taxonomic method most certainly does not imply the immediate classification of all manifestations with apparent cultural significance. It is only in those instances in which sufficient data are available, quantitatively and qualitatively, to create a problem of cultural differentiation that classification can serve to any advantage. In some provinces little in the way of detailed classification can logically be attempted at this time; in other provinces much can be accomplished toward detailed classification; and in all provinces a taxonomic method should be adopted before any serious attempts are made at classification. Naturally, this method should be standard for the largest area possible. Following an agreement as to method, the actual classification should be a slow, deliberate procedure, constantly experimental, subject to such major and minor corrections as newly accumulating data may dictate, and to a maximum of constructive criticism and resulting improvement. Such is the history of any scientific classification. It is the method of classification to be employed, not any specific classification, that offers an immediate, initial problem for which a solution is now being attempted.

The method now in tentative operation is a simple one, based solely on complexes of cultural factors. Time will not permit a comprehensive exposition of the method here; most of those who are directly interested in the subject are familiar with the details, which they indeed have helped to formulate. Four arbitrary divisions are made, differentiating between broadly influential types of manifestations and increasingly specialized, localized types.

The most general division is the *basic culture*, characterized by a few fundamental, or essential, determinants dealing with the primary adjustments of peoples to their immediate physical environments. Each basic culture is divisible into groups called *phases,* identified by complexes of important cultural limitations. As greater cultural detail is taken into consideration, a phase may be found to be manifested in readily distinguishable groups, all sharing the phase determinants, but each exhibiting an important modicum of more specific traits peculiar to itself alone. These subdivisions of a phase are termed *aspects.* Similarly, aspects may be subdivided into *foci,* each focus exhibiting peculiarities in the final analysis of cultural detail. An additional term completes the taxonomic framework; the manifestation of any given focus at a specific site is termed a *component* of that focus. This is in no sense a fifth type of cultural manifestation; rather, it is the fourth division as represented at a site and serves to distinguish [74] between a site which may bear evidence of several cultural occupations, each foreign to the other, and a single, specified manifestation at a site. Several components may be found to occur at a single site.

This method is comparable to a filing cabinet equipped with labeled drawers to facilitate the orderly arrangement of materials. I can best demonstrate its usefulness by example, illustrating from a field of my own experience, the Wisconsin field. Investigations at a large village site on the Grand River, in east-central Wisconsin, produced a quantity of culture-indicative materials. These materials offered a variety of details not characteristically encountered at sites in general throughout the state but duplicated at several adjacent sites. Thus a specific list of culture traits was found to be typical for several sites, establishing through its recurrence the fact that it was a true culture complex. At the time, for lack of better terminology, the complex was said to characterize the Grand River culture.

Later, similar but not identical complexes were encountered in two other widely separated districts, one in the Mississippi uplands of the state and the other on the shores of Lake Winnebago. In each of these two districts a complex of detailed traits was found to occur at site after site, but the complex for each district exhibited strong peculiarities of its own. These data seemed to warrant the use of such terms as the Mississippi Uplands and Lake Winnebago cultures.

It was at once apparent, however, that all three cultures, so called, were closely related divisions of a more-inclusive parent culture, which we finally called the Upper Mississippi culture. Thus we had an Upper Mississippi culture divided into three subcultures: the Grand River, Mississippi Uplands, and Lake Winnebago.

Comparative research soon demonstrated that similar manifestations were important in several adjacent states. There was a logically sound basis for including all of these manifestations within a single, large cultural division, which tentatively retained the name Upper Mississippi culture. But the subdivisions in Wisconsin were more like each other than like some of the more distant manifestations. This seemed to require a further subdivision in recognition of a type of culture between the very general Upper Mississippi category and the specialized local divisions which had been named Grand River, Mississippi Uplands, and Lake Winnebago. We began to call this the Wisconsin Upper Mississippi culture. We now had the Upper Mississippi culture, represented in Wisconsin by the Wisconsin Upper Mississippi culture, of which [75] three variants were known—the Grand River, Mississippi Uplands, and Lake Winnebago cultures.

It was apparent from the first that this entire cultural manifestation was not characteristic of the Woodland area in which it so largely occurred, but that it was strongly reminiscent, in many important features, of certain southeastern area manifestations. A detailed comparison of traits seemed to demonstrate the fact that Upper Mississippi was but a subdivision of an even larger cultural order, which was most frequently referred to as the Mississippi culture.

The form of description with which I have been attempting to portray a decidedly complex cultural picture has produced an increasingly grotesque and unwieldy structure and will serve, I believe, to illustrate the need for specific terminology. However, when our taxonomic framework is applied, the picture clarifies and a modicum of order appears out of chaos, permitting the student to observe the identically same facts devoid of confusion. The most general division is the Mississippi basic culture, identified by a complex of fundamentals. The Mississippi basic culture is subdivided into a number of phases, including the Upper Mississippi phase. This phase is represented in a district including Wisconsin by at least one variant, the Wisconsin aspect. Another aspect is apparent in a district including Nebraska. These and other aspects will eventually be found to account for Upper Mississippi manifestations in Iowa, the Dakotas, Minnesota, Illinois, Indiana, Ohio, and in any other states where they are found to occur. In Wisconsin there are three known foci of the Wisconsin aspect: the Grand River focus, the Mississippi Uplands focus and the Lake Winnebago focus. Each of these foci is represented by a number of known components (the occurrence of a focus at a single site), and additional components for the several foci are being discovered yearly.

To one unfamiliar with the status of archaeological information and problems in Wisconsin, such a term as "Grand River culture" is not intelligible; it is a title which carries with it no cultural conception whatever. On the other hand, such a term as "Grand River focus" categorically places in one's mind immediately the type of cultural manifestation referred to, and if "Wisconsin aspect of the Upper Mississippi phase" is added, the division at once takes its place in the culture scheme. Furthermore, one having had experience in some other division of the same phase is immediately in a position to review critically the data offered in substantiation of the classification submitted.

[76] It will be noted that whereas the statement has been made that our taxonomic method is independent of the distributional factor, certain geographical terms have nevertheless been employed in the specific classification used here for purposes of illustration. However, these terms are used without a sense of geographical limitation. For example, a component of the Grand River focus might be encountered on the Fox River or outside Wisconsin. Such terms are convenient, rather than being distributional in any limiting sense, although they may indicate the place of original discovery. In a similar sense, the geologist, without implication of geographical limits, speaks of the Niagara formation, which occurs not only in the Niagara Falls district but in Illinois, Iowa, Wisconsin, and elsewhere. In a similar sense, we speak of Neanderthal man, who has been found to have inhabited various districts remote from the Neanderthal of the Rhine province. In exactly the same sense, we speak of the Aurignacian culture in Africa, far from Aurignac, France. A culturally descriptive term might be considered more ideal, but reasonably short, practical terms that are truly descriptive of a culture complex are exceedingly difficult, in many instances utterly impossible, to devise.

If later it can be established that the Mississippi Uplands focus, Wisconsin aspect of the Upper Mississippi phase, is culturally identical to primitive Ioway Sioux, that fact will in no way disturb the cultural classification but will serve to establish a point of contact between archaeological-cultural and ethnological-linguistic classifications, and thus give the archaeological division a historical import. Or, if it later develops that the Wisconsin aspect of the Upper Mississippi phase is a later manifestation than the Elemental aspect of the Hopewellian phase, this will furnish the basis for a temporal classification complementary to the cultural classification.

One of the most difficult problems is that of identifying the determinants for a cultural division, particularly those more specific divisions, the aspects and foci, in which cultural details are important criteria for differentiation. This difficulty is partly due to the difference in complexity between objects such as simple bone awls and pottery vessels. Culture criteria available to the archaeologist are demonstrated by culture-indicative materials, for the most part artifacts. A simple type of artifact may serve as one element in a trait complex for one cultural division, and therefore may serve as a determinant for that division. This may be the case with a simple type of bone awl. However, when a comparatively complex type of pottery is characteristic for a cultural division, the question arises as to whether it [77] should be considered as a single determinant or as comprising a number of determinants. It certainly is more culturally indicative than a pointed fragment of bone. Single pottery traits, such as shell temper, loop handles, or cord-imprinted decoration, would seem to be at least as important as culture markers as a simple implement with a single differentiating trait. Thus, apparently, we may have a considerable variety of determinants exhibited by pottery alone. Probably other elements of cultural import will each supply more than a single determinant, as for instance, mound structure, burial methods, and house types.

The problem, then, narrows down to selecting, from the traits comprising a complex element, those traits having sufficient cultural significance to qualify as culture determinants. This implies a separate classification of the essential traits for any given complex cultural subject. For example, pottery should be classified under such essential heads as temper, texture, hardness, surface, color, shape, and decoration. Determinants in pottery for a cultural division could then be selected to cover these standard pottery traits. In the same way, the essential traits for burial methods and other complex subjects could be standardized through special classifications, with the result that the determinants for one cultural division would cover the same ground as, and carry similar weight to, those for another division of the same type.

However, the problem is not as simple as that. We have to deal with types of culture as different from each other as a basic culture and a focus. The basic culture is identified by determinants that express fundamental cultural trends. These fundamentals are quite different in character from the detailed material traits so important in determining highly specialized divisions such as foci. Thus, the character of a determinant will depend on the type of cultural division for which it serves as a determinant. The presence of horticulture might serve as a determinant for some sedentary basic culture as distinct from a nomadic, hunting basic culture, but it could not serve to distinguish between two subdivisions of a horticultural basic culture. In the opposite extreme, a specific motif in pottery decoration might serve as one determinant of a focus but not as a determinant for the aspect (or still less-specialized division), since it is peculiar to the focus.

One general axiom must guide the student in attacking this problem. Determinants must be characteristic for that division which they serve to identify. That being the case, determinants for a basic culture will be general in character and relatively few in number. For the more-specialized divisions, progressing from lesser to greater specialization, the determinants will be [78] an enriched edition of the determinants for the immediately preceding, more general division, as altered to include greater detail plus a considerable number of specific traits peculiarly characteristic of the more specialized division. The focus determinants, for example, would be the aspect determinants made richer in detail and augmented by additional traits peculiar to the focus and exhibiting the greatest cultural detail apparent for the entire basic culture.

To exemplify, employing for the sake of simplicity a single form of cultural expression, pottery, two of the determinants for the Mississippi basic culture might be stated as: (1) manufacture and use of pottery vessels and (2) unamalgamated type of temper.

The Upper Mississippi phase of this culture is characterized by pottery determinants more or less in the nature of primary limitations, as follows:

Temper—shell, cell, grit, or crushed pottery
Hardness—1 to 4, softer wares predominating
Texture—fine to medium coarse
Structure—compact and scaly to porous and flaky
Natural color—grays, drabs, and dull terra-cottas preponderating
Surface—generally smooth, rarely rough or polished
Thickness—walls ranging from medium thick to very thin
Shape—simple variety of wide-mouthed jars and bowls
Decoration—ornamentation in intaglio lines on outer surface between shoulders and rim

A more detailed list of traits is required as determinants for the Wisconsin aspect of this phase, although it will be noticed that a few of the latter remain identically the same as for the phase:

Temper—preponderatingly shell
Hardness—1 to 4, softer wares predominating
Texture—fine to medium coarse
Structure—compact and scaly to porous and flaky
Natural color—grays predominating
Surface—smooth to polished
Thickness—walls thin for larger vessels, commonly to such an extent that exterior intaglio lines have produced pronounced corresponding interior cameo reliefs
Shape—a. lips smooth and square or rounded
b. rim angles ranging from horizontal flaring to secondary contracting
c. neck, mere line of juncture between rim and body
d. mouth broad
[79] e. shoulders absent or round and unpronounced
f. base round; conoidal base never characteristic
g. handles either vertically placed loops in diametrically opposed pairs or similarly placed lugs; many vessels without handles
Decoration—a. straight-lined patterns predominating
b. notched or scalloped rims important
c. absence of decoration by pigmentation
d. motifs combined of geometric arrangements of lines and dots important
e. curvilinear motifs rare

The maximum in detail is required of pottery determinants for the focus. Those for the Mississippi Uplands focus of the Wisconsin aspect will serve to illustrate:

Temper—a. exclusively shell
b. abundant temper most common
Hardness—1 to 3
Texture—fine to medium coarse, finer wares relatively rare
Structure—flaky, porous wares predominating
Natural color—varying slightly from gray to dull buff, light grays predominating
Surface—characteristically smooth; pronouncedly smooth surfaces very rare; polished surfaces have not been encountered
Thickness—2 to 12 mm, as thin as 4 mm in places in very large vessels

Shape—a. for the most part simple variations of round-bodied, wide-mouthed jars with flaring rims
b. square lips predominating
c. rims most commonly broad with slight to pronounced flare; vertical rims rising from contracting neck rare; rims with secondary contraction rising from contracting neck rare
d. necks mere line of juncture between rim and body
e. mouths invariably broad
f. shoulders absent or but slightly apparent
g. base invariably round
h. handles round in cross section or broad, straplike loops, the latter predominating; approximately one-third of vessels equipped with handles
i. beakers and hemispherical bowls absent
[80] Decoration—a. present on majority of vessels
b. intaglios preponderatingly incised, trailed, or indented
c. patterns for most part carelessly designed and roughly executed
d. curvilinear motifs absent
e. repetition of a motif to form a symmetrical design rare
f. combination of lines and dots very common as patterns
g. corrugated, incised, or indented patterns on handles common
h. notched or scalloped rims usual
i. cord or other surface roughening does not occur

These series of pottery determinants are not submitted for close inspection as to their accuracy. As a matter of fact, some of the determinants, particularly for the more general divisions, are based on rather meager information, and all are subject to correction. However, I believe they are sufficiently accurate to serve the illustrative purpose for which they are here presented.

Thus, for pottery alone, we have 2 determinants for the basic culture, 9 for the phase, 19 for the aspect, and 26 for the focus. These numbers have no mathematical significance but illustrate the numerical increase in determinants from the more general to the more specific divisions and serve to justify the acceptance of the classification itself. They also illustrate the minimum degree of detail which I believe should be recognized in selecting determinants. It should be remembered, of course, that pottery is but one factor in a cultural manifestation and should never serve alone to determine a classification.

In pottery, I have purposely selected one of the more complex cultural subjects, since it offers a more difficult problem than some more simple product such as a projectile point. There are many subjects that may be adequately handled by a single determinant, discussed in increasing detail for each increasingly specific subdivision. It can be readily seen that a complex of determinants for any of the more specific cultural subdivisions may be composed of from 30 to 40 percent of pottery traits. I do not believe that this is an overemphasis but that it fairly represents the value of pottery, when pottery is present, as a culture indicator.

This method of classification is arbitrary, like all other scientific taxonomic methods. Under the blending influences of diffusion and cultural growth by invention, there can be no hard and fast natural division lines. This absence of sharp lines of [81] demarcation between classes applies equally to the subject matter of all natural sciences and cannot be advanced as a valid argument against classification. Our method is not nearly so arbitrary as the division of a continent into culture areas, which involves an inelastic temporal factor and a confusing and inaccurate distributional factor. The maximum degree of arbitrariness in our method is attained in the division of cultural manifestations into four rather than into some other number of culture types. The four divisions were finally agreed on by the authors of this method as satisfying all apparent needs for major subdivision. With the means provided for subclassification into specific groups under these culture-type heads, the major requirements of remotely separate fields seem to be satisfied, as reported by specialists in those fields.

In applying any taxonomic method, there is always the danger that an unleashed enthusiasm may induce the classifier to attempt to make the facts fit the method. The lure of being methodical at all costs is a constant threat to the wholly profitable use of any method. It is well to bear constantly in mind the rule that the classification is, and forever must be, subservient to the facts. It is convenience and orderliness in handling archaeological data that is required of the classification, not a flawless, natural regimentation of the facts.

Rather recently, it has become increasingly popular for American archaeologists to search out some site known or traditionally reputed to be one formerly occupied by a certain historic group of Indians and to examine data left there as illustrative of the former culture of that ethnic group. This has been termed the direct historical method. As a secondary method, employed to identify a known preliterate cultural manifestation as that possessed by an historic ethnic group, this method has important value. But as a primary method employed to identify culturally data encountered for the first time, without comparing it with other data archaeologically important in that province, it is a method productive of inaccuracy and error.

Granting that it is established that the Iroquois, let us say, occupied a definitely known site in historic times, there remains the strong possibility that the same site was occupied previously by inhabitants possessing a culture, or cultures, foreign to that of the Iroquois. Consequently, there is the important question of precisely which of the various culture indicators present are attributable to the Iroquois. Even where clear-cut stratification is present, there is no way in which the student not previously acquainted with the value of traits as culture determinants can be [82] certain that the culture detritus of any given stratum is not the result of cultural admixture. Such mixtures of culturally divergent materials often result from intermittent occupation of a site by culturally distinct groups or from the disturbance of deposits by human or natural agencies during or subsequent to occupation.

As a matter of fact, there is only one way in which the student can be certain that he is observing the products of a single cultural group in the culture detritus of a given site. This requires a knowledge that the traits found at this site belong to a culture complex definitely known to occur at other sites. Such information can develop only as the result of the use of the recurring complex method—the comparative study of culture-indicative materials encountered at various sites, leading to the knowledge that a certain complex of culture traits, adhering together as a unit, occurs repeatedly at a number of sites. Only when that information is available can the direct historical method be employed accurately to identify a previously established cultural division as a known historic division. The recurring-complex method alone can serve to determine a cultural division; subsequently, the direct historical method may serve to identify it historically.

The statements which I have made are admittedly controversial and have been presented in the hope of awakening critical thought on the subject of culture classification and of encouraging trial of a taxonomic method which, in the absence of some better method, promises to contribute materially toward introducing order into the existing chaotic status of culture concepts in middle North American archaeology.

May 1934.

Appendix 1

State Archaeological Surveys

Suggestions in Method and Technique

Prepared by the Committee on
State Archaeological Surveys
Division of Anthropology and Psychology
National Research Council

Clark Wissler, Chairman,
Amos W. Butler, Roland B. Dixon,
F. W. Hodge, Berthold Laufer

Issued in mimeograph form by the
National Research Council
Washington, D.C.

1923

Table of Contents

Introduction   435

The Archaeological Ideal [1]   437

Archaeology and History [2]   437

Classification of Materials [5]   440

Data on Private and Public Collections [6]   440

Filing Systems [7]   441

Mapping [7]   441

Publications [8]   441

Personnel and Supporting Organizations [8]   441

Collecting [11]   443

Locating Sites [12]   444

Plotting a Site [15]   446

The Examination of Graves, Cemeteries, and Village Sites [16]   447

Mound Exploration [19]   449

The Sounding Rod as an Aid in Field Exploration [22]   451

Summary [23]   452

Introduction

In this report the term archaeological survey is used in a broad sense to cover all aspects of the aboriginal Indian problem, and it is taken for granted that every state is interested in conserving and investigating its archaeological and historical resources. In order to deal with these resources intelligently and to make them of real service to the state, all archaeological and historical Indian sites must be located and classified. Just what kinds of materials are found within the state and where they occur must be determined. It follows, then, that an archaeological survey is for one thing an inventory of these resources. And for practical guidance in its conservation work the state needs such an inventory.

The making of such a survey is essentially a scientific procedure involving special techniques, the essentials of which should be acquired by all who undertake the work. As information of this practical nature is not readily obtainable, the accompanying suggestions are offered. They embody the latest improvements in archaeological technique and have been compiled by the committee from statements prepared by experienced American archaeologists.

[1]
State Archaeological Surveys

The Archaeological Ideal

In all archaeological research the ideal should be to record accurately all of the pertinent facts. What is found? Where? How related to topography? To the earth strata? And lastly, the spatial relations of all objects. All data should ultimately be visualized in a three-dimensional scheme, their places in the horizontal plane, and their relative depths. The former express the geographical distribution, the latter the time sequence. The geographical distribution is primary and is also the immediate objective of the survey.

Examples of stratification are rare, but when found they should be noted with the utmost care. They are also the most precious of finds, to be preserved whenever possible for future detailed study.

[2]
Archaeology and History

In all the states there are known sites of what were Indian villages during the period of colonization, and in many of the states there still remain remnants of Indian tribes once living and flourishing there. It is thus possible to connect the immediate pre-historic with the historic. The reconstruction of the original culture of these tribes at the time of their first meeting with the settlers is a most important problem. For example, the Menomini of Wisconsin when first discovered were residing about where they now are, so that an intensive study of that territory would enable one to identify the pre-historic sites, to determine their culture characteristics, and eventually to distinguish between the early and the late sites. A good example of this kind of work is to be found in Skinner's "Material Culture of the Menomini."[1] But for a more exhaustive study see "The

---

1. Skinner, Alanson. Chap. 7. Published in Indian Notes and Monographs, Museum of the American Indian, Heye Foundation. [Ed. note: *Museum of the American Indian, Heye Foundation, Indian Notes and Monographs* 20. 1921]

Mandans."[2] Many other similar studies could be cited, but all of them are still deficient in archaeological data and particularly in the use of such refined methods as are now available for the determining of time relations. All the states in the Union, particularly those in the Mississippi Valley, offer many such problems in the archaeology of known tribes, for throughout the length and breadth of that great area there lived, in pre-historic times, many Indians of different stocks and cultures. Whether the historic Indians were the same people or whether they were the descendants of those who held the country in pre-historic times, is, of course, a question in many cases. It is often possible, however, by the examination of known historic sites of given tribes to trace those tribes back to the pre-historic period. [3] When articles identical with those found on the historic sites occur on those of pre-historic origin, careful comparison with other sites in the locality will leave little doubt as to the identity of the people inhabiting the locality.

Two cultures may be looked for in the Mississippi Valley region and also northward toward the Great Lakes. The first of these is the Siouan culture, which is not at all well known, but which, judging by the remains found on the upper Missouri River and in Nebraska, especially on the Mandan village sites, was rich in bone and antler implements and possessed a high development of characteristic pottery. In the lower stretches of the Mississippi Valley, there is a second culture, associated with the mounds, which was exceedingly rich in many varied forms of pottery, including necked bottles, effigy jars, and painted ware. Neither of these cultures has received very full description, although the latter has been worked over to a considerable extent by Mr. Clarence B. Moore, whose splendid publications have been printed under the auspices of the Philadelphia Academy of Sciences. Mr. M. R. Harrington, in "Certain Caddo Sites in Arkansas,"[3] has produced an exceedingly valuable paper.

Two other cultures which have been identified in the east and which may be expected to occur in certain parts of the Mississippi Valley and Great Lakes region are the Algonkin and Iroquoian. The Algonkin culture has been shown to be a complex in which articles of stone predominate over those of other materials. In this complex are found polished slates,

---

2. Will, George F. and Spinden, Herbert J., Papers, Peabody Museum of American Archaeology and Ethnology, vol. 3, no. 4. [Ed. note: 1906]

3. [Ed. note: *Museum of the American Indian, Heye Foundation, Indian Notes and Monographs* 10. 1920]

banner stones, gorgets, tubes, and bird stones; platform pipes and Micmac pipes; pottery vessels possessing conical bases; long stone pestles; grooved axes; arrow points of many types, shapes, sizes, and materials, [4] especially the notched and stemmed varieties. Bone and antler work are weakly developed. While first identified in New York, this culture has since been found to have spread throughout New England, southern Ontario, through Pennsylvania, Ohio, Indiana, Illinois, Wisconsin, and westward as far as Minnesota. Its southern range has not yet been determined.

The Iroquois culture, on the other hand, is one in which the use of bone and clay predominates over that of stone. Iroquoian pipes are especially beautiful, and on their ancient sites vast numbers of handsome effigy pipes of clay have been obtained. Their pottery is characteristic; the jars have rounded bottoms, constricted necks, and usually, overhanging rims. Bone implements include, of course, awls, fishhooks, harpoons—usually unilaterally barbed—combs, spoons, bowls, beads, gorgets made of human skull, and many other articles. Triangular flint arrow points of small size seem to be the only stone arrowheads used by these people. Celts are found, but not the grooved ax which is a component of Algonkin culture. The long pestles noted in the former culture are absent; as are, for the most part, gorgets and all the polished slates.

It must not be thought, however, that the mere occurrence on a site of triangular arrowheads, to the exclusion of others, proclaims this site to be one of Iroquois origin. Several or more of the component units of these complexes must be present before the culture can be definitely identified. Taken separately, they are only indications or symptoms.

Overlapping of the various cultures is to be expected; both Algonkian and Iroquoian sites sometimes occur on the same spot, though, of course, they are not contemporaneous. But more than this, certain articles are found to be common to many different cultures. For example, the celt is found almost universally over eastern North America, and likewise certain [5] other types of implements, so that, as mentioned above, one must take into consideration not only one, but a number of units comprising any complex before a culture can be identified.[4]

These suggestions are offered merely as hints as to how the historical problem articulates with the archaeological one, thus making it clear that both historical and archaeological students should be interested in these surveys.

4. Alanson Skinner [Ed. note: see footnote 1]

Classification of Materials

Archaeological sites, or places at which archaeological data may be obtainable, naturally fall into definitive classes, for which a classificatory nomenclature has been agreed on. It is usual to classify such sites by the following scheme:

Agricultural plots, or fields	[6] Pits
Burial grounds	Quarries
Caches	Refuse heaps
Cairns	Reservoirs
Campsites	Rockshelters
Canals	Ruins
Caves	Salt fields
Embankments	Shell heaps
Fortifications, or forts	Shrines
Mines	Trails
Miscellaneous finds	Village sites
Mounds	Workshops
Pictographs	

For convenience in mapping, each of the above should have a symbol. (See Ohio Atlas and New Jersey Survey.)

Then when an archaeological site has been detected, its place in the above scheme should be determined and so reported on as to give full information on the following essential points:

1. Class and location
2. Extent, plotting, etc.
3. Description, archaeological characteristics
4. Notes on collections made
5. History of the site (former surveys, excavations, etc.)

Data on Private and Public Collections

The distribution of artifact types is important. Thus grooved stones of a specific type may occur in one section of the state and not elsewhere. If notes are taken of the types in collections and the [7] localities from which they come, such facts of distribution will ultimately appear. Incidentally, such inquiries will stimulate collectors and local students to give more attention to the precise locations of their finds.

Special attention should be given to pottery as well as to the several types of stone artifacts. Collectors often neglect sherds, or fragments, because they consider them of no value, when the fact is that all the distinguishing characteristics of the pottery for a given area can be determined from the small samples gathered from the surface or found in the ground. For each site, too, special note should be taken of the decorations on the pottery, the structural character of the sherds, and similar matters. Also, the presence or absence of pottery at a site is of itself of the greatest significance.

Filing Systems

The most practical method of filing original data is by the envelope system. The logical unit for the state is the county. Sites and other information can best be tabulated on cards, by counties, each site being given a serial number, which then stands as its index throughout. Here, or in the file, should be recorded all bibliographic references. For a published bibliography, see Harlan I. Smith's Michigan Bibliography, Michigan Geological and Biological Survey, Publication No. 10.

Mapping

Presumably for each county (except in a few of the eastern states) there is available a map showing all the section lines. All sites and finds should be carefully placed on such a map. Preferably all highways, [8] towns, and streams should be indicated on the base map, as these will greatly facilitate precise entries by the fieldworker. For examples of such mapping, see publications of the Ohio, New Jersey, and New York surveys.

Publications

The formulated data on sites, etc., seem best segregated under the heads of counties. The form followed by the New Jersey and the New York surveys is recommended.[5]

Personnel and Supporting Organizations

A state survey may be conceived of as a distinct state department—parallel to such other departments as the geological survey or the biological survey

---

5. See A. C. Parker, The Archaeological History of New York, New York State Museum Bulletin, Nos. 235–238, Albany, 1920; also, Alanson Skinner and Max Schrabisch, A Preliminary Report of the Archaeological Survey of the State of New Jersey, Bulletin 9, Geological Survey of New Jersey, Trenton, 1913.

—under the more or less independent direction of a single official. No state has so far fully realized this ideal. In Indiana and New Jersey, for example, the director of the geological survey has been charged with the archaeological survey. On the other hand, we find examples of state archaeologists as officers of more or less independent organizations to which the state gives support. Thus in Ohio we find the State Archaeological and Historical Society, which directs a state museum and a state survey, appointing a director and a staff for the same. A somewhat similar situation exists in Wisconsin. On the other hand, we have state museums supporting departments of anthropology which have undertaken intensive surveys, as in New York. [9] Thus without going into an exhaustive review for the several states, we find that the tendency in general is for such surveys to develop either at the hands of a permanent organization having under its control a museum, or as one part of an independent museum organization.

One notable fact is that, whereas state universities take a large interest in the geological, biological, and other scientific work of the state, they show no such tendency with respect to anthropological problems. The one exception is the University of California, which supports a well-organized staff of anthropologists. This staff, incidental to its teaching and research function, has carried on and is still carrying on a survey of the state. So taking it for granted that instruction in anthropology will soon be given in all the state universities, we may look forward to the time when these institutions shall lead in the researches which such a survey entails.

In the meantime, facing conditions as they are at present, the taking up of a survey brings the state agency involved face to face with the question of personnel.

Someone must necessarily be the active, responsible initiator of the work and must himself take up at least a part of the burden of fieldwork. It goes without saying that he must have the requisite training, the breadth of view, and the scientific qualification for research. The ideal condition would be for such a man to give his entire time to the work, half of which should be spent in fieldwork and the remainder in working up his data. A young man just completing his graduate work in anthropology and possessed with the requisite qualifications could safely be given such a directorship, under the general direction of some appropriate agency, such as the state geologist, the curator of the state museum, or the conservation commission.

[10] If, however, it is not feasible to provide for full-time service, the

assistance of some teacher of anthropology who can give his spare time should be sought. Although the number of men available is not great, there are still such men and institutions from whom aid and service can be anticipated. Further, in the event that this method is followed, it may be expedient to give parts of the work to different investigators according to their specific qualifications. In this way, a satisfactorily high research efficiency may be attained.

Among the incidental, but by no means inconsequential, results from a survey are the reaction and stimulus of its interested citizens. In each community may be found a few individuals who have more than a passing interest in the subject and who stand ready to cooperate under wise and efficient leadership. It seems probable that the survey in any given state will stand or fall according to the skill with which its leaders approach this great body of amateurs and enlist their support. Visits to a county by the state fieldworker must necessarily begin with calls on these local antiquarians and the study of private collections, and out of this contact should develop a relation of a permanent kind.

Appropriate questionnaires may be prepared for circulation among local students, for example, the circular issued by the Indiana Historical Commission.

While it is possible to canvass a state entirely by mail, the result will be far from satisfactory because the person in charge of the survey must himself see most of the sites and meet most of the correspondents before he can evaluate their communications or intelligently follow them up.

[11] In this connection, an early publication of the data for a group of counties will be serviceable in stimulating additions and corrections as well as in setting a pattern for reports from other localities.

Collecting

As a large part of the data for the survey will be gathered by enthusiastic collectors for the localities in which they reside, collectors should acquaint themselves with the methods experience has shown necessary to the proper recording of data. By taking such precautions everyone interested can contribute to our knowledge of the past; and so it comes about that the making of an archaeological collection, when properly done, is a real service.

In the first place, a collector should give chief attention to one locality or section. A desultory collection is too scattering to be of scientific value, but one confined to a restricted area will stand as a distinct unit and an index to the culture of its pre-historic inhabitants. One of the most satisfactory collections ever noted by the writer was from a single farm of 320 acres, the precise locations in which were recorded for each specimen.

Although a number of suggestions are offered here as to the exploration of mounds, graves, and village sites, it is our firm conviction that the ordinary, untrained collector would do better not to attempt the excavation of a given site. On the contrary, let him write to the director of his state survey and perfect some kind of arrangement whereby the site may be properly explored. Much valuable archaeological evidence is lost because men who have had no training in excavation attempt that difficult and technical work.

[12] Finally, in an archaeological survey it is important to list all of the village sites, mounds, caverns, and other pre-historic remains in the state. Independent of what remains have been found in these places, it is of the greatest importance that their location should be known and properly recorded. Such sites are quite as important in archaeology as the specimens found in them. Thus collectors can be of the greatest assistance to the survey if they will report such sites and prevent their exploration by ignorant persons.

Locating Sites

What to look for is in the main obvious, but some of the less noticeable of archaeological materials call for special methods. Among these are village sites, camping places, and graves. In this connection, the following statement should be noted:

> In considering the Indian village and campsites of the Mississippi Valley and Great Lakes Region, we may first note certain general characteristics not only well-nigh universal throughout the entire region under consideration but throughout all the vast territory of the Mississippi River and probably also the western part of the region. Two things were absolutely requisite to every aboriginal community. One was the presence of fresh water, the other, situation in a sheltered spot, preferably on the northern bank of a river, and if possible on a sunny knoll. The searcher for Indian sites may therefore, as a rule, ignore all localities where the soil is not light and dry and that are not located within easy reach of an everlasting supply of pure, fresh water. Important village sites may be expected at the forks of important rivers, or where tributaries join the main waterway. [13] They will generally be on the first terrace above the river and not on the floodplain,

where they are open to inundation in the spring of the year. Along the Great Lakes, where the shores are sandy, sites occur on the second ridges, i.e., the sand dunes which are separated from the cold blasts which come from over the water by another ridge between them and the lake. This is true all along the west shore of Lake Michigan from a point a few miles north of Milwaukee practically up to the Straits of Mackinack.

Campsites resemble villages but are very much smaller in area and more sparsely covered with indications of former occupation. They are also found scattered along the main waterways and are quite prevalent in the interior of the country in remote places. On them the searcher is less apt to find rare or unusual specimens.

The most prominent criteria of the camp or village sites are, first, quantities of burnt stone that have seen use in aboriginal fires. In locations where the light sand is apt to be blown away by the prevailing winds, sometimes the burnt stones marking the fireplaces of the wigwams may be found in the original circles. Black earth, i.e., earth full of carbonized animal matter and charcoal, is also a very sure criterion of former Indian occupation. This discolored earth occurs in large patches, ordinarily known as kitchen middens, and is, as a rule, full of burnt stone, flint chips, implements, fragments of charcoal, animal bones, decayed unio, or freshwater clamshells, and potsherds, all of which, even when found without the accompanying black earth, are invariable indications of former Indian occupation.

In some localities, especially in the southern part of the area, Indian villages are often marked by low circular mounds which were erected as foundations for the wigwam. In the northern part of the territory under discussion are found somewhat similar mounds which have a central [14] depression. These are the remains of earth houses which have fallen in and decayed in former years. The domiciliary mounds just mentioned are not apt to have any particular objects in them unless they have been made use of as a secondary burial place. The fallen dirt houses are often full of camp debris and sometimes contain specimens of particular interest. The old earth-house sites are especially abundant from Wisconsin westward to Minnesota on the upper waters of the Missouri River, being found abundantly in the states of North and South Dakota, and in Nebraska.

Many Indian village sites are often marked by the occurrence of caches or fire pits. The Indians frequently dug bowl-shaped holes in the ground for many different purposes. Sometimes these occupied the center of the wigwam and in the bottom of them a fire was built, the depth of the hole preventing the sparks from flying upward and setting the lodge on fire. Such fireplaces, in the course of years, gradually filled with ashes and accumulated camp debris, such as potsherds, discarded or broken implements, lost articles, and the bones of animals. They were cleared from time to time by throwing fresh earth over the foul-smelling debris, or the ashes and garbage were scraped out and left along the sides. Often in winter when the ground was frozen and it was hard for Indians to dig with primitive tools, the bodies of their dead were buried in the fireplace and the lodge removed. Sometimes pits were dug outside the wigwam at a little distance, to receive the camp garbage, and very often similar holes lined with mats or bark were used to store wild rice and corn or other articles until they should be required for use. When the ground has not been plowed, especially in the forested regions, traces of these pits may often be identified as small, round dimples in the soil measuring from two to three feet across and a few [15] inches in depth. When dug open, however, the disturbed earth will often be found to run down for several feet, marking the outline of the pit. Usually this earth is readily distinguishable from that of the surrounding undisturbed virgin soil because of its darkness and mixed color. Sometimes traces of the mats or bark with which the pits were lined are plainly visible and occasionally pieces of some value, such as perfect pottery vessels and entire implements, are found in these places, where they were stored away or lost. They are far more apt to yield articles of interest than are the middens or the surface soil of the village, and because of their depth, perishable materials are more apt to be preserved. They rank next in importance to graves in the estimation of an archaeologist as repositories for valuable specimens.[6]

Plotting a Site

Whereas it is sufficient to locate small sites on a county map by numerals and letters, the more important of them call for plotting on a large scale. In every case, excavations should be preceded by plotting and the establishment of levels and sectional lines in order that the depths and transverse locations of all finds may be precisely recorded. To do this accurately requires some technical training; one without such training should seek the advice of a person experienced in surveying or building.

Cemeteries and village sites are usually on level ground so that all one need do is to run a base line, taking care to have it level and to record its

6. Alanson Skinner [Ed. note: see footnote 1]

position by a compass. From this base, run lines parallel [16] and at right angles, in the same plane, thus marking off squares or rectangles by which all finds can be accurately located.[7]

Before digging, plot the lines on section paper to scale, and letter and number so as to make identification sure. Subsequently, all trenches and pits should be drawn in accurately. Remember that digging destroys the evidence forever; hence, the record should be as complete and accurate as possible.

The Examination of Graves, Cemeteries, and Village Sites

Allow us to repeat here that the person interested in the Indian remains of a given locality should proceed carefully and cautiously in his work. The essentials of technique are, in the main, as follows: graves are found singly or in groups, and there are seldom surface indications. No grave should be explored unless it can be done thoroughly. That is, photographs taken as the work proceeds, the skeleton, whether whole or fragmentary, carefully dug into relief by use of hand trowels, and notes written as to the position of all objects. The bones should be carefully preserved. They will seldom break unless carelessly handled. If a bone is decayed, dig its entire length under it and take it out adhering to the clay and wrap it up carefully. If all the fragments of the skull are there, save them all, as the skull can be restored later. Where one grave is found there may be others, and a trench should be run in the direction in which the cemetery exists, a ground plan made, and all graves numbered. If it is a village site, one should look for the ash pits. [17] Ashes have a wonderful preservative quality, and carbonized food, corn, seeds, cloth, mattings, and so forth are frequently found. The ashes and black soil of fire pits should be most carefully examined.

One of the best published statements of detailed procedure will be found in Arthur C. Parker's "An Erie Indian Village and Burial Site" (Bulletin 117 of the New York State Museum, Albany, New York), from which we quote the following:

> *Method of Excavating in the Village Section.* The village section was staked out in parallel and adjacent trenches 16 feet wide. Excavations were commenced at the wire fence 20 feet from the shoreline. A sectional trench 3 feet wide was dug and the dirt thrown back. This left a cross section of

---

7. Consult Parker, A. G., in Bulletin 117, New York State Museum, Excavations in an Erie Indian Village and Burial Site at Ripley, Chautauqua Co., N.Y., Albany, 1907.

the trench exposed with the 3 feet of floor serving as a working space. The archaeologist examined this cross section and if indications pointed to the probable presence of objects, he troweled into the bank, allowing the earth to fall to the floor until it had filled when it was removed by a laborer. If the indications pointed to a barren spot, the workmen spaded ahead until signs of disturbance again appeared, and the section was again examined. When a pit was discovered, a clean working space was made and the pit vertically exposed at one side. The pit filling was then troweled from top to bottom, great care being taken not to break the specimens that might come to light with any trowel stroke. As the work progressed, measurements of the pit were taken and all the important specimens labeled and placed in trays for subsequent numbering. The refuse material such as animal bones, potsherds, flint chips, and rude implements was placed in labeled bags. A diagram of the pit was drawn and the details of its excavation recorded in the trench book. Trenching was continued until the trench became [18] barren, at which point another trench was worked.

Every pit, pocket or posthole was charted, the varying character of the soil and the manner of its disturbance was noted, and it is possible for anyone familiar with our methods to take a specimen from the collection and after examining its number and referring to the records, point out on the map or on the actual site itself exactly where that object was found.

To ensure accuracy in field records, three of a different kind were made, so that any circumstance omitted in one might be found in one of the others. The first record was made in a trench book and written as the actual work progressed; the second record was made on data slips and supplemented the trench book in the matter of measurements, locations, and positions of trenches, pits, and objects, and added the details of the particular thing described on the slip; the third was a survey record, in which every pit, grave, or trench cutting was charted to a degree of mathematical exactness. All these records are supplemented by drawings, diagrams, maps, and photographs.

Method of Excavating Graves. The burial section was staked out in the same manner as the village section. The workmen removed the disturbed topsoil for a distance of 3 feet, leaving a working space of 3 feet by 16. Excavations were continued until signs of deeper disturbance appeared. These "signs" were foreign substances in the regular strata, such as fire-burned stone, flint chips, charcoal, and lumps of clay. Earth of the character here found once disturbed is never as compact again as originally, and even if there were no intruding substances in the sand its very looseness as distinguished from the rather compact sand surrounding it was a sign of its disturbance. The topsoil over the grave was removed and its outline ascertained. The superincumbent earth was removed for a foot, and a depth of 6 inches below explored for signs [19] of the grave bottom, and if not found the earth for another 6 inches was shoveled out with great care—the shovel scooping up the earth rather than spading into it. The trowel was used again to dig down, and the process repeated until the skull or pottery-vessel top was reached. The soil was then removed carefully with trowels. The skeleton and grave bottom were cleaned with fine pointing trowels and finally swept with a brush, care being taken not to move any bone or other object in the grave. A diagram of the grave and its contents was made and the exact position of these objects ascertained by means of a compass and tape. The dimensions of the grave, its number and position in the trench, and the character of the soil and other items of importance were recorded in the field book. If the burial was of sufficient interest, photographs from one or more positions were made. The skeleton when removed was wrapped in excelsior or cotton and placed in a labeled box, but not finally packed until dry. The objects found in the grave were placed in a tray with a proper label and afterward marked with the serial field number, this number being distinguished from the museum serial by prefixing the letter "F." Data slips numbered to correspond with the specimens were filled out and gave all the necessary details. Any information not found on the slip may be found in the field record. The various records thus countercheck each other.

The reader would do well to read the whole of the work from which the preceding extract is quoted.

Mound Exploration

Mounds make a strong appeal to the historical interests universal in man. No one looks upon a mound without experiencing a desire to dig into it. We have dwelt elsewhere on the inadvisability of careless and reckless digging. In fact, one of the greatest services a local [20] student can render is to discourage all such tampering with pre-historic remains. Then, when he himself feels ready and competent to undertake the investigation of a mound, he should note carefully the following recommendations.[8]

In the first place, anyone with a general knowledge of the topography of the country will be able to distinguish between a mound and a small

8. From notes supplied by Warren K. Moorehead.

hill. For the most part, a mound is round or conical, and small hills are seldom in this form. If in doubt, the only solution to the problem would be to dig into the mound from the edge until rewarded by finding the floor of the sacred place for which the mound site was used before the mound was erected over it. If this floor is not perceptible to the untrained eye, it may be detected by the general character of the soil. The usual way to excavate is to begin on one side of the mound on a level with the surrounding country and carry forward the excavation taking down every part of the mound. In due time the floor of the mound will be found, and as soon as it is found it can readily be traced by carefully watching the indications always found on the floor.

As you proceed, you will find little masses of very dark earth, probably caused by the decay of skins, etc. Upon striking these, go carefully, as they indicate the immediate presence of skeletons, and with the skeletons will frequently be found specimens. Sometimes the dark masses are deposits of ashes in which little is found. Do not pick into soft masses or try to dislodge skeletons with the shovel and pick. The hand trowel is better for that kind of work.

About the skeletons look for beads. They are at the wrist and neck. The soft, frail shells must be handled carefully. Do not try to get the [21] earth off from them but preserve them as they are and clean them several weeks later. About the arms copper bracelets may be found. These are green in color, having oxidized through age.

The pottery is often soft when first taken out. Set it aside carefully and do not try to take out the earth until it is dry. Whatever pottery you secure, pack carefully in excelsior or sawdust in a strong box. Do not put stones or heavy things with the pottery. The shells and fragile objects should be packed in sawdust (or better still in tissue paper) in cigar boxes.

Too great care cannot be exercised when taking the earth away from about the bones. Do not use shovels, as you may throw aside stone pipes or ornaments. Save the skulls entire if possible, as the skeletons of prehistoric peoples are needed by anatomists for study.

Keep your work to yourself, since finds excite people, and many visitors interfere with operations. In case you do not have sufficient leisure or funds properly to explore a place or site, do not undertake it at all. Graves, mounds, and caves unexplored are of more real value to American archaeology than when either partially or superficially explored. It frequently happens that one digs in a site, secures a few objects, and abandons the site. It becomes known in the community that "Indian relics" were discovered. Curiosity-seekers flock to the place and soon ransack it. Therefore it is better to leave a site unexplored, unless as has been stated, the work can be properly carried to a successful end.

[22] **The Sounding Rod as an Aid in Field Exploration**

For locating objects and burials below the surface, as well as detecting former disturbances of the soil, the use of the "sounding rod" is recommended. The following statement by Reginald Pelham Bolton should be noted:

> This implement consists of a slender steel rod, 3/16 inch in diameter and about 3 to 4 feet in length, provided with a wooden handle such as is commonly used on bench tools. The end is ground to a point, and the tool is used to penetrate the soil, giving indication of its density and of the existence of objects below the surface. Its material should be spring steel, and care must be exercised not to buckle it by undue pressure.
>
> It was developed as a means of avoiding much heavy labor involved in digging trial holes and to discover the presence of shells and waste debris in Indian, colonial, and military sites, and it has proved most effective and informing. Mr. W. L. Calver, who originated the plan, has, with the writer, used this tool for upward of 20 years in exploration in and around New York City. The writer has tried various modifications such as grooving the point, using a triangular rod, and extending the length, but the simple form above described has been found to answer all practical requirements. Its continued use is rather hard on the palm of the hand unless the handle be made well rounded, or a glove is worn. In soft and wet soil it can be thrust down with a single motion, but in dry ground it should be forced down in a series of short advances, and should be turned slightly on the down thrust. Practice gives considerable sense of the character of object with which it may come in contact. Thus wood or roots can be distinguished from stone, and such objects as shells or bone are recognizable by their penetration, while it becomes possible to [23] recognize glass and crockery.
>
> It is very helpful in determining buried stones, lines of brickwork, or hard floors. The size of a stone can be outlined on the surface by the position of the holes pierced around it, a wall can be followed by successive proddings, and a hard surface like a floor can be traced and its level decided by a series of equidistant penetrations.
>
> In excavating or in trenching, the rod is thrust in sidewise to determine

the lay of debris in any direction and is very useful in giving advance notice of delicate objects such as pottery, glass, etc. The rod can be mounted inside a hollow cane for convenience in traveling.

Summary

Finally, for convenience and reference these recommendations may be formulated as rules. The first and most important rule is, *do not dig until you have in mind the technique.* With the technique in mind, observe the following rules:

1. Photograph (or draw accurately) the site or mound before commencing excavation.
2. Stake off the spot (or mound) in squares of 3 or 5 feet each.
3. Small hand trowels or broad, dull knives, and whisk brooms are indispensable. Ordinary large digging tools need no explanation.
4. In case of a mound, run a trench tangent to its base as previously stated, to at least three-fourths of the diameter. Dig down slightly below the original surface, or the "floor." In some mounds there is a "sod line," or dark streak at the base; in others, a hard, burned floor. In many others you cannot determine bottom positively and must continue on down until the undisturbed clay or gravel is reached. Note with extreme care the face of the cut. Scale it down in narrow sections.
[24] 5. Throw the earth behind and keep a clear space of 4 or 5 feet between the earth and the front wall, or face, of the trench. When through, the excavation will be nearly filled, and little damage will have been done to the structure. Mounds should not be opened by means of an irregular pit sunk from the summit (or center).
6. For village sites and grave groups, rules 4 and 5 must be somewhat changed. Long, narrow trenches sunk down as far as charcoal and ashes occur must be run. Throw earth behind as you proceed. Excavate all ash pits carefully, as interesting objects are frequently found in them.
7. Enter all finds on a map or ground plan, and note in the squares (by numbers or letters) the skeletons or objects, and so forth, found.
8. Photograph skeletons or objects in situ.
9. Number or letter the objects or crania (or entire bones) and also designate the mound or site so that it and its contents may not become confused with the results of explorations in other monuments.

10. Keep a careful field catalogue or diary and retain the same series of numbers or letters in the packing boxes and so forth.

11. Pack specimens for transportation so that there is no danger of breakage.

12. Provide shellac or a light solution of glue or other good preservatives for bones, pottery or soft substances, as well as strong packing boxes, cigar boxes, paper, excelsior, cotton, and string.

Appendix 2

Number 93 Price 25 cents

Reprint and Circular Series of the National Research Council

Guide Leaflet for Amateur Archaeologists

Issued under the Auspices of the
Committee on State Archaeological Surveys

Division of Anthropology and Psychology
National Research Council

National Research Council
Washington, D.C.
1930

Guide Leaflet for Amateur Archaeologists

In 1920 the National Research Council organized the Committee on State Archaeological Surveys to encourage systematic study of fast-vanishing Indian remains. In the 10 years of its existence, the committee has assisted in the formation of research organizations in various states, has sought to systematize and unify methods of investigation, and through publications, conferences, and visits by its chairman, has endeavored to keep all workers in the field informed of the progress of archaeological research throughout the United States.

The activities of the committee have been purely advisory. It has not sought to control the actions of any group or state but has freely offered its help and advice in the advancement of scientific work. It now seeks to extend its services to amateur archaeologists and to all who are interested in the early history of our country. In presenting this booklet, the committee hopes to enlist the active cooperation of all intelligent laymen in the preservation of archaeological sites. It seeks to give information which will enable the local investigator to carry on work according to the most approved methods, so that he may assist in unraveling the story of human development on the American continent.

It is evident to everyone that the great majority of our Indian remains have already been destroyed. This has been due in part to the fact that many pre-historic sites have been occupied by white settlers who have found it necessary to level Indian mounds and earthworks in order to utilize the land for farm purposes, for city development, or to make way for roads. However, the greatest destruction has been wrought by curio hunters who have dug into the mounds in search of relics, without realizing that they were destroying valuable historical material. To open an archaeological site without knowing how to preserve the record is equal to tearing pages out of a valuable book, a book which can never be rewritten.

In each state there are some people who are interested only in securing specimens which they can sell for personal gain. They care nothing for history or science and are not disturbed by the fact that their ruthless methods destroy materials of great interest to their fellow citizens. This leaflet is not addressed to such. Their activities will cease only when public opinion is strong enough to make their work unprofitable. Today no scientific institution and no well-informed person will purchase archaeological material which is not accompanied by a full record. When intelligent local collectors take the same attitude, the work of these [4] commercial "pot hunters" will cease. An Indian relic without data is as worthless as an unidentified postage stamp or a bird's egg. The pages which follow seek to show how amateur archaeologists may assist in recovering the prehistory of our country and at the same time help to preserve the existing Indian sites for future generations.

It is well known that some of our Indian tribes were nomadic. They were wanderers who made their camps near to favorable hunting grounds and who moved to new sites whenever whim or necessity dictated. Other Indian groups were dependent chiefly on agriculture, and these made permanent settlements which were occupied for long periods. But exhaustion of soil, hostile raids, epidemics, and other causes led to their abandonment and the establishment of new camps. Thus it sometimes happened that a single campsite was occupied several times, and the record of these periods of occupation can now be read by careful excavation. In some places it is possible to carry back the record through successive stages of development from historic to ancient times. Examples of such stratification are rare and should be noted with the utmost care. Through them we can trace the movements of peoples, the growth of culture, and the effects of environment on man in America.

But such a story can be obtained by neither the careless digger nor by those who are interested only in beautiful specimens. It can only be revealed by those who preserve every evidence of this early life. Every potsherd, every implement of bone or stone, no matter how crude or fragmentary, every animal bone or vegetable product, becomes an important part of the record. Nothing should be discarded until it has been made the subject of careful study. Even the scattered surface finds have great value if their location is recorded, for when their distribution is plotted on a map they tell of migrations, of trade routes, and of local development.

In some places, the Indians built great earthworks, fortresses, and pyramids. In others they constructed mounds of earth in the form of birds and animals—the so-called effigy mounds. In some localities they buried their dead in graves dug in the earth or surrounded them with stone slabs. In other places they placed the corpses on the surface and raised over them mounds of earth, some of considerable size; still others constructed mounds in which they placed the dead. Many different methods of preparing the body were employed. Sometimes it was laid out full length on its back. Again it was placed on its side with hands and feet drawn close up to the body. In some instances cremation was practiced, while still other groups placed the dead on platforms until the flesh had vanished, then tied the bones into bundles and placed them in the mounds. All these methods are of extreme interest to the student, [5] and the record of their presence may go far toward identifying the Indian groups in question.

It not infrequently happened that a mound was originally built by a people practicing one method of burial, but was later used by incoming tribes. Such intrusive burials are most instructive in deciphering the sequence of cultures.

In the southern, eastern and far western states, Indians living near to the sea lived largely on shellfish, and during long periods of occupancy built up great refuse piles in which are found animal bones, broken bits of pottery, and other objects which help to reveal the life and habits of the builders.

Cave dwellings are for the most part restricted to the southwestern part of the United States, yet important sites have been discovered in the Mississippi Valley and elsewhere.

Within recent years, reports of finds of early man have been current. These range from the finding of utensils associated with the bones of animals now extinct to the discovery of arrowheads and similar objects lying in undisturbed gravels at points where river erosion or excavation has exposed successive strata. Still other important sites are ancient mines and quarries from which the Indians obtained their flint and in some cases copper.

How to Obtain the Record

No single collector can hope to obtain a representative exhibit from the whole country, nor would such a collection be desirable, for upon the death of the owner it is almost certain to be scattered and its scientific value lost. However, each local archaeologist can become a specialist in his own locality. He can gather the most accurately recorded collection from that area. He can obtain information which when added to that of his fellow workers will ultimately reveal the pre-history of America, and he can have the satisfaction of knowing that he has assisted in preserving pre-historic monuments for future generations.

The survey

In many sections of the country it is possible to obtain plat books which give locations of farms, roads, lakes, and other features which may serve as guides in the field. If these are not obtainable, township or section maps may be used, but here it is necessary to transfer from county maps, streams, roads, and other information by which it is possible definitely to locate a site. On such a map first place all existing Indian sites, then those whose former existence can be definitely determined, [6] and finally the approximate location of doubtful sites. In order that all work may be uniform, the symbols shown in Figure 1 are suggested.

Indian trails which can be located from old land surveys, maps, or county histories should be drawn in with blue pencil but only so far as they can be definitely and accurately identified.

Should there be several mounds so close together as to make it impossible to place them on the map, this can be indicated by placing a number at the lower-right-hand side, e.g., for eight circular mounds: O_8. If further identification becomes necessary in describing, letters can be placed above the figures, e.g., $O\ ^A/_8$.

For describing particular sites, squared paper should be used, and the exact location and size of each mound should be noted. Thus each square might be considered as 5 feet, and the group of mounds $O\ ^A/_8$ might be shown as in Figure 2.

[7] In such a case the use of a tape and compass is necessary to place the mounds in their exact relationship to one another.

Surface collecting

When mapping the Indian remains in a township, it is desirable to make surface collections and to locate the material with relation to the nearest mound, village site, and so on. Such surface material should be carefully numbered and entered in the catalogue. Never depend on your memory alone for locating specimens.

Village and campsites are often located by the profusion of broken pieces of pottery on the surface. Black earth containing charcoal and burned animal bones is also a good indication of former occupation. In places, low circular mounds reveal the foundations of wigwams, while low mounds with central depressions may be the remains of earth lodges.

	Now Existing	Formerly Existing Definitely Located	Reported
Round or conical mound	○	⊗	?
Elongated or elliptical mound	⬭	⬭⊗	?
Effigy mound	⋁	⋁××	⋁?
Village site	△	△×	△?
Earthwork or fortification	▭	▭×	▭?
Quarry	⬆	⬆×	⬆?
Burial ground (not a mound)	† †		
Rock shelter or cave showing human occupancy	⊐		

Figure 1

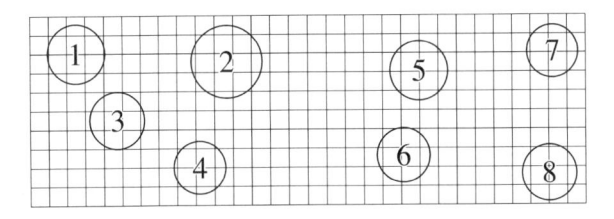

Figure 2

Survey of collections

In nearly every section of the country, private collectors will be found. These may be farmers who have preserved only the specimens found on their property, or they may be those who have collected materials from several townships. In all cases where the owners have any knowledge of

Figure 3

the locality from which their collections came, it is desirable to make a record of their specimens. For this purpose it is not necessary to draw in or photograph every piece. First of all, separate the arrowheads into classes.

Then with a lead pencil, trace in the outline of one of each class and state the number of such pieces in the collection. Or place one of each type on a suitable background, photograph them, and indicate the number of each. Thus if three classes of arrowheads are found, they might be indicated as in Figure 3.

[8] A similar method should be followed for stone axes, hammer stones, and so on. It is desirable to photograph pottery, but if this is impossible, make drawings, and always indicate the style of decoration if any is present. Also state if the pottery is sand or shell tempered. Pictures and descriptions of potsherds are also desirable. With such information it will ultimately be possible to learn the distribution of types of utensils. Local archaeologists can render service of great value if they will obtain the data indicated and make them available to the Committee on State Archaeological Surveys or to the local institution whose name appears on the last page of this leaflet.

Excavation

Every amateur who desires to carry on excavation should first of all receive instruction from a trained archaeologist. The ability to see the record in the ground frequently depends on training and experience. A beginner, with the best of intentions and with every attempt at care, will often miss stratification lines or fail to recognize the difference between disturbed and undisturbed deposits.

Your state university or museum, any member of the Committee on State Archaeological Surveys of the National Research Council, and particularly the institution furnishing these instructions will gladly assist you. *You are urged not to excavate without this instruction unless it becomes necessary to save the record of a site which is about to be destroyed.* In such a case, the following methods should be followed (the letters refer to the points and lines so designated on Figure 4):

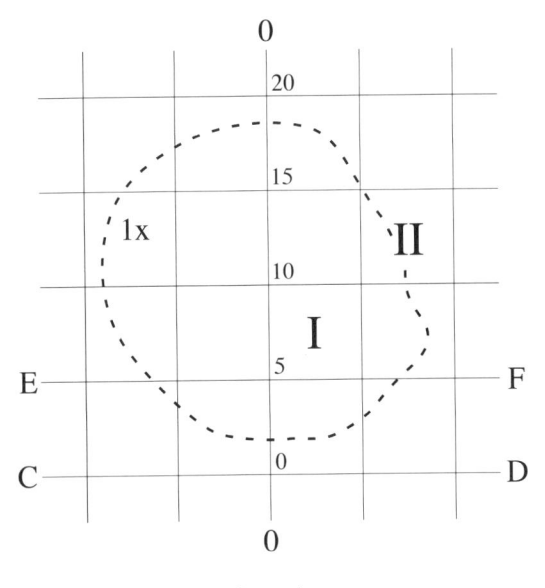

Figure 4

[9] Run a line across the north and south axis of the mound, as line 0–0. Five feet to the east run another line parallel to 0–0 and continue these 5-foot lines until you are well outside the mound. Now do the same on the west side of 0–0. Then, beginning on the south, well outside the mound, run an east and west line C–D. Five feet to the north run another such line, E–F, and continue this procedure until you have gone beyond the northern limits of the mound. Now place stakes at each point of intersection of the lines, and your whole site will be divided into 5-foot squares. Before starting work, you should make a map of the squares, such as in Figure 4. Along the line C–D sink a trench to a depth of about 2 feet below the surface or disturbed soil. Now carry this trench forward much as you would cut a loaf of bread. Always keep a straight face to the cut, throwing the dirt behind you so as to leave an open space.

As you enter the mound, you may find evidence of a prepared or hard-beaten floor, or of the undisturbed ground on which the mound was erected. You should be constantly on the watch for fire lines or evidences that the mound was built in two or more different periods. If the primary mound stood for years and grass and other materials accumulated on the surface and then at a later time more earth was heaped on it, this will probably be indicated by a dark or humus line. All evidences of this character should be carefully noted, and your record should indicate the situation for each square. Likewise, every find of a stone implement, pottery, or skeleton should be accurately placed in your plan and should receive further notice in your field notebook. By following the plan indicated in Figure 4, it is an easy matter to place every object found in its exact place on the map.

Thus such a square as the one marked "I," which begins on the 5-foot line E–F and lies east of the zero line 0–0, can be written: I=5E0 (i.e., it begins on the 5-foot line, east of the zero line), while square II=10E5 (i.e., it begins on the 10-foot line, 5 feet east of the zero line). If an object is found at 1x, it can be written in your notebook as 12.5-W-7, which indicates that it lies 12 feet and 5 inches north of the line C–D, and 7 feet west of the line 0–0. You should also note in your book how far below the present surface and how high above the floor of the mound the object lies. Each time an east and west line is encountered, as E–F, you should measure the height of the mound from the floor at each stake. By following such a method, you will have a complete record of the mound, its composition, and its contents. In all excavations test pits should be sunk from time to time below the level of your work, to be [10] sure that you are not overlooking some more ancient site. Village sites and cave deposits should be staked for excavation in like manner.

Utensils

A pick and shovel can be used for the preliminary trench, but when entering the mound it is necessary to use other tools. A mattock with a short handle can be employed for shaving down the face of the cut from top to bottom, until objects of interest are encountered, at which point smaller tools, trowels, dull knives, orangewood sticks, whisk brooms, and smaller brushes become necessary.

Preservation of Material

Never remove a specimen by pulling it out. Always expose the object fully by cutting away material above and on either side of it. If it appears to be associated with other objects or with a skeleton, allow it to remain in place until all are uncovered and photographed. Pottery, human, and animal bones are sometimes so soft when encountered that they cannot be removed without injury, but exposure to the air for a few hours often hardens them considerably. Very fragile bones can be strengthened by spraying them with a very thin solution of shellac. Often it is desirable to cut below a fragile object and slip in a thin piece of wood or tin on which it can be removed. When working around bones and similar materials, remove the soil by means of thin knives, orangewood sticks, or brushes. Any object which is worth uncovering is worth preserving. *Unless you are willing to give this time and care to preserving the record, you should not attempt excavation.* Preserve all fragments of pottery and bone; they may be capable of restoration later. Each specimen should be numbered and entered in a notebook. Since tags are easily lost, it is wise to mark each specimen with a 6-H (hard) pencil. Wrap each specimen separately in paper and attach tags to them. When many potsherds are found together, they may all be placed in a box and properly labeled. Never place pottery, arrowheads, and heavy stone specimens in the same box. Copy all your notebooks, drawings, and pictures in duplicate, and send one copy to your local institution or to the Committee on State Archaeological Surveys for interpretation and safekeeping. Your interests will be protected, and you will be given full credit for any information used.

Mention has been made of the possibility of finding evidences of early man in places where excavations or stream cutting is exposing the strata of the rock. In all such localities the face of the cut should be carefully studied and if human bones or stone utensils are found at considerable depths or associated with extinct animals, your state institution or the Committee on State Archaeological Surveys should be notified at once.

Last but not least, every collector should make provision for the care and disposition of his collection in case of his death. The amateur collector has made himself custodian of information of great historical interest, and he should guard it against loss or scattering.

The foregoing instructions are far from complete, especially the pages dealing with excavations. Opening a pre-historic site is a task which should only be undertaken in an emergency. Use your influence to preserve all mounds and village sites until you can have assistance or advice from a trained archaeologist. The Committee on State Archaeological Surveys and your local organization are anxious to aid you in recovering and preserving the story of man in America.

Index

Abihka, 285, 288
Aboriginal Sites on Tennessee River, 273–74
Academy of Natural Sciences of Philadelphia, 256, 272, 438; *Journal of,* 35
Academy of Science of St. Louis, 7, 8, 9; *Journal of Proceedings,* 8
Acolapissa, 232; sites, 271
Adair, James, 271
Adena aspect, 396
Adena culture, 47, 52, 289, 295, 351, 354, 355, 356, 358, 360, 385, 395, 396, 408
"A Form for Collection Inventories," 24
Ais, 232
Akokisa, 233
Alabama, 228, 229, 232, 264, 288, 290, 292, 293, 294
Alabama Anthropological Society, 36, 157, 261, 321
Alabama Museum of Natural History, 36, 121, 219, 256
Alabama State Department of Archives and History, 69 (n. 24)
Alamo, 226
Algonquian (Algonkin), 14, 15, 53, 160, 230, 231, 238, 243, 289, 295, 296, 297, 298, 346, 354, 363, 364, 365, 383, 387, 388, 390, 397, 405, 406, 438, 439
Amateur archaeological societies, x, xi, xii, xiii, 4, 32, 37
Amateur archaeologists: guidelines for, 177–78
American Anthropological Association, 2, 3, 4, 18, 20, 25, 57, 65, 223, 340

American Anthropologist, xi, xii, 11, 31, 267, 276, 317, 318
American Antiquity, 11, 408, 409
American Association for the Advancement of Science, 6, 18, 20, 64
American Institute of New York, 188
Americanist archaeology, ix, xii, xiii, xiv, 4, 12, 16, 28, 34, 39, 46, 54, 62, 65, 66
American Museum of Natural History, 3, 4, 21, 68 (n. 21), 321
"Ancient Monuments of the Mississippi Valley," 115
Andaste, 15
Anderson, E. A., 277
"An Introduction to Nebraska Archaeology," 30
Anthropological Society of St. Louis, 6, 7, 10, 11, 17
Antillean culture, 243
Apalachee, 225, 226, 232, 284, 285
Archaeological method, xi, xii, 23
Archaeological record, x, 22, 33, 66; basis of understanding, 45; identifying cultures in, 45; preservation of, xii, 94; shallow time depth of, 16; viewing of in ethnological terms, 29
Archaeological Survey of Illinois, 95
Archaeological Survey of Iowa, 190
Archaeological Survey of Missouri, 190
Archaeology, value of, 117–18, 132, 183, 186
Archaic Algonkin, 380, 399, 404
Archaic pattern, 361, 362, 364
Architecture: prehistoric, xii

Arikara, 231, 289
Arkansas, 265
Arkansas Museum of Natural History and Antiquities, 19
Arthur, Gabriel, 372, 378
Artifacts: analysis of, 24, 36, 37, 124, 125; comparison of, 116; distribution of, 440; European trade, 109, 266, 271, 291, 347, 366, 377, 378; exhibition of, 108, 119; faked, 307; importance of, 103; interpretation of, 126, 127, 189, 305; preservation of, 37, 319; repositories for, 119
Aspect, 55, 56, 60, 61, 62, 353, 382, 386, 392, 400, 403, 404, 405, 406, 419, 422, 423, 425
Assanai, 276
"A Suggested Classification of Cultures," 57
Atakapa, 232, 233, 265, 290, 294; sites, 271
"A Tentative Archaeological Culture Classification for Upper Mississippi and Great Lakes Areas." *See* Midwestern Taxonomic Method
Autosse site (Alabama), 148
Avoyel, 232, 265; pottery, 266; sites, 271
Ayllon expeditions, 223
Aztalan site (Wisconsin), 113, 175
Aztec, 98

Badia, 287
Barrett, Samuel A., 58, 139, 164, 166
Barrows, Albert L., 41
Bartram, James, 271
Base, 55, 60, 383, 403, 404, 405
Basic culture, 60, 341, 343, 383, 386, 387, 392, 400, 403, 407, 418, 422, 423, 425
Basketmaker, 384
Baskets, 260; of the Ozark bluff dweller, 253, 254
Baum site (Ohio), 360
Bayogoula, 232; sites, 271
Bidai, 233
Big Mound (Missouri), 91

Big Sioux focus, 349, 394
Biloxi, 238, 259, 290, 291, 366, 367, 368, 373, 376, 377, 379
Birmingham Museums Association, 219
Bishop Museum, 50
Black, Glenn A., 48, 356, 394, 403, 413
Blackduck focus, 350, 351
Black Minqua, 372
Blackmore Museum, 8
Black Sand component, 352, 353
Black Sand culture, 47, 297
Black Sand site (Illinois), 353
Blood Run site (Iowa), 347
Blom, Frans, 21
Blue Island group, 393
Blue Island site (Illinois), 353
Bluff lands: acquisition of, 136–38
Boas, Franz, 67 (n. 2), 68 (n. 21)
Bolton, Reginald Pelham, 451
Brame, J. Y., 284
Brannon, Peter A., 17, 18, 36, 69 (n. 24), 151, 157, 159, 252, 256, 321
Breech blocks, 157
Britten, M. H., 322
Broken Kettle site (Iowa), 349
Browman, David, 10
Brown, Calvin S., 21
Brown, Charles E., 13, 17, 18
Bureau of American Ethnology, xi, xii, 3, 9, 10, 16, 28, 30, 32, 34, 35, 36, 40, 41, 43, 46, 47, 48, 50, 193, 194, 195, 196; *Annual Report of,* 33, 322; *Bulletin,* 9, 368
Burial complex, 395, 396
Burials, 124, 161, 176, 245, 246, 266, 278, 346, 350, 351, 352, 356, 357, 361, 362, 363, 391, 396, 397, 400, 401; at Oneota sites, 347; gravel-kame, 359; in stone graves, 275, 360; intrusive, 461; in urns, 148, 149, 150, 238, 244; kinds of, 237–38; of the Moundville culture, 257; of the Ozark bluff dweller, 253, 255
Bushnell, David I., Jr., 10, 297

Bushnell, David I., Sr., 10, 139
Butler, Amos W., 3, 17, 18

Caddo, 226, 228, 231, 238, 239, 241, 243, 244, 247, 256, 268, 276, 286, 287, 289, 294, 296, 297, 298, 365, 381, 389; pottery, 267; sites, 266
Caddo Confederacy, 271
Cahokia culture, 47
Cahokia Mounds (Illinois), 6, 21, 22, 26, 87, 97, 130, 135, 140, 161, 175, 181, 246, 353
Calusa, 225, 226, 232
Calver, W. L., 451
Canasauga, 285
Cane, 269, 273
Capiché, 266
Carnegie Corporation, 65
Carnegie Foundation, 10
Carnegie Institution, 18
Casqui(n), 154, 286, 288, 293
Castalian Springs site (Tennessee), 274
Catawba, 229, 231, 285, 287, 290, 367, 374, 376, 377
Caulfield, Henry S., 21, 22, 88, 94
Caves, 106
Cayuga, 15
Central Basin aspect, 342
Central Basin pattern, 405, 406
Central Basin phase, 344, 345, 354, 356, 358, 394, 396, 398, 405; in Illinois, 351, 352; in Indiana, 355; in Iowa, 346; in Michigan, 359
Ceramic Repository for the Eastern United States, 18; Notes from, 393
"Certain Caddo Sites in Arkansas," 438
"Certain Culture Classification Problems in Middle Western Archaeology," 58, 340
Chaco Canyon (New Mexico), 385
Chakchiuma, 232, 259, 286, 292
Chambers, Moreau B., 19, 261, 262, 263, 294
Cherokee, 229, 231, 234, 238, 239, 244, 263, 285, 287, 289, 291, 293, 365, 392, 393
Chiaha, 285
Chicago World's Fair, 21, 192
Chichen Itza, 101
Chickasaw, 227, 228, 229, 232, 237, 238, 259, 263, 264, 286, 292, 293, 294, 368, 377
Chief Deer Foot, 111
Chief Joseph, 182
Chisca, 285, 295
Chitimacha, 227, 232, 238, 239, 265, 290, 294; sites, 271
Chiwere Sioux, 348
Choctaw, 43, 148, 152, 224, 227, 228, 229, 232, 233, 237, 238, 244, 248, 259, 261, 264, 284, 286, 292, 293, 294, 295, 365, 368, 377, 400, 401; pottery, 261, 262; sites, 261
Chongue, 369, 373, 375
Chowanoc, 230
Chronological ordering, xii, xiii, 39, 43, 50, 53, 65, 260; of New York prehistory, 14–16; prehistoric, 47; relative, 31
Chunky stones, 246, 258
Citico Mound (Tennessee), 272, 273, 274
Civil Works Administration, 63
Cladistics, 57
Clans, 238, 294
Clark, William, 182
Clark's Point site (Indiana), 357
Classes, 25, 401, 402, 417, 426; cultural, 418
Classification, 24, 25, 55, 417, 418, 426; paradigmatic, 24; systems of, 54, 55
Cliff dwellings, 106, 112, 113, 186
Clothing, 237; of the Ozark bluff dweller, 253
Coahuilteco, 233, 294, 296
Coastal-environments program, 40
Coastal focus, 362, 363, 364
Cofitachequi, 285
Colburn, W. B., 322
Cole, Faye-Cooper, 21, 31, 35, 37, 38, 39,

48, 58, 60, 62, 100, 102, 111, 299, 320, 384, 392, 403
Coles Creek site (Mississippi), 389
Coligoa, 286
Coligua. *See* Coligoa
Collins, Henry B., 36, 37, 39, 42, 43, 44, 46, 245, 259, 267, 389
Columbia University, ix, 38, 67 (n. 2), 311
Comanche, 281
Committee on Archaeological Nomenclature, 25
Committee on Basic Needs in American Archaeology, 63
Committee on State Archaeological Surveys (CSAS), ix, xi, xii, xiii, xiv, 3, 6, 7, 14, 21, 22, 32, 35, 36, 45, 63, 66, 87, 88, 94, 141, 162, 176, 322, 339, 341, 408, 409, 412, 415, 464, 467; and training amateur archaeologists, 20, 37; *Circular* series, 58, 69 (n. 22); focus of, 4, 12, 64, 65, 190, 191, 459; enlargement of, 17; establishment of laboratory, 18; Missouri work, 9, 10
"Comparative Research and Publication," 313
Compartment vessels, 250
Complex, 380, 383, 390, 398, 401, 417, 419, 427, 439
Component, 55, 61, 387, 392, 401, 404, 405, 407, 409, 419, 420
Conference on Midwestern Archaeology, 14, 17, 20, 23, 48, 50, 52, 58, 60, 63, 65, 66, 68 (n. 21), 87, 180, 196
Conference on National Parks, 102
Conference on Southern Pre-History, 31, 32, 36, 37, 48, 65, 68 (n. 21), 219, 322
Congaree, 231
Connor place site (Tennessee), 273
Cooper, William J., 21
Coosa, 285
Copper, 246, 247, 258, 263, 267, 272, 274, 283, 344, 390, 450
Corn Mother, 239

Costehe, 285
Coweta, 285
Coxe, Daniel, 369, 372, 373, 375
Cox, P. E., 192, 196, 272, 273
Creek, 152, 159, 225, 227, 228, 229, 232, 233, 237, 238, 239, 242, 263, 264, 285, 286, 287, 288, 290, 292, 294, 365, 400, 401
Creek Confederacy, 111, 231, 259, 293
Crow, 231, 290
Crowley's Ridge, 154
Cults, 239
Cultural classification, 52, 69 (n. 22), 340
Cultural complexity, 26
Cultural determination, 26
Cultural development, 235
Cultural lineage, 30
Cultural relationships, 340, 385, 386, 395
Culture, 49, 221, 341, 385, 386, 403, 404, 417; areas, 45, 46, 59, 60, 242, 243, 265, 267, 392, 416, 426; classification, 45, 393, 400, 402, 415; comparative study of, 48, 339; dynamics, 299; history, xiv, 29, 46, 106; identification of, 439; reconstruction, 316; sedentary, 222; study of, 116; taxonomic scheme of, 343; types, 58, 426
Culture-area concept, 59
Culture-indicative materials, 422, 427
Cusabo, 224, 232
Cushing, Frank H., 195
Cutler, Mannassah, 100

Dakota, 231, 290, 291, 350, 373, 374, 376
Danforth, C. H., 6, 10, 11
Davis, Edwin H., 12, 115, 271
Dawes, Rufus, 21, 99, 181
Day, Edmund, 20
Deadoses, 233
de Anasco, Juan, 155
Deasonville site (Mississippi), 267, 389
DeJarnette, David L., 256
de Laudonniére, Renaud, 224
Delaware, 365

de Leon, Alonso, 226
de León, Ponce, 223
Dellinger, Samuel C., 19, 36, 252, 278, 319, 322, 408
de Luna, Don Tristan, 224
de Montigny, Dumont, 227
de Pinedo, Alonso Alvarez, 223
Design elements, 148, 149, 150, 151
de Soto, Hernán, 41, 148, 150, 151, 152, 153, 154, 155, 156, 157, 158, 236, 255, 270, 274, 283, 284, 285, 286, 287, 288, 292, 294, 295
Determinants, 55, 62, 341, 386, 388, 389, 393, 399, 401, 406, 418, 422, 423, 425, 426
De Tontí, Henri, 266, 375
de Vaca, Cabeza, 224
Deuel, Thorne, 48, 58, 60, 61, 62, 358, 386, 392, 402, 403, 411
Dhegiha, 290, 348, 376
d'Iberville, Sieur, 227
Dickson, Don F., 21
Dickson site (Illinois), 353
Diffusion, 44, 222, 411, 426
Direct historical approach, 30, 31, 33, 40, 41, 43, 44, 45, 47, 52, 54, 59, 60, 64, 398, 416, 427
Division of Mound Exploration, 30, 32–33, 34
Dixon, Roland B., ix, 3, 4, 17, 26, 30, 68 (n. 21), 190, 377
Dorsey, J. O., 229, 290, 374
Douglass, A. E., 110
Duck River site (Tennessee), 159, 408
Dunbar, William, 269
Dunlap, Knight, 20, 21, 196, 322
Du Pratz, Le Page, 227, 268, 292, 368
Durant's Bend site (Alabama), 150

Earle, S. L., 220
Ear plugs, 246
East Tennessee Archaeological Society, 36, 272, 274, 275
Effigy Mound culture, 141, 289

Effigy mounds, 87, 95, 175, 185, 243, 245, 460
Effigy Mounds aspect, 342, 349, 401
Elemental aspect, 421
Elrod site (Indiana), 357
Eno, 231
Erie, 15, 231, 365, 372, 374, 377
Eskimo, 15, 131
Ethnology, 4, 7, 28, 29, 112, 188, 221, 259, 265, 385, 417
Etowah culture, 26, 159
Etowah site (Georgia), 157, 159, 221, 241, 242, 245, 247, 272, 274, 294, 355
Eufaula, 293
Evans site (Tennessee), 274
Excavation, xiii, 14, 127, 186, 188, 314, 444; methods of, 37, 38–39, 123, 124, 126, 300–2, 320, 446–47, 448–49, 452, 466–67; of a mound, 450, 452, 465–66; permits for, 126, 127; stratigraphic, 34, 39
"Excavations at an Erie Indian Village," 122, 447

F°14 site (Illinois), 353
Fanning site (Kansas), 348
Father Davion, 227, 369
Father Massanet, 226
Faunal remains, 107, 278
Federal Emergency Relief Administration, 63
Federal-relief archaeology, 37, 64
Field methods, xii, 37, 105, 108, 109; excavation, 123; surveying, 122
Field Museum of Natural History, 3, 121
Fieldwork, x, 36, 112, 163; lack of standards in, xi
Fisher, Alton K., 50, 51, 52, 53, 55, 57
Fisher group, 393
Fisher site (Illinois), 353, 385
Florence site (Alabama), 257
Focus, 55, 56, 60, 61, 62, 353, 386, 392, 400, 401, 402, 404, 406, 419, 422, 423, 425

Folsom people, 388
Ford, James A., xii, 19, 27, 28, 40, 47, 66, 261, 262, 263, 294, 389, 390
Fordyce, John R., 158, 284, 286
Form, xiii, 25, 50; conflation with function, 25; units based on, 53
Fort Ancient Aspect, The, 53, 62
Fort Ancient aspect, 60, 393, 394, 401, 407
Fort Ancient culture, 26, 47, 52, 159, 289, 296, 355, 358, 360, 385, 387, 390, 393, 396, 406; sites, 399, 406
Fort Ancient site (Ohio), 97, 130, 378
Fort St. Peter site (Mississippi), 262
Fort Toulouse, 228
Fowke, Gerard, 9, 10, 28, 68 (n. 21), 140, 394
Fox Farm site (Kentucky), 360
Franciscan missionaries, 225
Francisco of Chicora, 224, 292
Franz-Green Mound (Indiana), 346
Freeland, W. E., 21, 91, 113
Funkhouser, W. D., 145, 245, 297
Futrall, John C., 21, 68 (n. 20)

Game animals, 236; of the Ozark bluff dweller, 254
Geological Survey of Illinois, 135
Geometric enclosures, 359
George Washington University, 28
Gibson, Jon, 35, 40
Gillet's Grove site (Iowa), 347
Gilmore, Melvin R., 254
Glendora site (Louisiana), 266
Glenwood aspect, 394
Glenwood culture, 47, 349
Globe Conference, 54
Glover Mound (Kentucky), 145, 146
Goddard, Charles F., 363
Gordon culture, 145
Gordon-Fewkes site (Tennessee), 353, 355
Grand River culture, 141, 419, 420, 421
Grand River focus, 342, 348, 401, 420, 421
Great Lakes, xiii, 48, 174, 175, 229, 230, 289, 295, 298, 339, 412, 438, 444
Great Lakes culture, 295, 404

Great Northern Trail, 296
Great Serpent Mound (Ohio), 87, 97
Greenman, Emerson F., 24, 25, 48, 166, 170, 171, 172, 354, 359, 396, 406
Griffin, James B., 46, 47, 48, 53, 58, 60, 62, 64, 65, 66, 386, 395, 398, 399, 400
Grigra, 232, 259
Ground Slate phase, 364
Groups, 402
Guacata, 232
Guasili, 285
Guernsey, E. Y., 48
Guthe, Carl E., ix, 6, 17, 18, 19, 35, 48, 50, 58, 63, 64, 65, 141, 196, 299, 322, 364, 384, 401, 402; on archaeological surveys, 14; on survey summaries, 11

Haida, 400
Halbert, Henry S., 261
Hale, George E., 1
Handbook of American Indians, 284
Harrington, M. R., 249, 254, 278, 296, 297, 438
Harrison, William Henry, 182
Harvard University, ix, 3, 4, 16, 17, 18, 34, 35, 41, 58, 113
Hasinai, 226
Hayden, Ferdinand V., 8
Head form, 229, 230, 297, 362, 363, 364; deformation, 276
Henry, Joseph, 8
Hewett, Edgar, 10
Hidatsa, 290
Hilder, Frank, 8
Hinsdale, W. B., 139, 398
Historical period, 31, 109, 374
Hitchiti, 232, 285, 290, 292, 294
Hiwassee, 285
Hodge, Frederick W., ix, 3, 17, 26, 28, 31, 109, 111, 112
Hohokam, 381, 383, 384, 403
Hohokam classification, 382, 384
Hollywood Mound (Georgia), 250, 294
Holmes, William Henry, 16, 45, 242, 278, 279, 322

Hooton, Ernest A., 377
Hopewell aspect, 383, 391
Hopewell culture, 52, 159, 175, 274, 289, 295, 319, 354, 356, 358, 385, 387, 399, 405, 407, 409; geographical limits of, 408; in Georgia, 26; in Illinois, 26, 47, 352, 353, 358; in Ohio, 26, 27, 28, 46, 47, 143, 144, 342, 352, 357, 359, 391, 394, 395, 396, 407; in Tennessee, 26; in Wisconsin, 141, 143, 144, 358; mounds, 397; origin of, 160–61; pottery, 241, 267, 352, 354; sites, 61
Hopewellian phase, 405, 421
Hopewell Mounds (Ohio), 159, 243, 358, 359, 387
Hopewell phase, 344, 345, 346, 364, 388, 389, 410, 411
Hothliwahali, 286
Houma, 232, 265, 292; pottery, 266; sites, 271
House remains, 263, 344
Howell, W. H., 41
Huguenots, 224
Huron, 231, 365, 371, 372, 374
Hutchins system, 165, 166

Ibitoupa, 232, 259
Illinois, 230, 365, 372
Illinois aspect, 388
Illinois Bluff culture, 47
Illinois Indian culture, 159
Illinois State Academy of Science, 57
Illinois State Geological Survey, 21
"Importance of Systematic and Accurate Methods in Archaeological Investigation, The," 26
Index markers, 39
Indiana Academy of Science, 3–4, 6
Indiana Historical Commission, 443
Indiana Historical Conference, 6
Indiana Historical Society, 48
Indianapolis Conference. See Conference on Midwestern Archaeology
Indian Knoll site (Kentucky), 357, 361
Initchigami, 230

"Interest of Scientific Men in Pre-History, The" 221
International Congress of Americanists, 6
Intrusive Mound culture, 363
Invention, 426
Iowa, 291, 367
Ioway, 348
Ioway Sioux, 59, 160, 416, 421
Iroquois, 14, 15, 58, 117, 231, 243, 244, 289, 296, 297, 298, 319, 354, 358, 359, 361, 364, 365, 368, 371, 372, 374, 378, 393, 394, 406, 415, 427, 438; pottery, 439
Isle Royale, 174

Jackson, Andrew, 157
Jeaga, 232
Jennings, Jesse, 34
Jesuit missionaries, 225, 227, 371
Jesup North Pacific Expedition, 68 (n. 21)
Jones, Walter B., 36, 220, 221, 256
Jonesville site. See Troyville site
Joutel, Henri, 226, 228
Judd, Neil, 31, 35, 37, 39, 303, 322
Jumano, 381
Juniata, 377

Kansas, 290
Karankawa, 233, 278, 281, 294
Kasihta, 285
Kaskaskia site (Illinois), 300
Kaskinampo, 286, 293
Kathio site (Minnesota), 346, 350
Kehoe, Alice, 57
Kellogg, Vernon L., 2
Kelly, A. R., 58, 300
Keno Place site (Louisiana), 266
Kethlas, 239
Keyes, Charles R., 17, 18, 48, 296, 355, 394
Kichai, 231
Kidder, A. V., ix, 16, 17, 18, 67 (n. 15), 190, 384
King Philip, 182
Kinietz, 411, 412

Knapp, Thomas M., 21
Kniffen, Fred B., 40, 271, 322
Knockers, 9, 10
Koasati, 229, 232, 285, 286, 288, 290, 293
Koroa, 232, 233, 259, 265, 286, 368, 369; pottery, 266; sites, 271
Kroeber, Alfred L., 4, 16, 31, 41, 47, 53, 58, 68 (n. 21)

"Laboratory and Museum Work," 303
Laboratory methods, xii, 133, 305–6, 309–12, 319
Lake Michigan culture, 47
Lake Michigan phase, 341, 346, 349, 351, 394, 398, 408, 409
Lake Winnebago culture, 419, 420
Lake Winnebago focus, 342, 348, 420
Lamoka focus, 361
Langford, George, 301, 385
Larto Lake Mounds (Louisiana), 267
La Salle, René-Robert Cavelier, Sieur de, 156, 157, 226, 227, 233, 295, 368, 370, 373, 411
Latham, Roy, 362
Laufer, Berthold, 3, 6, 68 (n. 21)
Laura Spelman Rockefeller Memorial, 20
Laurel Mound (Minnesota), 350
Lawson, John, 228, 367
Lederer, John, 228, 367
Leechman, D., 313, 319
Leidy, Joseph, 8
Leighton, M. M., 21
Lemley, Harry J., 21
Lewis, Meriwether, 182. *See also* Clark, William
Lewis, Theodore H., 12, 408
Lilly, Eli, 48, 413
Lineages, 238
Linguistics, 53
Linguistic stocks, 230, 232, 242, 284, 289, 364, 366, 416
Linton, Ralph, 35, 44, 45, 46, 221, 222
Little Sioux focus, 394
Liverpool focus, 352
Louisiana aspect, 388

Louisiana State Geological Survey, 140
Louisiana State University, 40, 271
Lower Mississippi phase, 380, 381, 405

Madisonville focus, 353, 406
Madisonville site (Ohio), 366, 377, 378, 379
Maize, 222
Mammoth Cave, 146
Mandan, 290, 349, 367
"Mandans, The," 437–38
Marksville aspect, 391
Marksville complex, 389
Marksville component, 388, 389
Marksville focus, 391
Marksville site (Louisiana), 28, 46, 61, 242, 267, 390, 405
Marquette, Jacques, 227, 370, 372, 374, 378
Martyr, Peter, 224, 292
Mason, J. Alden, 21
Mason, Otis T., 45
"Material Culture of the Menomini," 437
Maya, 98, 114, 230, 281
McKenzie site (Tennessee), 274
McKern Method. *See* Midwestern Taxonomic Method
McKern, W. C., xiii, 48, 49, 50, 51, 52, 53, 55, 56, 57, 59, 60, 61, 63, 65, 139, 141, 144, 163, 289, 296, 340, 348, 355, 386, 389, 390, 394, 395, 398, 403, 408
Meherrin, 231
Menard Mounds (Arkansas), 286
Menendez, Pedro, 224
Menomini, 437
Menomini material culture, 346, 349
Mercer, Henry C., 16, 34
Merwin, B. W., 377, 378
Merwin, R. E., 377
Mesa Verde National Park, 134
Metz, Charles L., 377
Mexico, 222, 233, 234, 276, 289
Miami, 365, 366, 378, 379
Middle Mississippi, 66, 294, 349, 351, 353, 356, 394, 408, 411

Middle Mississippi complex, 353
Middle Mississippi phase, 342, 380, 381, 392, 394
Middlesex focus, 363
Midwest, xiii, xiv, 32, 51, 53, 54, 60, 65, 69 (n. 22), 175, 179, 184, 185, 188, 297, 395, 398, 412
Midwestern Taxonomic Method, xiii, xiv, 51, 52, 54, 56, 57, 60–61, 62, 65, 393
Migration, 44, 222, 289, 367, 411, 460; legends, 292, 377; through Texas, 280, 281
Mill Creek culture, 47, 290, 291, 349, 350, 394
Miller, Townsend, 163
Mills, William C., 12, 13, 17, 26, 159, 161, 399
Milwaukee Public Museum, 48, 50, 58, 141, 162, 163
Mines, 107, 175
Mississippi basic culture, 342, 343, 351, 355, 394, 397, 402, 404, 420
Mississippi culture, 387, 388, 390, 391, 392, 409, 410, 420
Mississippi Department of Archives and History, 19, 261
Mississippi Geological Survey, 21
Mississippi pattern, 61
Mississippi River, 21, 87, 233, 286, 287, 288, 344, 366, 368, 370, 375, 376, 379, 407, 411, 444; changes in river channel, 270; survey along, 408
Mississippi Uplands culture, 348, 419, 420
Mississippi Uplands focus, 420, 421, 424
Mississippi Valley, xiii, 17, 18, 27, 36, 39, 44, 47, 48, 53, 60, 62, 66, 95, 96, 97, 98, 121, 128, 131, 160, 164, 176, 184, 192, 193, 219, 245, 246, 247, 265, 266, 267, 276, 321, 322, 339, 382, 383, 384, 386, 388, 415, 438, 444, 461; oxbow lakes in, 270
Missouri, 291, 348, 367
Missouri Archaeological Society, 9
Missouri Historical Society, 8, 9, 140
Mitchigamea, 369
Mobile, 232, 286, 287
Moneton, 367, 372
Monks Mound (Illinois), 135
Mooney, James, 195, 224, 367
Moore, Clarence B., 35, 140, 150, 151, 155, 159, 242, 246, 256, 257, 261, 266, 272, 273, 278, 279, 295, 357, 361, 438
Moorehead, Warren K., 6, 12, 21, 22, 27, 28, 35, 37, 87, 135, 140, 161, 162, 221, 273, 319, 320, 321
Morgan, Lewis Henry, 26
Morley, Sylvanus G., 18
Morris, Earl, 10
Moscoso, Luys, 156, 286
Mosopelea, 231, 366, 370, 373, 375, 376, 377, 378, 379
Mound Bottom site (Tennessee), 272
Mound builders, 100, 117, 131, 132, 138, 143, 187, 191, 280, 388, 398; culture, 158, 159, 319; myth of, 30, 33, 34, 40
Mounds, 98, 102, 112, 131, 133, 137, 139, 150, 176, 189, 192, 243, 265, 278, 281, 302, 346, 350, 391, 408, 449, 460, 461; at Marietta, 100, 101; building of, 26, 27, 161, 269, 355, 390; destruction of, 96, 97, 100, 270, 274, 459; functions of, 245; identification of, 450; in Kentucky, 360; looting of, 104; made of burned rock, 279; made of stone, 355, 356, 394; of the Moundville culture, 257; preservation of, 92–93, 103, 106, 162, 176, 270; rebuilding of, 113, 124; shell, 159, 243, 245, 270, 361; structural analysis of, 136, 142; temple, 152; value of, 99
Moundville culture, 36, 256, 257, 258, 259, 263, 293
Moundville site (Alabama), 151, 159, 219, 220, 221, 224, 241, 243, 245, 247, 250, 256, 257, 262, 293
Mugulasha, 232
Museum of the American Indian, 3
Museums, x, xi, xii, xiii, 304, 306; exhibits in, 307–8

Muskhogean, 150, 160, 224, 231, 232, 233, 259, 284, 285, 287, 288, 290, 292, 294, 295, 296, 319, 365, 368, 411
Muskogee, 42, 231, 232, 287, 292, 293, 294, 295
Myer, W. E., 272, 274, 275

Nacoochi site (Georgia), 294
Nahyssan, 367
Napochi, 224, 293
Narvaez expedition, 224
Natchez, 227, 228, 229, 232, 233, 236, 238, 239, 245, 248, 259, 264, 265, 268, 286, 287, 288, 290, 292, 294, 296, 365, 369, 389, 406; pottery, 261, 262, 266, 267; sites, 261, 398
Natchez site (Mississippi), 262
Natchitoches, 266; sites, 271
National Academy of Sciences (NAS), 1
National monuments, 134
National Museum of Canada, 313
National Park Service, 58, 133, 134
National Research Council (NRC), 14, 18, 41, 47, 63, 65, 87, 88, 94, 138, 175, 184, 190, 196, 256, 322, 339, 341, 412, 413, 415, 459, 464; Archives, xiv; *Bulletin,* xiv; creation of, 1; Division of Anthropology and Psychology, ix, 2, 3, 4, 12, 20, 21, 221, 322; organization of, 2; *Report and Circular Series,* xiv; role in Americanist archaeology, ix
Nebraska aspect, 394
Nebraska culture, 47, 349
Nelson, Nels, 4, 16, 21, 171, 384
Neutrals, 231, 359, 365, 372, 374
Newcomb site (Indiana), 357
New Galena Mound group (Iowa), 346
New York State Museum, 12, 120
Nicholls Mound (Wisconsin), 142, 143
"Notes on State Archaeological Surveys," 10
Nottoway, 231
Nowlin Mound (Indiana), 354, 356
Numerical taxonomy, 57

Occom, Samson, 182
Offagoula. *See* Ofo
Ofo (Ofogoula), 231, 259, 291, 366, 367, 368, 368, 369, 372, 373, 374, 375, 376, 377, 379
Ohio complex, 389
Ohio State Archaeological and Historical Society, 12, 17, 24, 442
Ohio State Museum, 22, 87, 166, 399
Okelousa, 232; sites, 271
Old Algonkin pattern, 389
Old Ocmulgee town (Georgia), 111
Old Trading Trail, 153
Omaha, 231, 290
Oneida, 15
Oneota aspect, 348, 350
Oneota focus, 342, 346, 347, 349, 393, 394; sites, 348
Onondoga, 15
Opelousa, 271
O'Regan site (Iowa), 347
Oriental Research Foundation, 185
Orient focus, 362
Orr, Ellison, 349
Osage, 231, 290, 291, 367, 368
Osotci, 232
Oto, 291, 348, 367
Ouachita, 266
Ouesperies, 369, 372, 375
Owasco aspect, 363
Owasco focus, 364

Pacaha, 154, 286, 294
Pafallaya, 286
Palmer, George Herbert, 133
Pamlico, 230
Pan American Union Bulletin, 11
Pardo, Juan, 225, 295
Parker, Arthur C., 12, 13, 14, 16, 22, 23, 128, 361, 447
Parks, 101–2, 113, 131, 132, 134, 138, 181, 189
Pascagoula, 232
Patiri, 233
Pattern, 55, 60, 61, 341, 404, 405, 406

Pawnee, 231, 254, 289, 349, 406
Peabody, Charles, 25, 34
Peabody Museum (Harvard), 3, 8, 10, 17, 121, 377
Peacock, Charles K., 36, 271
Pearce, James E., 36, 276
Pecos Conference, xiii, 54, 380, 382, 384, 385
Pecos Pueblo (New Mexico), 18, 383
Pedee, 231
Petroglyphs, 107
Phase, 55, 60, 62, 343, 380, 382, 386, 387, 392, 400, 401, 402, 403, 404, 405, 406, 407, 409, 419, 425
Phillips, Philip, 62, 66
Phillips Academy, 12, 17, 25, 113
Phyletic history, 46
Phylogenetic relations, 55, 57
Pictographs, 255
Pierrette, Rosa, 368
Pits, 254, 257, 397, 446, 447, 448
Plains, 54, 60, 69 (n. 22), 182, 221, 231, 289, 296, 322, 365, 383, 406
Pleasant Creek Mound group (Iowa), 346
Poffenberger, A. T., 221, 322
Point Peninsula focus, 363
Ponca, 290
Pot hunters, 96, 106, 108, 122, 124, 193, 270, 273, 321, 459, 460
Pottery, 26, 45, 105, 283, 422, 423, 425, 426, 438, 440, 450; classification, 342; decoration, 248–50, 262, 266, 267, 277, 279, 347, 350, 354, 409; Gulf Coast, 267; killed, 151, 243, 267, 396; of the Moundville culture, 258; of the Southeast, 241; tempering, 53, 146, 243, 344, 347, 356; use in chronological ordering, 43, 44, 260, 262; use in cultural determination, 104
Powell, John Wesley, 30, 33, 194, 195
Powell Mound (Illinois), 102
Powhatan, 238
Powhatan Confederacy, 230
Prather site (Indiana), 356
Pre-Algonkin, 361

"Pre-historic Ethnology of an Archaeological Site," 316
"Pre-Historic Southern Indians, The," 241
"Preservation of Antiquities," 319
Projectile points, 24, 426; classifying, 166–71
Protohistoric period, 40, 271, 347
Publication, 314–15, 320, 321
Pueblo area, 106, 279, 280, 381, 383, 384
Putnam, Frederic W., 12, 17, 34, 39, 68 (n. 21), 161, 366, 377, 378
Putnam, Rufus, 100

Quarries, 107
Quapaw, 233, 286, 290, 291, 355, 367, 368, 369, 373, 374, 375
Quinipissa, 271

Raleigh Colony, 228
Rathgen, Friedrich, 319
Read, M. C., 272, 273
"Recent Field Work in Southern Archaeology," 36
Red Cedar component, 388
Red Cloud, 182
Rediscovering Illinois, 62
Red Ochre phase, 351, 354, 356, 358, 395
Red River focus, 388
Reed flutes, 255
"Relation of the Southeast to General Culture Problems of American Pre-History, The," 41, 283
Relative dating, 52
Relic hunters. *See* Pot hunters
"Report of the Committee on Archaeological Nomenclature," 25
"Report on the Mound Explorations of the Bureau of Ethnology," 33
Robbins place site (Tennessee), 272
Roberts, Frank H. H., 48
Rochester Municipal Museum, 124–25
Rock-shelters, 106, 272, 273, 346, 398, 461; in Kentucky, 361; of the Ozark bluff dweller, 252, 253

Russell, Richard, 40
Ryan, Frank, 317

Sa-cah-gah-wea, 182
Santa Elena, 225
Santee, 231, 237
Sauer, Carl, 40
Sauk and Fox, 365
Saville, Marshall, 17
Sawokli, 292
Schulte, Dorothy, 339
Scioto aspect, 388
Scioto Valley, 26, 28, 159, 160, 161, 358
Scioto Valley focus, 61, 405
Seashore, C. E., 3, 12
Seip component, 388
Seip focus, 388
Seip Mounds (Ohio), 358, 359, 387
Seltzertown Mound group (Mississippi), 262
Seminole, 226, 229, 232
Seneca, 15, 372, 373
Seriation, 39, 53
Setzler, Frank M., 27, 28, 46, 47, 48, 61, 62, 64, 242, 243, 263, 267, 284, 354, 390, 394, 404
Shakori, 231
Shawnee, 230, 231, 232, 239, 294, 355, 366, 371, 372, 376, 378, 379; sites, 412
Shell: currency, 240; gorgets, 246, 247, 279, 283
Shell-heaps, 106, 245, 265, 391
Shetrone, Henry C., 12, 22, 87, 133, 137, 161, 393, 399
Shuman, 289
Signal Butte culture, 47
Sigma Xi, 100
Similarity: formal, 47, 53; homologous, 31, 46; morphological, 55, 57; typological, 46
Sioux, 53, 153, 182, 224, 228, 231, 237, 238, 259, 287, 290, 291, 296, 297, 298, 346, 365, 366, 367, 368, 373, 376, 378, 379, 438
Sites, 14, 267, 302, 385, 397, 409, 419, 437; as sight-seeing attractions, 92, 97, 118; classifying, 435, 440; destruction of, xii, 4, 64, 87, 180, 275, 459; historic, 99; kinds of, 461; locations of, 105, 444, 445; looting of, x, 96, 111; mapping of, 440, 446, 462; occupied by historic tribes, 271, 410, 411, 412, 426, 427, 438; preservation of, 22, 275; recording, 163, 164; stratified, 349; surveys of, 4, 13; value of, 99; village, 396, 438
Sitting Bull, 182
Skeletal remains, 107, 162, 251, 252, 259, 264, 278, 302, 311, 351, 447, 449, 450, 466; postmortem perforation of, 360
Smith, Harlan I., 316, 317, 321, 377, 440
Smithsonian Institution, 8, 30, 32, 40, 98, 115, 193, 194, 272, 274, 308, 352, 408
Society for American Archaeology, 12, 64, 65
Society of American Indians, 12
"Some Resemblances in the Ceramics of Central and North America," 280
Sounding rod, 451
Southeast, xii, 32, 34, 35, 36, 37, 39, 40, 41, 42, 44, 45, 46, 53, 65, 219, 221, 222, 223, 228, 229, 234, 235, 239, 240, 241, 242, 243, 252, 260, 263, 264, 288, 297, 365, 383, 389, 390, 396, 400, 411, 412
"Southeastern Indians of History," 13, 41, 223
Southern culture, 392
Southwest, xiii, 16, 17, 36, 39, 53, 54, 106, 110, 113, 150, 190, 194, 223, 242, 253, 260, 269, 276, 289, 297, 380, 381, 382, 383, 385, 386, 396, 403, 412
Spanish knives, 157–58
Species concept, 45, 46, 56
Spier, Leslie, 4, 16, 384
Spoon River Mississippi material, 47
Squier, Ephraim G., 12, 115, 271. *See also* Davis, Edwin H.
Stallings Mound (Georgia), 361
Starved Rock site (Illinois), 97

"State and Local Archaeological Surveys," 133
State Historical Society of Iowa, 17, 48, 133
State surveys, 10, 11, 13, 21, 23, 117, 128, 130, 166, 180, 270, 275, 276, 300, 321, 444; operation of, 118–21, 271, 440, 441, 442, 443
Sterns, Frederick H., 18
Stevenson, James, 195
Stevenson, Matilda, 195
Steward, Julian, 54
Stirling, Matthew W., 21, 28, 29, 30, 32, 35, 36, 45, 47, 241
St. Louis Society of the Archaeological Institute of America (AIA), 8, 9, 10
St. Louis University, 21
Stone Fort site (Tennessee), 272
Stratification, 26, 105, 106, 123, 281, 301, 427, 437
Stratigraphic revolution, 17, 34
Stratigraphy, 16, 17, 45, 241, 317, 350, 357
Strong, William Duncan, 30, 36, 37, 39, 60, 64, 289, 290, 296, 319, 320, 321, 348, 349, 394, 406
Suma, 289
Superposition, 16, 39
Surface collection, 462
Survey, xiii, 14, 24, 435, 462; of collections, 464; records, 164, 191; summaries, 11, 12
Susquehanna, 231, 238, 365, 372, 374, 375, 377
Swanton, John R., 36, 39, 41, 42, 46, 48, 53, 66, 68 (n. 21), 223, 241, 252, 259, 265, 283, 298, 316, 364, 366, 400, 411

Taensa, 232, 233, 239, 265, 365, 370, 375; pottery, 266; sites, 271
Takesta, 232
Tali, 285
Talikwa, 285
Talisi, 286
Talladigi, 285, 288
Tampico phase, 351, 352, 408, 409
Tanico, 286
Taposa, 232, 259
Tasqui, 293
Tawasa, 232, 286, 294
Taxonomic units, xiii
Taylor, Jay L. B., 21
"Technical Methods in the Preservation of Anthropological Museum Specimens," 319
Tecumseh, 182
Tellico, 285
Teotihuacan (Mexico), 174
Terminology, 24, 25, 49
Terry, R. J., 196
Thomas, Cyrus, 30, 33, 34, 45, 279, 322
Throop, G. R., 21, 187, 196
Thruston, Gates P., 272
Time, xii, xiii, 45, 50, 53, 54; depth, 16
Timucua, 224, 225, 226, 238, 284, 287, 290, 294, 296, 365
Tiou(x), 232, 259
Tlingit, 400
Tobacco pipes, 247, 248, 255, 258, 262, 263, 267, 268, 283, 344, 352, 363, 364, 439
Tohome, 232
Tolu site (Kentucky), 353, 355, 411
Tomahitans, 371, 372. *See also* Yuchi
Tonkawa, 233, 281, 287, 294
Trade goods, 109
Trails, 152, 156, 165, 166, 175, 462. *See also* Great Northern Trail; Old Trading Trail; Warrior's Path
Traits, 242, 243, 383, 385, 392, 395, 396, 419, 422, 427; analogous, 57; ancestral, 57; cultural, 45, 51, 52, 144, 248; determinant, 399, 402; diagnostic, 55, 62, 400, 401, 402, 409, 410; fundamental, 407; general versus specific, 400; homologous, 57; Hopewell, 61, 144; inherent, 62; linked, 55, 62, 399, 401, 402; Mexican, 222; Mississippi, 61; overlapping, 53, 245; presence/absence of, xiii; shared, derived, 57; use of in the Midwestern Taxonomic Method, 55; Woodland, 61, 387, 390

Transitional materials, 406
Travertine deposits, 317
Treaty of Greenville, 182
Tree-ring dating, 110, 269, 317, 385
Trempealeau aspect, 342, 343
Trempealeau focus, 383, 388
Trench runs, 125
Troyville site (Louisiana), 267, 269, 270
Tukabahchee, 292
Tula, 155, 156, 286
Tulane University, 21
Tulla. *See* Tula
Tunica, 227, 232, 233, 241, 259, 265, 286, 288, 290, 294, 368, 369, 372, 374, 389; pottery, 262, 266; sites, 261, 262, 271, 398
Tunica Oldfields site (Mississippi), 286, 294
Turner site (Ohio), 358, 359, 391
Tuscarora, 228, 231, 238
Tuskegee, 232, 285, 288, 293
Tutelo, 290, 291, 367, 373, 374, 376, 379
Tylor, Edward B., 26
Types, 24, 25, 44
Type site, 409

University of Arkansas, 19, 21, 36, 68 (n. 20)
University of California, ix, 4, 10, 28, 31, 40, 50, 58, 442
University of Chicago, ix, 21, 38, 48, 57, 58, 185, 301, 352, 408
University of Illinois, 6, 58, 113, 135, 300; *Bulletin*, 181
University of Indiana, 6
University of Iowa, 3
University of Kentucky, 21, 145
University of Michigan, ix, 6, 18, 24, 48, 62, 254, 318, 411; Museum of Anthropology, 18, 24, 62, 139
University of Minnesota, 48
University of Missouri, ix
University of Pennsylvania, ix, 21; Museum, 16

University of Texas, 36, 276, 278
University of Wisconsin, 36
Upper Mississippi culture, 47, 58, 350, 351, 353, 357, 358, 359, 393, 402, 403, 411, 415, 419, 420
Upper Mississippi Oneota, 59, 416
Upper Mississippi phase, 342, 343, 380, 381, 401, 408, 420, 421, 423
Upper Republican aspect, 394
Upper Republican culture, 47
U.S. Department of the Interior, 133
U.S. Geological Survey, 8, 165, 194, 233
U.S. National Museum, xii, 28, 31, 32, 34, 36, 37, 40, 43, 46, 48, 121, 284, 303, 305, 308; accession procedure of, 309–10; annual report of, 320; preservation procedures at, 312

Vaillant, George C., 279, 280
"Value to the State of Archaeological Surveys, The," 22
Variation: distribution of, xiv; formal, xiv; identification of, 25
Vessel shape, 250–51, 278, 344, 347, 350, 351
Vine Valley aspect, 362, 363, 364
Virginia Colony, 228
Volsine Chiki, 368

Waccamaw, 231
Walker, Winslow, 36, 39, 40, 41, 46, 265
Warrior's Path, 289
Washa: sites, 271
Washington University (St. Louis), 10, 21
Wateree, 231
Weapemeoc, 230
Webb, William S., 21, 37, 145, 245, 255, 297, 319, 355, 407, 412, 413
Wedel, Waldo, 30, 60, 348, 394
Weer, Paul, 48, 413
Wheeler, H. E., 219, 220, 221, 322
Whelpley, Henry M., 6, 17, 139
Wichita, 231, 289

Wickliffe site (Kentucky), 353
Wilford, Lloyd A., 48
Willey, Gordon R., 38, 47, 62
Williams Mound (Kentucky), 145, 146
Winnebago, 231, 290, 291, 348, 366, 367
Winyaw, 231
Wisconsin Archaeological Society, 163
Wisconsin aspect, 342, 353, 420, 421, 424
Wisconsin Historical Society, 185
Wisconsin Upper Mississippi culture, 420
Wissler, Clark, ix, 3, 4, 6, 17, 21, 35, 37, 45, 67 (n. 2), 133, 181, 190, 221, 241, 283, 313, 320, 321, 322; and the role of the CSAS, 9, 10, 11
Willoughby, Charles C., 161, 366, 377, 378, 388, 389, 391
Wolf, Abraham, 418
Woodland area, 420
Woodland basic culture, 341, 342, 343, 344, 349, 355, 357, 359, 360, 380, 383, 387, 389, 390, 391, 392, 394, 396, 397, 399, 401, 402, 403, 404, 405, 408, 410; in Illinois, 351, 352; in Indiana, 355; in Iowa, 346, 350; in Michigan, 359; mounds, 397; sites, 397
Woodland pattern, 342, 408, 409, 411
Woodward, Henry, 295
Works Progress Administration, 63

Yale University, ix, 4
Yamasee, 225, 226, 228, 229, 232, 285
Yazoo, 232, 259, 265, 368, 369, 398
Yuchi, 232, 233, 239, 285, 287, 295, 296, 371, 372

Zuñi, 381